D0821663

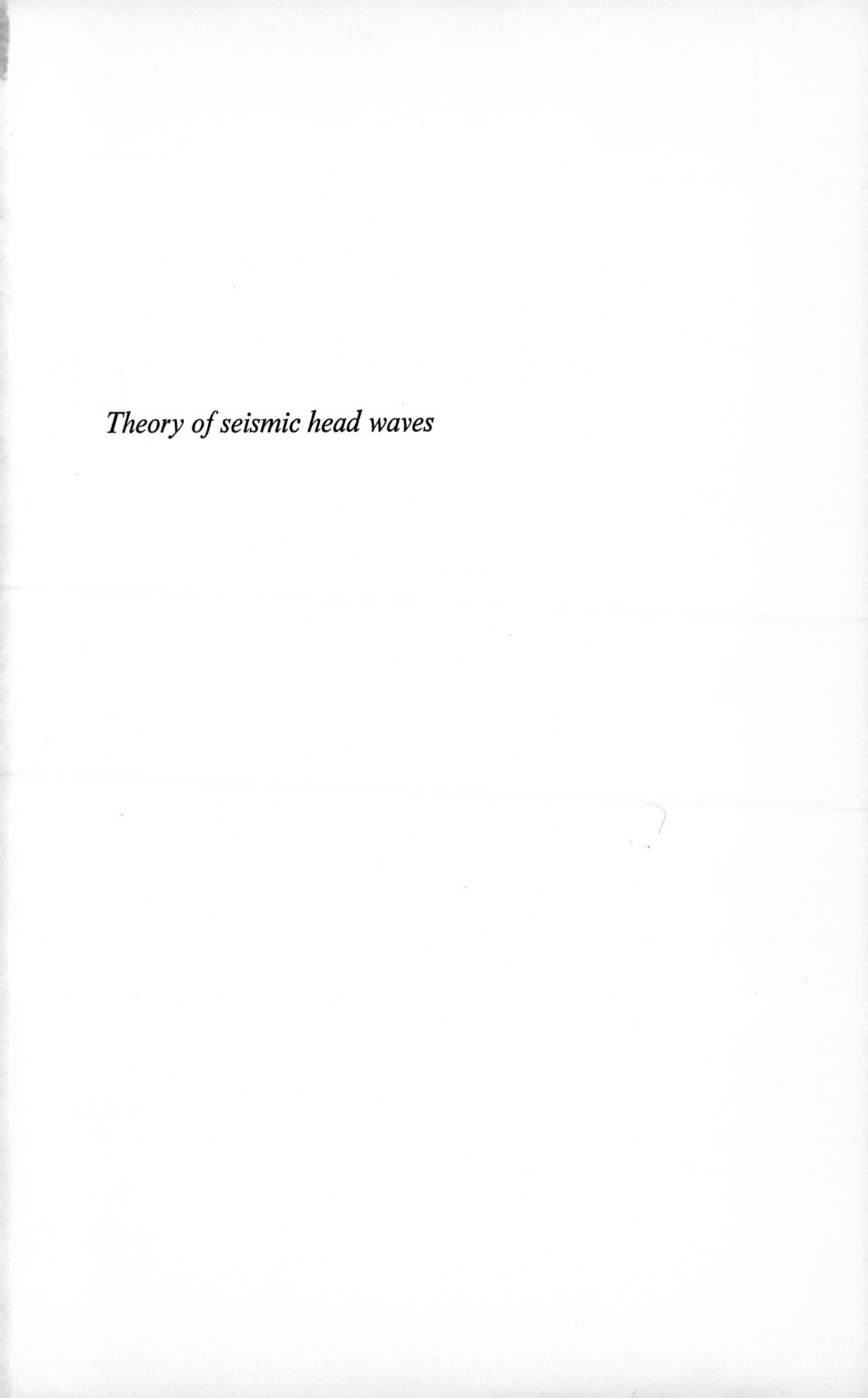

Theory of seismic head waves

Vlastislav Červený

and

Ravi Ravindra

Theory of seismic head waves

University of Toronto Press

Printed in Great Britain for

University of Toronto Press, Toronto and Buffalo

ISBN 0-8020-1678-2

Microfiche ISBN 0-8020-0049-5

LC 76-151364

AMS 1970 Subject classifications 86.00, 69.00

To the memory of canoe trips in
Kejimkujik and Little Harbour,
Nova Scotia

CONTENTS

PREFACE xi

LIST OF IMPORTANT SYMBOLS xv

1

INTRODUCTION 3

1.1 Brief historical remarks 3
1.2 A simple picture of head waves 8

2

THE RAY METHOD 11

2.1 Ray series 11
2.2 The phase function and coefficients of the ray series 17
2.2.1 Determination of the phase function 18
2.2.2 Determination of coefficients of the ray series 26
2.2.3 Recapitulation of the general procedure 38
2.3 Behaviour of waves at an interface 39
2.3.1 Discussion of interface conditions 39
2.3.2 Waves moving away from an interface 53
2.4 Formulae for the zeroth approximation 57
2.4.1 Continuation of the displacement vector along the ray 58
2.4.2 Influence of an interface 59
2.4.3 Displacement vector of reflected and refracted waves at an arbitrary point of the ray 70
2.4.4 Arbitrary number of interfaces 71
2.5 Point source 74
2.6 Final expression for R-waves for a point source for some types of media 76
2.6.1 Velocity a function of depth 77
2.6.2 Homogeneous layers 85
2.6.3 Constant velocity gradient inside layers 86
2.6.4 Curved and dipping interfaces 88
2.6.5 Generalization for an impulse source 93

3

HEAD WAVES FROM A SINGLE INTERFACE 97

3.1 Description of source and media 97
3.2 Mathematical procedure 98
3.2.1 Source of P or SV waves 98
3.2.2 Source of SH waves 113
3.3 Summary of results 116
3.3.1 P and SV head waves 116
3.3.2 SH waves 122
3.3.3 Accuracy of the formulae for head waves 122
3.4 Classification of head waves 123
3.5 Head wave coefficients 127
3.5.1 Formulae for head wave coefficients 127
3.5.2 Relationship between head wave coefficients and coefficients of reflection and refraction 131
3.5.3 Formulae connecting different head wave coefficients 132
3.5.4 Numerical values of head wave coefficients 133
3.6 Amplitudes of head waves 146
3.6.1 Comparison of amplitudes of different types of H and R waves 150
3.7 Impulse source and construction of theoretical seismograms 163
3.8 Interference waves 166
3.9 General expressions for head waves 179

4

HEAD WAVES IN MULTILAYERED MEDIA 188

4.1 Harmonic source 189
4.2 Pulse source 199
4.3 Particular types of head waves 200
4.3.1 Symmetric head waves 200
4.3.2 Partially symmetric head waves 201
4.3.3 Fundamental head waves 203
4.4 Homogeneous layers 204
4.5 Constant velocity gradient inside layers 205
4.6 Arbitrary velocity–depth functions 206
4.7 Kinematic and dynamic analogues and multifold head waves 208
4.8 Amplitudes of fundamental head waves and comparison with amplitudes of R-waves 212
4.8.1 Overburden composed of homogeneous layers 213
4.8.2 Overburden with inhomogeneous layers 217

4.9 Construction of theoretical seismograms 223

5
CURVED AND DIPPING INTERFACES IN THE OVERBURDEN 225

5.1 General curved interfaces 225
5.2 Plane dipping interfaces 234

6
INTERFERENCE HEAD WAVES 235

6.1 The C-wave: Zero-order expressions 240
6.2 The interference head wave C^+ 245
6.3 Numerical examples 247

7
HEAD WAVES BY WAVE METHODS 251

7.1 Introduction 251
7.2 The one-interface problem 252
7.2.1 Formal solution 253
7.2.2 The contour of integration 257
7.2.3 The reflected wave 265
7.2.4 The head wave 271
7.2.5 The interference reflected-head wave 276
7.2.6 Exact numerical integrations 280
7.2.7 The reflected wave in the critical region: Numerical examples 280
7.3 Layered media 287
7.3.1 Exact ray expansion 289
7.3.2 Normal mode expansion 291
7.3.3 Direct evaluation 292

APPENDIX: Some useful vector identities 295

REFERENCES 297

INDEX 307

PREFACE

So-called refraction arrivals have been used extensively in near earthquake studies, deep crustal explosion seismology, and geophysical prospecting for the past fifty years or so. However, the attention of workers has been largely confined to studies of the kinematic characteristics of these arrivals, which have been called by various names including "head waves." Only recently have the wave dynamic characteristics (such as the amplitude, frequency spectrum, and wave-form) of these signals, which are usually the first to appear on a seismogram, attracted more attention. Many new analytical and computing techniques have been employed in these studies aimed at obtaining information about the dynamic characteristics and preparing theoretical seismograms for realistic media.

It is our intention to present here a consistent treatment of the theory of head waves starting from the relatively simple problem of one interface and moving on to a much more realistic situation in which there are a number of layers, some of which are not necessarily homogeneous, horizontal, or bounded by plane interfaces. Some complexities found in nature, such as absorption effects and anisotropy of the medium, are not taken into account.

We shall confine our attention mainly, though by no means exclusively, to the ray method, which is conceptually clearer and simpler than others. Until the mid-fifties, the problem of head waves was always attacked by wave methods, and formal solutions had then to be made tractable by approximate methods. In the last decade, however, many workers – particularly in the USSR – have successfully used the ray theory (asymptotic series solution) based on the work of Luneburg, Keller, Babich, Alekseyev, and others to study several problems in the theory of head waves. This method has its own limitations, of course, and these will be brought out here by comparing the results with those obtained by other techniques. In particular, the ray method is not generally applicable when interference effects become important, as in thin layers of thickness less than or comparable with the wavelength. In this work, we shall concentrate mainly on non-interfering head waves propagating in different types of thick layers. The ray method is quite suitable for such a study and will be used extensively. To provide an appreciation of the inaccuracy involved in using this method, we shall compare some of the results with those obtained by more exact methods in the last chapter, where wave methods are discussed in some detail. A major portion of the book is devoted to a study of the one-interface problem, since the solution of this problem is the most important prerequisite to the solution of all other problems in the theory of head waves, particularly when using the ray theory where one more or less necessary assumption is that layering, when present, is of dimensions greater than several wavelengths to ensure the validity of treating any boundary as an isolated interface between two half-spaces.

All the relevant theoretical considerations regarding the ray method are outlined in chapter 2, preliminary to its application in subsequent chapters. For various reasons, we have thought it advisable to give the details of this method in the most general form possible within the scope of this book in the second and third sections of chapter 2. We hope the reader will not be daunted by the formidable appearance of the formulae in these sections. He may find it advisable in the beginning to skip to section 2.4 after reading section 2.1 and quickly glancing over 2.2 and 2.3 and refer back as the need arises. Some of the details in sections 2.2 and 2.3 have been included for the sake of completeness and are not absolutely essential to the understanding of the rest of the book.

We have tried to be as consistent as possible in the use of symbols without sacrificing clarity. The most important ones have been listed at the front of the book. The reader should be very careful, however, when comparing our results with those published earlier. The use of symbols differs considerably from author to author. Also, unfortunately, misprints or errors seem to have crept into even some well-known works.

Although we have referred only to head waves in the title of the book, and they are our main concern here, all related waves such as reflected and refracted waves are also discussed in detail. We hope that readers interested in any aspect of seismic wave theory will find something of value in this book. The methods used are quite general and can be applied to many problems in the theory of elastic wave propagation. Similar techniques have been used extensively in the last two decades in the study of electromagnetic waves. We lay no claim to originality for a great deal of the material; indeed, we have not hesitated to borrow complete sentences if they appeared to us to convey the ideas clearly and succinctly.

We have assumed that the reader is acquainted with seismology and elastic wave theory, and is proficient in the use of vector algebra. For convenience, the most useful vector identities have been brought together in the appendix.

The first-named author was on leave of absence from Charles University,

Prague, as a Senior Killam Research Fellow during the preparation of this book. The other author is grateful for the support of the National Research Council of Canada. Drs. C. Chapman, C. Keen, and F. Hron made several useful suggestions. Staff of the Geophysical Institute of Charles University prepared many programmes and graphs, and Dalhousie University Computer Center was generous with time on their IBM360-50 computer. The manuscript was improved enormously by the editing of the staff of the University of Toronto Press.

V.Č.
R.R.

Dalhousie University,
Halifax, Nova Scotia, Canada
January 1969

LIST OF IMPORTANT SYMBOLS

$E_{\parallel}$	Plane of incidence for a single interface, or plane of the ray in multilayered media
$E_{\perp}$	Plane perpendicular to $E_{\parallel}$
f	Frequency
H-wave	Head wave
$\mathbf{i}, \mathbf{j}, \mathbf{k}$	Unit vectors along x, y, and z axes, respectively
k	Wave number
m, n, p	Indices characterizing a head wave from a single interface
$\mathbf{n}_P, \mathbf{n}_{SV}, \mathbf{n}_{SH}$	Unit vectors tangent to a ray, perpendicular to a ray but in the plane of $E_{\parallel}$, and perpendicular to the plane $E_{\parallel}$, respectively
r^*	Critical distance
r_1, r_2	Principal radii of curvature of a wave front
$r_{\parallel}, r_{\perp}$	Radii of curvature of normal sections of a wave front, associated with $E_{\parallel}$ and $E_{\perp}$
R_1, R_2	Principal radii of curvature of an interface
$R_{\parallel}, R_{\perp}$	Radii of curvature of normal sections of an interface, associated with $E_{\parallel}$ and $E_{\perp}$
R-waves	Reflected (refracted) waves; also, multiply reflected and refracted waves
R_{mn}	Coefficients of reflection and refraction
s	Arc length along a ray
$S(\omega)$	Spectrum of an impulse
t	Time variable
$\mathbf{t}_1, \mathbf{t}_2, \mathbf{t}_3$	Unit vectors tangent, principal normal, and binormal to the ray, respectively
T	Radius of torsion of a ray; or period
$\bar{v}$	Velocity of propagation of head waves along the interface on which the head wave is generated
$\mathbf{W}$	Displacement vector for harmonic waves
$\mathbf{W}_k$	kth amplitude coefficients of the ray series
$\mathscr{W}$	Displacement vector for transient waves
x, y, z	Ordinary Cartesian coordinates
α	Velocity of compressional waves; $\alpha = [(\lambda+2\mu)/\rho]^{1/2}$

β	Velocity of shear waves; $\beta = (\mu/\rho)^{1/2}$
γ	Ratio of shear and compressional velocities
Γ_{mnp}	Coefficient of head waves
ϑ, θ	Angles between the ray and the normal to an interface
θ^*	Critical angle
λ, μ	Lame's parameters
Λ	Wavelength
ρ	Density
$d\sigma$	Cross-sectional area of an elementary ray tube
τ	Phase factor; or arrival time
ω	Angular frequency

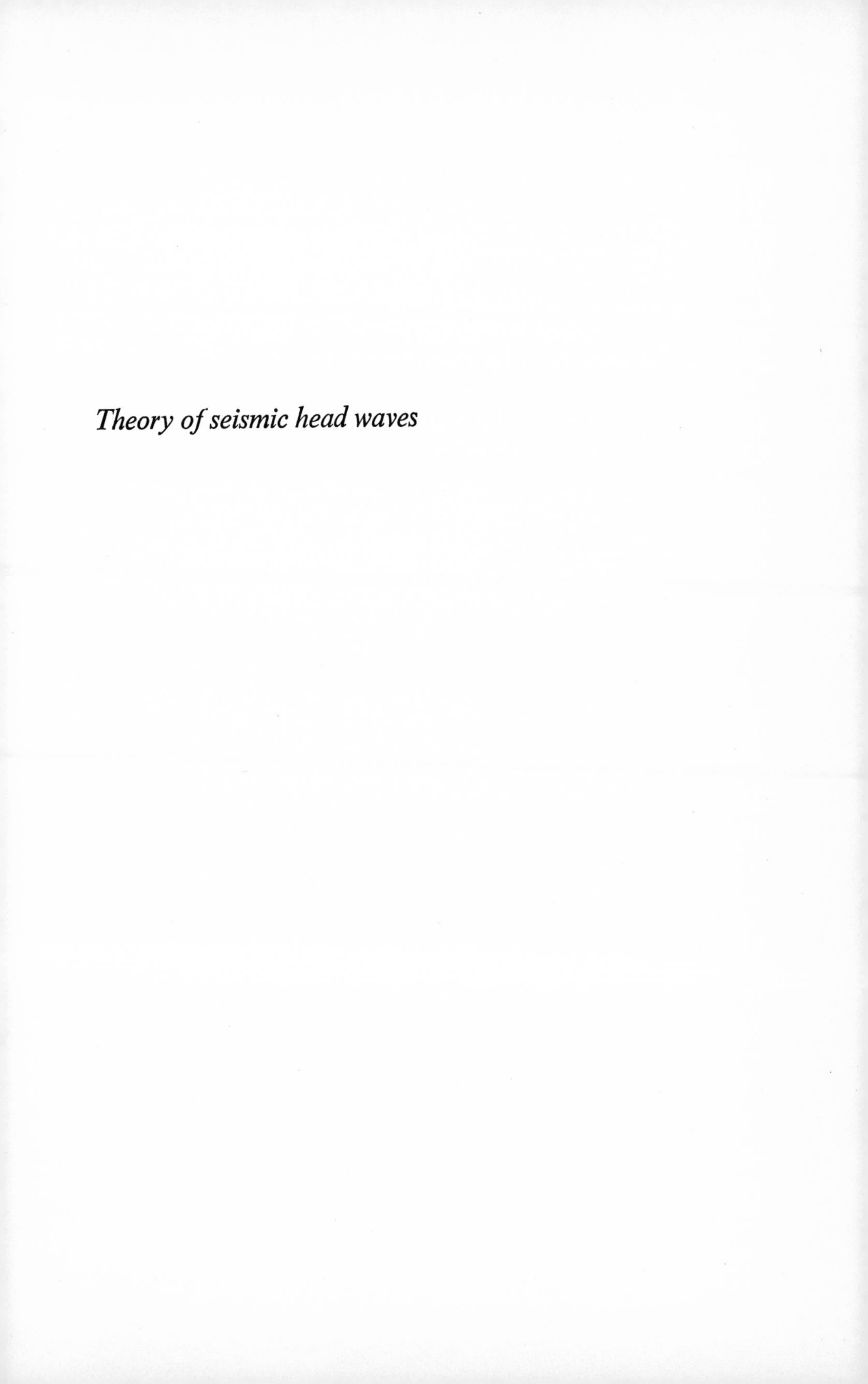

Theory of seismic head waves

1
INTRODUCTION

1.1
BRIEF HISTORICAL REMARKS

Head waves have been used very extensively in near earthquake studies, geophysical prospecting, and explosion seismology for the last fifty years or so (see, for example, McConnell *et al.*, 1966; Musgrave, 1967). It is usual in seismic prospecting to call these waves *refracted* or *refraction* waves (also *refraction arrivals*) and the method based on them the *refraction method*. Mohorovičić (1910) seems to have been the first to have used head waves in an earthquake record to deduce continental crustal layering. He observed that for several near-distance earthquakes, the first arrivals of seismic energy constituted a straight line on a time–distance graph. From these purely empirical observations, he surmised correctly that in the first arrivals seismic energy must travel from the source M_0 to the receiver M along a path M_0O^*AM (as shown in figure 1.1) such that the velocity of propagation along the segments M_0O^* and AM is given by a constant, α_1, and along O^*A by another constant, α_2.

It was assumed that O^*A was the critically refracted ray path along which seismic energy was carried in apparent conformity with the laws of geometrical optics. These laws, however, predict that the segment O^*A should carry no energy. Also, on the basis of geometrical optics, it is impossible to explain how energy will return to the first medium once it has entered the second. It appears, then, that although it is difficult to see from elementary considerations of the principles of geometrical optics how the refraction paths can be of significance for the transfer of energy, the empirically established linear time–distance curves still demand the postulation of these paths. To resolve this difficulty, it seemed evident that a close analysis of the details of the whole process from the wave point of view was required. Consequently, almost all theoretical papers about head waves until 1955 were based on wave (integral) methods. Most of these papers considered only the simplest seismological situation of a point source and a plane interface between two homogeneous media. Details of the methods used cannot be given here, but they are well known from a number of books and papers (see, for example, Honda and Nakamura, 1953, 1954; Ewing *et al.*, 1957; Brekhovskikh, 1960; Cagniard, 1962; Zvolinskiy, 1965; Grant and West, 1965; Petrashen, 1965; Pod"yapol'skiy, 1966a; Müller, 1967, 1968). We shall mention only the most important of these methods.

Jeffreys (1926) was the first to obtain a term which could be interpreted as a wave with appreciable energy, which "appears to have travelled along the interface with the velocity of sound in the lower medium." He considered a point source in a fluid layer. Later, Jeffreys (1931) considered a similar problem for SH-waves. Muskat (1933) obtained approximate expressions for head waves for the case of two semi-infinite *elastic* layers by the method of Lamb (1904) and Sommerfeld (1909). The Sommerfeld integral method was used also by Wolf (1936), Joos and Teltow (1939), Ott (1942), Heelan (1953), and others. These authors devoted their attention mainly to approximate formulae for head waves (valid for high frequencies and at epicentral distances sufficiently far removed from the critical distance). Head waves in the critical region were first studied by Brekhovskikh (1948, 1949, 1960) for *liquid* media, and by Červený (1957a, 1966a) and Smirnova and Yermilova (1959) for *solid* media. Other references to work on head waves in the critical region will be given in chapter 7.

Cagniard was the first to give an exact solution in a computable form. The approach developed by him during the 1930s was based on the Laplace transform rather than on the Fourier transform which was essential to the Sommerfeld method. Cagniard's approach has been further developed and applied to problems more complicated than those he considered. The interested reader should consult the English translation by Flinn and Dix (Cagniard, 1962) of Cagniard's classical work. There he will also find an

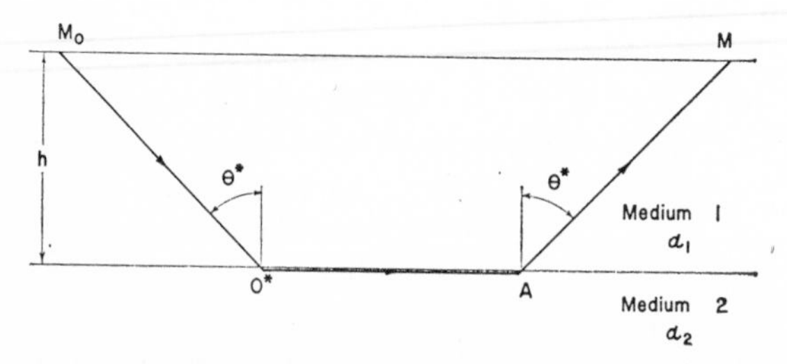

FIGURE 1.1 Schematic ray representation of *refraction arrivals* (head waves) from the source M_0 to the receiver M ($\alpha_2 > \alpha_1$, $\sin\theta^* = \alpha_1/\alpha_2$).

exhaustive bibliography of papers dealing with exact solutions of transient elastic wave propagation problems up to 1961.

An independent method, developed by Smirnov and Sobolev (Sobolev, 1932, 1933; Smirnov and Sobolev, 1933) and briefly described in Petrashen (1965), was used for the study of head waves by Zaitsev and Zvolinskiy (1951a, 1951b), Zvolinskiy (1957, 1958), and others. Zvolinskiy (1958) also found formulae for the transient reflected-head waves in the critical region.

The investigation of head waves from a plane interface between two homogeneous media is of basic importance in the theory of head waves. However, solutions of this problem cannot be used directly in seismology because real structures are considerably more complex. A natural extension of this simple problem is to study non-interfering head waves in a medium consisting of homogeneous thick layers, with all interfaces plane and parallel to each other. A large number of papers dealing with the properties of non-interfering seismic body waves propagating in thick-layered media (including head waves) are based on Petrashen's method, described in Petrashen (1965), where other references are also given. In the early 1950s, Petrashen and his co-workers derived approximate formulae for arbitrary head waves propagating in thick-layered media in a form suitable for numerical calculation and later also published extensive tables of head wave coefficients and of some other relevant functions (Petrashen, 1957b). These results and tables have been used widely in seismic prospecting and crustal reflection experiments in the USSR.

Similar problems were investigated by many other seismologists and mathematicians. The wave field generated by a point source in an *n*-layered medium can be expanded into a series of integrals, which correspond to various types of multiple reflections. These integrals can be calculated exactly (by exact ray theory) or approximately. Individual integrals contain contributions corresponding to head waves of various types. Exact ray theory can, in principle, be used for thin layers also, but it becomes very cumbersome in this case. It has been used mainly to study waves propagating in thick-layered media (see more details in section 7.3). Extensive work using this theory has been done by Pekeris and his co-workers (Pekeris, 1955a, 1955b, 1960; Pekeris and Lifson, 1957; Pekeris and Longman, 1958; Pekeris *et al.*, 1963, 1965); see also Spencer (1965a), Müller (1967), and Bortfeld (1967). For numerical evaluation of individual integrals (corresponding to multiple reflections), some modifications of Cagniard's method can also be used. Pod"yapol'skiy (1959a) used the method of Smirnov and Sobolev to obtain expressions for multiple reflections (including head waves); he has also developed a procedure for exact calculation of the corresponding integrals (Pod"yapol'skiy, personal communication). Červený (1968) and Smirnova (1968) have found expressions for head waves in a thick-layered medium, valid in the critical region.

The situation becomes considerably more complicated if some layers in the medium are *thin*. Head waves from a thin high-velocity layer are of considerable importance in seismic prospecting. These waves, which have a distinct interference character, cannot be investigated by ray theory, but require methods for studying a wave field *as a whole*. Some situations have been considered recently; the reader is referred to Rosenbaum (1964, 1965), and to the papers of Petrashen, Krauklis, and Molotkov (relevant references are given in Petrashen, 1965, and Krauklis, 1968). Petrashen *et al.* have also investigated so-called *screened head waves*, i.e. head waves propagating along some deeper interface screened by a thin high-velocity layer in the overburden (the velocity in the thin layer is higher than the velocity with which the head wave propagates along the deeper interface). In spite of the fact that the ray of the screened head waves cannot be constructed, they may be quite important in seismic prospecting.

Generally, any medium in which velocity depends only on depth can be simulated by a combination of thin, homogeneous parallel layers with plane boundaries. The formal solution for a wave field generated by a point source in an n-layered medium is known (see section 7.3); however, numerical evaluation of the corresponding integrals is very complicated. The situation becomes simpler if we are interested in calculating only the first arriving body phases (which often correspond to compressional head waves). This procedure has been used recently to study head waves from transition layers (Fuchs, 1968a; Bessonova and Michota, 1968; C. Keen, personal communication). C. Keen has used the method of λ-spectra, developed by Phinney (1965), which seems to be very suitable for considering compressional head waves registering as first arrivals. Head waves from transition layers have also been studied by Nakamura (1964), Tsepelev (1961, 1968), and Hirasawa and Berry (1969). These head waves have, of course, a clear interference character.

All the papers mentioned above (except the last three for transition layers) have dealt only with homogeneous, plane-parallel layers, for which the formal solution is known. In reality the structure is often more complex – the medium can be inhomogeneous, and the interfaces curved. We shall give some references to these more complicated situations later.

In the late 1950s it was shown that practically all the approximate results for non-interfering head waves, yielded earlier by wave methods, can be derived by the simple generalization of geometrical optics by means of the so-called ray method (see chapter 2). In the frequency domain, the ray method looks for a solution in the form of an asymptotic series in inverse powers of frequency. This asymptotic series is known as a ray series. Similar series can also be written directly for impulse waves.

Friedrichs and Keller (1955) obtained formulae for acoustic head waves, generated on a plane interface between two fluids, by the ray method. This method was first applied to the problems of *elastic* wave propagation by Karal and Keller (1959), Babich and Alekseyev (1958), and Chekin (1959), probably independently. The first investigators to obtain expressions for head waves from an interface between two solid media using the ray method were Alekseyev and Gel'chinskiy (1958, 1961); see also Yanovskaya (n.d.). A computation matrix formula for head wave coefficients, based on the method of Alekseyev and Gel'chinskiy, was given by Yanovskaya (1966).

The ray method is not restricted to situations for which the formal solution is known, and, therefore, it allows us to find expressions for (non-interference) head waves in many important cases not yet tractable by integral wave methods (such as problems involving curved and dipping interfaces in the overburden, an inhomogeneous overburden, and incident waves of arbitrarily shaped wave front). For situations where both the approximate solution found by integral methods (using the method of steepest descent) and the ray solution are known, they turn out to be the same (see chapter 7). However, the ray method is only an approximate method and has a number of serious limitations. It can be used only for non-interfering head waves and when the frequency is high. If a head wave interferes with other waves or is formed by a superposition of many waves, it cannot generally be used. These limitations apply, for example, to the critical region (where the head wave interferes with the reflected wave) and to thin-layered media. In these cases, a close analysis of the details of the process of propagation of head waves must be made from the wave point of view, using more complicated methods for numerical evaluation of the integrals. In some situations it is also possible to use modifications of the ray theory such as the method of parabolic equations (known also as the boundary layer method); see Yanovskaya (n.d.), where other references are given.

Finally, we should like to mention a few important papers on head waves propagating along a curved interface or along a boundary of an inhomogeneous medium (see figure 3.41). Head waves are very sensitive to curvature of the interface and to inhomogeneity of the substratum if the velocity changes in the direction perpendicular to the interface. For example, a very small positive gradient of velocity below the interface, which has practically no influence on the time-distance curve of head waves, can increase their amplitudes by as much as a few orders of magnitude (see chapter 6). The head waves generated in the above situations (called here interference head waves and damped head waves) were investigated by Chekin (1964, 1965, 1966), Buldyrev and Lanin (1965, 1966), and Lanin (1966, 1968).

Head waves have also been studied experimentally in laboratory models and in real media. We shall not discuss the results of these investigations in this book; the reader is referred to Musgrave (1967), Yepinat'yeva (1960), O'Brien (1957a, 1957b, 1963, 1967), Berzon *et al.* (1962, 1966), and Donato (1963a, 1963b, 1964) where he will find many other references also.

Head waves have been given different names by various investigators; some of these are *Mintrop* waves (Salm, 1934), *conical* waves (Cagniard, 1962), *interface generated waves* (Jardetzky, 1952), *lateral waves* (Brekhovskikh, 1960). These names have been determined by the chief feature ascribed to the wave in question by the investigator. Throughout this work, we shall adhere to the name *head waves*, a name which is now used in most theoretical papers. The German and the Russian equivalents are *Kopfwellen* (Schmidt, 1939) and *Golovnyye volny* (Petrashen, 1959), respectively.

1.2
A SIMPLE PICTURE OF HEAD WAVES

A simple picture of head waves can be given by using the Huyghens principle, which was first applied to these problems by Merten in 1927 (according to Thornburgh, 1930). Ansel (1930), Dix (1939), and Jardetzky (1952) also used wave-front diagrams to show the physical conditions for the generation of different types of waves in layered media. The generation of head waves from the point of view of the Huyghens principle is also described in many books on seismology and the propagation of elastic waves; therefore, we shall describe this only briefly.

The characteristic physical conditions under which head waves are generated may be obtained in the simplest case of two semi-infinite homogeneous liquid media. We assume that a point source of spherical compressional waves lies at a point M_0 at a distance h from a plane interface $z = 0$ and that the velocity of propagation in the second medium (α_2) is higher than that in the first medium (α_1); see figure 1.2. The cylindrical coordinates of an arbitrary point are r, z, φ, where r is the epicentral distance. We shall consider only the two-dimensional section r, z with $\varphi = \text{const}$. Assume that the source starts to emit waves at the time $t = 0$. For $t < h/\alpha_1$, i.e. before the incident wave impinges on the interface, there is only the incident wave. The wave front of this wave is a sphere with centre at M_0 and radius $R_0 = [r^2+(z-h)^2]^{1/2}$ proportional to t (i.e. $t = R_0/\alpha_1$). For $t = h/\alpha_1$, the wave front of the incident wave reaches the interface at the point O and is tangent to it. As t increases further, reflected and refracted waves appear, as each point of the interface struck by the incident wave becomes a source of disturbance according to the Huyghens principle. The wave fronts of these waves for $h/\alpha_1 < t < h/(\alpha_1 \cos \theta^*)$ are given in figure 1.2b. The wave fronts of the incident, reflected, and refracted waves are connected at the point P on the interface, which moves

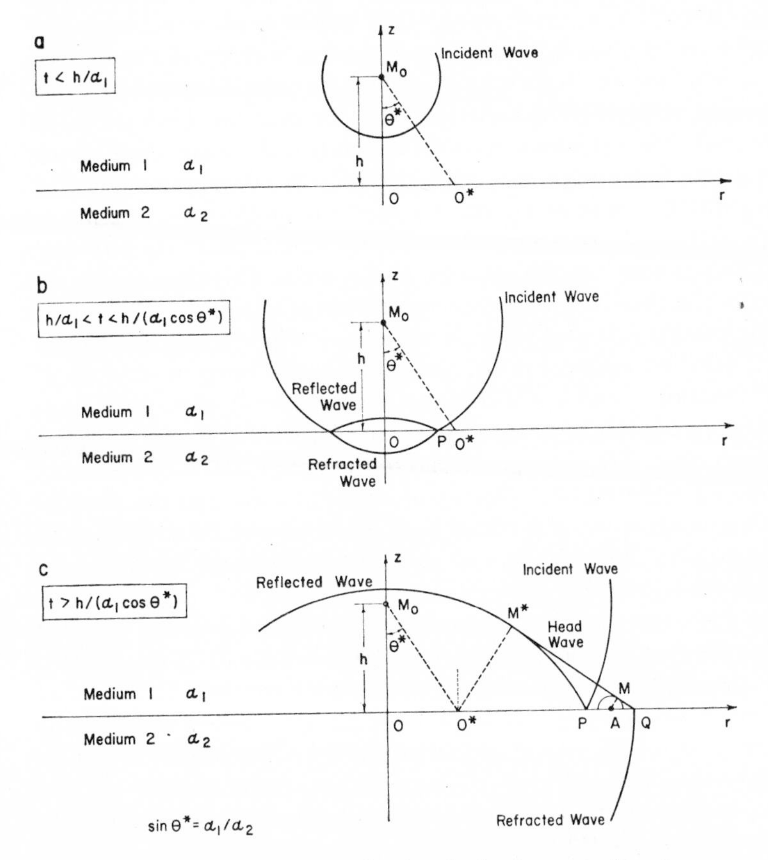

FIGURE 1.2 Wave front diagrams showing the development of head waves in fluid media when $\alpha_2 > \alpha_1$. Diagram a shows the incident wave before it impinges on the interface; b shows the incident, reflected, and refracted waves before critical incidence; c shows these waves and the head wave after critical incidence.

along the interface (with increasing t) from O to larger epicentral distances. We can easily calculate its velocity α_P:

$$\alpha_P = [1/(dt/dr)]_{z=0} = [\alpha_1/\{d[r^2+(z-h)^2]^{1/2}/dr\}]_{z=0} = \alpha_1/\sin\theta(P),$$

where $\theta(P)$ is the angle of incidence, i.e. the angle between the ray of the incident wave at P (PM_0) and the z-axis, $\cos\theta(P) = h/[r^2(P)+h^2]^{1/2}$. As $\sin\theta(P)$ increases with the distance r, α_P is a decreasing function of r. While $\alpha_P > \alpha_2$, the situation remains the same, all three fronts being connected at P. However, for $t = h/(\alpha_1 \cos\theta^*)$, where $\theta^* = \sin^{-1}(\alpha_1/\alpha_2)$ is the critical angle, the point P reaches the point O^*, and there $\alpha_P = \alpha_1/\sin\theta(O^*) = \alpha_1/\sin\theta^* = \alpha_2$. For larger t, $\alpha_P < \alpha_2$, and the refracted wave propagating from the point O^* as a centre of disturbance in the second medium (with velocity α_2) will be more advanced than the incident and reflected waves. At the point P, only the wave fronts of incident and reflected waves are then connected, the wave front of the refracted wave reaching point Q, $r(Q) > r(P)$, which propagates along the interface with the velocity α_2. The ray of the refracted wave arriving at point Q is M_0O^*Q (see figure 1.2c), and the wave front of the refracted wave is perpendicular to the interface at this point. Therefore, points on the interface, such as A, will become new centres of disturbances propagating back into the first medium with velocity α_1. These disturbances form the *head wave*, the envelope of the elementary circles being its wave front. As the velocities α_1 and α_2 are constant, the wave front of the head wave is a straight line (in three dimensions it is the frustrum of a cone). The ray of the head wave, AM, is perpendicular to the wave front. It can be easily seen that the angle between the ray of the head wave and the z-axis is again θ^*. The whole ray of the head wave is given by M_0O^*AM; the head wave propagates along M_0O^* and AM with the velocity α_1 and along O^*A with the velocity α_2.

As we have seen, head waves are generated at the interface only for $t > h/(\alpha_1 \cos\theta^*)$ at distances larger than $r(O^*) = h\tan\theta^*$. If we take into account the angle of the ray AM, θ^*, we can easily conclude that the head wave reaches any point $M(r, H)$ only if $r(M) > r^* = (h+H)\tan\theta^*$ at the time $t(M) = (h+H)/(\alpha_1\cos\theta^*)+[r(A)-r(O^*)]/\alpha_2$. The distance at which head waves begin to exist, $r = r^*$, is called the *critical distance*. At this distance, the wave fronts of the reflected and head waves are tangential to each other (see point M^* in figure 1.2c).

For two solid media, the discussion of head waves becomes more complicated as many more head waves can exist. However, in principle, the mechanics of generation of head waves is the same as for the simpler case of fluid media. More details will be given in chapter 3 (see figure 3.7).

2
THE RAY METHOD

2.1
RAY SERIES

It has been recognized for some time that the results of geometrical optics should follow from the wave equation. However, there have been mathematical difficulties in linking the two. The reader is referred to the excellent work by Kline and Kay (1965, pp. 3–21, 255–83) for an account of the historical development of the asymptotic series solution of the wave equation which yields the geometrical optics solution for high frequencies. The essential features of the method were developed by Luneburg during 1944–8 (see Luneburg, 1964) for the purpose of studying the propagation of discontinuities in electromagnetic fields. Further extensions were made especially by Friedlander (1946), Kline (1951, 1954), Keller *et al.* (1956), Keller (1958, 1963), and others to account for diffraction, which is not described by geometrical optics.

Karal and Keller (1959) in the USA, and Babich and Alekseyev (1958) and Chekin (1959) in the USSR, were the first, probably independently, to apply the series expansion to the problem of elastic waves (see also Alekseyev *et al.*, 1961; Yanovskaya, n.d.). This method is not restricted to equations with separable variables and can thus be applied to general inhomogeneous elastic media, whereas the vector wave equation of elasticity can be separated in only a few cases (see Hook, 1961, 1965; Gupta, 1966b). The series solution allows us to include as many terms as are suitable for the problem at hand; for example, for some seismic body waves, the first (leading) term in the series is often sufficient, as will be seen later.

We shall consider the application of the series solution to the propagation of elastic waves in an inhomogeneous, perfectly elastic and isotropic medium. We use ρ to denote the density of the medium and λ and μ its Lame's parameters. The linearized equation of motion is (see Karal and Keller, 1959; and after some simple modifications also the formulae given in Ewing *et al.*, 1957, p. 329; Grant and West, 1965, p. 44):

$$\rho \frac{\partial^2 \mathbf{W}}{\partial t^2} = (\lambda+\mu)\nabla(\nabla\cdot\mathbf{W})+\mu\nabla^2\mathbf{W}+\nabla\lambda(\nabla\cdot\mathbf{W}) + \nabla\mu\times(\nabla\times\mathbf{W})+2(\nabla\mu\cdot\nabla)\mathbf{W}, \tag{2.1}$$

where $\mathbf{W}$ is the particle *displacement* vector and t is the time.

We assume that a time-harmonic solution of (2.1) can be expressed in inverse powers of the frequency ω, i.e., that

$$\mathbf{W} = \exp[i\omega(t-\tau)]\sum_{k=0}^{\infty}(i\omega)^{-k}\mathbf{W}_k, \tag{2.2}$$

where τ and $\mathbf{W}_k$ are independent of ω and t. We shall call the expansion (2.2) a *ray series*, τ a *phase function*, and $\mathbf{W}_k$ ($k = 0, 1, 2, \ldots$) the *amplitude coefficients* of the ray series or simply the *coefficients* of the ray series. The moving surfaces of constant phase, $t = \tau(x, y, z)$, will be called *wave fronts* and the orthogonal trajectories of these surfaces *rays*. It follows that each ray has the direction of $\nabla\tau$ at its intersection with the wave front $t = \tau(x, y, z)$. We shall, of course, assume that $\nabla\tau \neq 0$, for otherwise (2.2) would not represent a travelling wave. τ and $\mathbf{W}_k$ ($k = 0, 1, 2, \ldots$) are unknown functions of the coordinates x, y, z. They can be determined by inserting (2.2) into (2.1), when some initial conditions are known (see section 2.2).

It should be noted that our approach is not completely rigorous mathematically; we shall make no attempt to investigate the question of the existence and validity of the asymptotic series (2.2). We shall assume that, for the problems to be studied in this book, the series (2.2) exists and is asymptotic to the exact solution of the equation of motion (2.1). The reader interested in the detailed mathematical treatment of the problems connected with the existence and validity of the asymptotic expansion (2.2) is referred to Kline and Kay (1965) and to other references given there. According to Yanovskaya (n.d.), the condition for the validity of the asymptotic expansion (2.2) is that the function $\tau(x, y, z)$ must be analytic. We cannot construct the ray series when τ is not analytic, because in the process of determining the coefficients of the ray series we must differentiate them along wave fronts and rays, and this is impossible if $\tau(x, y, z)$ is a non-analytic function. Thus, in the neighbourhood of the points where the phase function τ is non-analytic, the ray expansion (2.2) is not valid.

However, the condition that the function τ must be analytic does not tell very much to a practising seismologist. What does this condition mean in terms of seismic measurements on the surface of the earth? With a slight oversimplification, we can say that the ray expansion is not valid in the vicinity of those points where the time–distance curve of the wave has end points, cusps, tangent points with the travel time curve of another wave, or, generally, discontinuous derivatives. This situation exists, for example, in the vicinity of boundary rays (which separate the shadow zone from the illuminated one) or caustic and critical rays (where the time–distance curves of reflected and head waves are tangential to each other). The dimensions of the *neighbourhood* of these singular points where the ray method fails or gives very inaccurate results depend on the frequency. For high frequencies the method cannot be used only in a narrow region about these singularities, but for low frequencies the regions of inapplicability of the formulae are considerably larger. To investigate the wave field in these regions, other methods have to be used. When we know the formal solution of the problem at hand, we can use wave methods, as we do in chapter 7 to study the wave field in the critical region, which is of considerable importance in the theory of head waves. However, the formal solution has not yet been found for many situations which are of basic importance in seismology. In these cases, we can use modifications of the ray theory or other approximate methods, such as the parabolic equation method (boundary layer method). We shall not discuss these other methods here, as we shall not use them in our investigation of head waves; a short outline with many references is given in Yanovskaya (n.d.).

Inasmuch as the solution (2.2) of the equation of motion (2.1) is expressed as an asymptotic expansion in *inverse* powers of frequency, it yields best results for high frequencies. We shall assume that the velocity gradients are always considerably smaller than the frequency, and that the radii of curvature of the interfaces, if they exist, are substantially larger than the wavelength. Unfortunately, it is not possible to derive a precise formula connecting the frequency and degree of accuracy of the results for a given number of terms in the ray series. The accuracy of ray formulae depends not only on the parameters of the media, but also on the direction of the ray and on other conditions. The simplest method is to compare the ray results with exact calculations for situations in which the exact solution is known, and then to generalize the results of the comparison for more complicated cases.

Except in the close neighbourhood of singular points, the error which arises in keeping only the first few terms in the series tends to zero as the frequency becomes higher. In many cases it is sufficient to consider only the first (leading) term in (2.2). Then

$$\mathbf{W} = \exp[i\omega(t-\tau)]\mathbf{W}_0, \tag{2.3}$$

where τ and $\mathbf{W}_0$ are independent of ω and t. We shall call this solution the *zero-order solution* (or *zeroth approximation*). The zero-order solution is just the solution which can be obtained according to the principles of geometrical optics. The subsequent terms are corrections to this solution.

In studies of seismic body waves, formula (2.3) is often quite sufficient for finding their most important kinematic and dynamic properties. For example, for reflected and refracted waves (hereafter called R-waves) propagating in multilayered media, a great deal of information can be obtained by the use of this equation. As will be shown in chapter 3, R-waves propagating along an interface can generate head waves. However, to determine the properties of these head waves it is not enough to know the zeroth approximation of R-waves; we need to know the first two terms of (2.2). We shall call the solution which retains the first two terms the *first-order solution* or the *first approximation*, i.e.

$$\mathbf{W} = \exp[i\omega(t-\tau)][\mathbf{W}_0+(i\omega)^{-1}\mathbf{W}_1], \tag{2.4}$$

where τ, $\mathbf{W}_0$, and $\mathbf{W}_1$ are independent of ω. As we shall see, we need not consider any higher terms for R-waves if we confine our attention to the leading term of head waves.

It should be noted that the ray series may also be written in the form

$$\mathbf{W} = \exp[i\omega(t-\tau)]S(\omega)\sum_{k=0}^{\infty}(i\omega)^{-k}\mathbf{W}_k, \tag{2.2$'$}$$

where $S(\omega)$ is some known function of frequency and τ and $\mathbf{W}_k$ are the same as in (2.2). If (2.2) is a solution of the linearized equation of motion (2.1), then (2.2′) must also be a solution of that equation. Assume now that $S(\omega) = (i\omega)^{-n}$, where n is a positive integer. Then we can write

$$\mathbf{W} = \exp[i\omega(t-\tau)]\left(\frac{\mathbf{W}_0}{(i\omega)^n}+\frac{\mathbf{W}_1}{(i\omega)^{n+1}}+\ldots\right),$$

where τ and $\mathbf{W}_k$ ($k = 0, 1, 2, \ldots$) are again the same as in (2.2). From this follows an important conclusion: the *zero-order* ray formulae, which will be derived in this chapter, apply to the *leading* term of the ray series. It does not matter whether the leading term contains a factor $(i\omega)^{-n}$ or a factor 1 (as in (2.2)). In other words, when the coefficients $\mathbf{W}_k\,(x, y, z)$ in the ray series (2.2) vanish identically for $k = 0, 1, 2, \ldots, n-1$, then the *zero-order* ray formulae will apply to the coefficient $\mathbf{W}_n$, which is then the leading term. In all these cases, it is possible to take the factor $(i\omega)^{-n}$ out of the series and rewrite the ray series in the form (2.2′) (or, omitting the factor $S(\omega)$, in the form (2.2)). Therefore, without losing any generality, we shall consider the ray series for the harmonic waves in the form (2.2) and assume that $\mathbf{W}_0 \neq 0$ throughout this chapter. We must only remember that the zero-order formulae apply generally to the *leading* term of the ray series.

If we know the solution of (2.1) for harmonic waves, we can easily generalize it for impulse waves by using the method of Fourier transforms. However, the spectrum $S(\omega)$ of the impulse must be of high frequency, i.e. $S(\omega) = 0$ for $0 \leqslant \omega \leqslant \omega_0$ where ω_0 is large. We denote the impulse solution by $\mathscr{W}$, and we can express it as

$$\mathscr{W}(x, y, z, t) = \frac{1}{\pi} \operatorname{Re} \int_0^\infty S(\omega) \mathbf{W} \mathrm{d}\omega, \tag{2.5}$$

where $\mathbf{W}$ is the steady state solution of (2.1). It is clear that $\mathscr{W}$ is also a solution of (2.1). From (2.2) and (2.5), we find that

$$\mathscr{W}(x, y, z, t) = \frac{1}{\pi} \operatorname{Re} \sum_{k=0}^{\infty} \mathbf{W}_k f_k(t-\tau), \tag{2.6}$$

where

$$f_k(t-\tau) = \int_0^\infty (i\omega)^{-k} S(\omega) \exp[i\omega(t-\tau)] \,\mathrm{d}\omega \tag{2.7}$$

and the $\mathbf{W}_k$ remain the same as for the harmonic solution (2.2). We can easily see from (2.7) that

$$\frac{\mathrm{d}f_k(t-\tau)}{\mathrm{d}(t-\tau)} = f_{k-1}(t-\tau) \qquad (k = 1, 2, \ldots), \tag{2.8}$$

i.e. that all the functions f_k can be successively determined from the first one, f_0.

Generally, without invoking the use of Fourier synthesis at all, we can write the ray series in the form

$$\mathscr{W}(x, y, z, t) = \sum_{k=0}^{\infty} \mathbf{W}_k f_k(t-\tau), \tag{2.9}$$

where the functions $f_k(t-\tau)$ satisfy the relation (2.8). This form of the ray series was used by Friedlander (1946), Babich and Alekseyev (1958), Yanovskaya (n.d.), and others. It includes the solutions of (2.1) which are discontinuous along wave fronts. It follows from (2.8) that the order of discontinuity of the function f_k is less than that of f_{k-1} by one. The most important term in (2.9) is the first, as the function f_0 is of the highest order of discontinuity on the wave front. One example of the succession of functions f_k is given in figure 2.1. It is assumed that f_0 is the unit function: $f_0 = 0$ for $t < \tau$, $f_0 = 1$ for $t > \tau$ (this is the well-known Heaviside function). From this it follows that $f_k(t-\tau) = 0$ for $t < \tau$, $f_k(t-\tau) = (t-\tau)^k/k!$ for $t > \tau$ (see figure 2.1). If we are interested in the propagation of seismic body waves, it is often quite sufficient to consider only one or two terms of (2.9), as these terms represent the highest order of discontinuity on the wave front. The subsequent terms in the ray series change more slowly across the wave front, and therefore they are not so important. The situation is thus very similar to that for harmonic waves for high frequencies.

In theoretical studies concerning the existence and validity of the ray series and some other related problems, the ray series in the form (2.9) (with f_k discontinuous on the wave front) has many advantages in comparison with (2.2) or (2.6). In the sufficiently close neighbourhood of the wave front (for sufficiently small values of $t-\tau$), the ray series (2.9) is convergent if the function τ is an analytic function (see Babich, 1961),

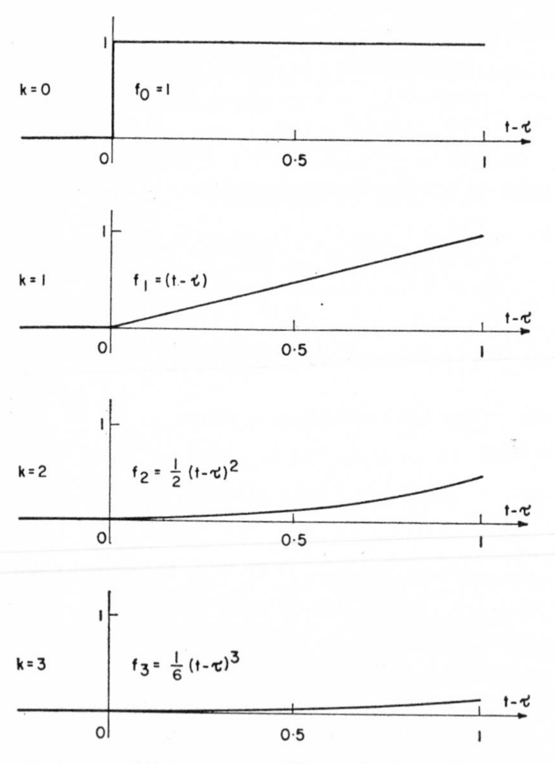

FIGURE 2.1 Succession of functions $f_k(t-\tau)$ in the ray series (2.9) when $f_0(t-\tau)$ is a Heaviside function.

whereas the ray series (2.2) for harmonic waves is not convergent but asymptotic. However, in practical applications, we can use any form of the ray series and we always obtain the same values of τ and $\mathbf{W}_k$ ($k = 0, 1, 2, \ldots$). In this book we shall first find the solution for the harmonic case and then apply the Fourier transform to the results. We shall, therefore, consider a time harmonic solution in the form of the ray series (2.2) (with $\nabla\tau \neq 0$ and $\mathbf{W}_0 \neq 0$).

2.2 THE PHASE FUNCTION AND COEFFICIENTS OF THE RAY SERIES

In this section we shall assume that λ, μ, ρ and all their derivatives are continuous functions of the coordinates (the case of interfaces at which λ, μ, ρ or their derivatives are discontinuous will be considered in section 2.3). We also assume that the above parameters do not change rapidly in short distances, and that the phase function τ in the asymptotic series (2.2) is an analytic function. (As previously, $\nabla\tau \neq 0$, $\mathbf{W}_0 \neq 0$.)

To find the formulae for τ and $\mathbf{W}_k$ ($k = 0, 1, 2, \ldots$) in the ray series (2.2), we insert (2.2) into (2.1). Using formulae of vector analysis (see Appendix), we obtain, after some simple mathematics,

$$\sum_{k=0}^{\infty} (i\omega)^{-k}\{(i\omega)^2\mathbf{N}(\mathbf{W}_k) - i\omega\mathbf{M}(\mathbf{W}_k) + \mathbf{L}(\mathbf{W}_k)\} = 0, \tag{2.10}$$

where

$$\begin{aligned}
\mathbf{N}(\mathbf{W}_k) &= -\rho\mathbf{W}_k + (\lambda+\mu)(\mathbf{W}_k \cdot \nabla\tau)\nabla\tau + \mu(\nabla\tau)^2\mathbf{W}_k, \\
\mathbf{M}(\mathbf{W}_k) &= (\lambda+\mu)\{\nabla(\mathbf{W}_k \cdot \nabla\tau) + \nabla\tau(\nabla \cdot \mathbf{W}_k)\} \\
&\quad + \mu\{2(\nabla\tau \cdot \nabla)\mathbf{W}_k + (\nabla^2\tau)\mathbf{W}_k\} + \nabla\lambda(\mathbf{W}_k \cdot \nabla\tau) \\
&\quad + \nabla\mu \times (\nabla\tau \times \mathbf{W}_k) + 2(\nabla\mu \cdot \nabla\tau)\mathbf{W}_k, \\
\mathbf{L}(\mathbf{W}_k) &= (\lambda+\mu)\nabla(\nabla \cdot \mathbf{W}_k) + \mu\nabla^2\mathbf{W}_k + \nabla\lambda(\nabla \cdot \mathbf{W}_k) \\
&\quad + \nabla\mu \times (\nabla \times \mathbf{W}_k) + 2(\nabla\mu \cdot \nabla)\mathbf{W}_k.
\end{aligned} \tag{2.11}$$

Since (2.10) must hold for any value of ω, the coefficient of each power of ω must vanish. Therefore,

$$\begin{aligned}
&\mathbf{N}(\mathbf{W}_0) = 0, \\
&\mathbf{N}(\mathbf{W}_1) - \mathbf{M}(\mathbf{W}_0) = 0, \\
&\mathbf{N}(\mathbf{W}_k) - \mathbf{M}(\mathbf{W}_{k-1}) + \mathbf{L}(\mathbf{W}_{k-2}) = 0 \qquad (k \geqslant 2).
\end{aligned} \tag{2.12}$$

This is a recurrent system of partial differential equations which enables us to determine τ and successively all $\mathbf{W}_k$ ($k = 0, 1, \ldots$), starting with $\mathbf{W}_0$. We shall call (2.12) the *basic system* of partial differential equations in the ray theory.

It will be shown in the following that the *phase function* τ can be determined from the first equation in (2.12), $\mathbf{N}(\mathbf{W}_0) = 0$. This equation also determines the direction of the vector $\mathbf{W}_0$, although its modulus can be determined only from the second equation. Generally, when we know $\mathbf{W}_{k-2}$ and $\mathbf{W}_{k-1}$, we can determine $\mathbf{W}_k$ from the $(k+1)$th and $(k+2)$th equations in (2.12).

If we introduce $\mathbf{W}_{-1} = \mathbf{W}_{-2} = 0$, then $\mathbf{M}(\mathbf{W}_{-1}) = \mathbf{L}(\mathbf{W}_{-1}) = \mathbf{L}(\mathbf{W}_{-2}) = 0$, and we can write the basic system of partial differential equations in the ray theory, (2.12), in a more compact form:

$$\mathbf{N}(\mathbf{W}_k) - \mathbf{M}(\mathbf{W}_{k-1}) + \mathbf{L}(\mathbf{W}_{k-2}) = 0 \qquad (k = 0, 1, 2, \ldots). \tag{2.12'}$$

This notation enables us to find the *general expressions* for $\mathbf{W}_k$ for any $k \geqslant 0$, where it is understood that $\mathbf{W}_{-1} = \mathbf{W}_{-2} = 0$.

2.2.1 *Determination of the Phase Function*

As mentioned above, the phase function τ can be determined from the first equation in (2.12), viz., $\mathbf{N}(\mathbf{W}_0) = 0$. Taking the scalar product of $\mathbf{N}(\mathbf{W}_0)$ with $\nabla\tau$, we find that

$$[-\rho + (\lambda + 2\mu)(\nabla\tau)^2](\mathbf{W}_0 \cdot \nabla\tau) = 0. \tag{2.13}$$

The cross-product of $\mathbf{N}(\mathbf{W}_0)$ with $\nabla\tau$ yields

$$[-\rho + \mu(\nabla\tau)^2](\mathbf{W}_0 \times \nabla\tau) = 0. \tag{2.14}$$

When we require that $\mathbf{W}_0 \neq 0$ and $\nabla\tau \neq 0$ (see section 2.1), only one of the products $\mathbf{W}_0 \times \nabla\tau$ and $\mathbf{W}_0 \cdot \nabla\tau$ can be zero. Moreover, one of these products *must* be zero, as both the expressions $-\rho + (\lambda + 2\mu)(\nabla\tau)^2$ and $-\rho + \mu(\nabla\tau)^2$ cannot vanish at the same time. Therefore, the system of equations (2.13) and (2.14) has two solutions:

$$\text{(a)} \quad -\rho + (\lambda + 2\mu)(\nabla\tau)^2 = 0, \qquad \mathbf{W}_0 \times \nabla\tau = 0, \tag{2.15}$$

then $\mathbf{W}_0 \cdot \nabla\tau \neq 0$; and

$$\text{(b)} \quad -\rho + \mu(\nabla\tau)^2 = 0, \qquad \mathbf{W}_0 \cdot \nabla\tau = 0, \tag{2.16}$$

then $\mathbf{W}_0 \times \nabla\tau \neq 0$.

Thus, in inhomogeneous media (where λ, μ, ρ and all their derivatives are continuous functions of the coordinates) there exist two independent families of wave fronts, corresponding to two waves. The phase function τ of the first wave is given by the first equation in (2.15) and the phase function of the second wave by the first equation in (2.16). The direction of $\mathbf{W}_0$ for the first wave is parallel to the ray (as $\mathbf{W}_0 \times \nabla\tau = 0$) and for the second wave perpendicular to the ray (as $\mathbf{W}_0 \cdot \nabla\tau = 0$).

To determine what the equations for τ, (2.15) and (2.16), mean from the physical point of view we can rewrite them in the form

$$(\nabla\tau)^2 = 1/\alpha^2, \qquad \text{where } \alpha = [(\lambda+2\mu)/\rho]^{1/2}, \tag{2.15'}$$

for solution (a), and

$$(\nabla\tau)^2 = 1/\beta^2, \qquad \text{where } \beta = (\mu/\rho)^{1/2}, \tag{2.16'}$$

for solution (b). Equations (2.15′) and (2.16′) are well-known *eikonal equations.* Each determines the phase function τ, in the appropriate case, in terms of its value on some initial surface. The right-hand side of the eikonal equation is the inverse of the square of the velocity of propagation of the wave front along a ray (i.e. along the orthogonal trajectory of the family of wave fronts). In the first case, this velocity is α, and in the second it is β, as defined above.

These expressions for α and β are standard in seismology. In a homogeneous, isotropic infinite medium (without boundaries and interfaces), the equation of motion separates into two different wave equations, characterizing the propagation of compressional (P) and shear (S) waves, with velocities $\alpha = [(\lambda+2\mu)/\rho]^{1/2}$ and $\beta = (\mu/\rho)^{1/2}$, respectively. The compressional and shear motions are not coupled to each other in this case. The compressional wave is also known as the longitudinal, irrotational, or dilatational wave, and the shear wave as the distortional, transverse, equivoluminal, or rotational wave. These terms describe the main features of the two kinds of waves in homogeneous media, i.e. the irrotational character of compressional waves and the equivoluminal character of shear waves (see Ewing *et al.*, 1957, pp. 8–10; Grant and West, 1965, pp. 30–1).

In *inhomogeneous* media, however, the equation of motion cannot generally be separated into two wave equations characterizing the propagation of equivoluminal and irrotational waves, since these motions are then coupled to each other. But, for *high frequencies* (or when we consider *propagation of discontinuities* in the wave field), there exist two independent wave fronts (at any given time) as long as λ, μ, ρ and all their derivatives are continuous. One wave front propagates with the *local velocity* of compressional waves α, and the other propagates with the local velocity of shear waves β. The equations of the wave fronts, $t = \tau(x, y, z)$, in the two solutions are not coupled to each other. Since only one value of τ, for each of the two solutions, holds for the whole ray series* (see (2.2)), we can write the ray series (2.2) for each case and determine the coefficients $\mathbf{W}_k$ *separately* in the two cases. This does not mean, however, that the wave motion connected with the wave front which propagates with the velocity α is purely longitudinal. As a matter of fact, we shall show in the following that the higher coefficients in the ray series (for $k \geqslant 1$) have non-zero transverse components in addition to the principal component which is longitudinal. Only the zeroth term of the ray series has no transverse component, in this case. Similar considerations apply to the wave motion connected with the wave front which propagates with the velocity β. It is not purely transverse, because the higher coefficients in the ray series have nonvanishing components along the ray (i.e. only the zeroth term is purely transverse).

Following Karal and Keller (1959) and Yanovskaya (n.d.), we shall call the wave motion connected with the wave front which propagates with the velocity α (with τ given by (2.15′)) the *compressional wave* (or P-wave) and the second motion (with τ given by (2.16′)), the *shear wave* (or S-wave). The compressional waves, as defined above, may not be purely longitudinal and may incorporate transverse components in the higher terms of the ray series. Similarly, the shear waves may not be purely transverse. Some seismologists might object to our terminology. As a partial recompense, our intention in this book is to distinguish clearly between the terms *compressional* and *longitudinal*; they are equivalent only in the zeroth approximation of ray theory. Similarly, we shall distinguish between the terms *shear* and *transverse*. The velocities of propagation of wave fronts α and β will be called the *velocity of compressional waves* (or compressional velocity) and the *velocity of shear waves* (or shear velocity), respectively. The same convention has been used in most papers on the ray theory of propagation of elastic waves.

We must remember that all our considerations are only asymptotic, applicable when the frequency is high. The ray series (2.2) cannot describe the complicated process of wave motion in inhomogeneous media completely and exactly. For example, those parts of the wave motion which decrease exponentially with frequency (e.g., inhomogeneous waves) are automatically excluded from consideration, since they are always smaller than any term of the ray series (2.2) for $\omega \to \infty$. Strictly speaking, from a wave point of view, *velocity* has no general and precise mathematical

*Let us emphasize that the value of τ (in either of the two cases), which is given by the eikonal equation as a solution, is not confined to the zeroth term of the ray series. The same value holds for the whole ray series, including all the higher terms.

meaning in non-homogeneous media (see Gupta, 1965), because it may depend on gradients of elastic parameters as well as on frequency. In general, a heterogeneous medium is necessarily a dispersive medium (see Ravindra, 1968, 1969), and only for high frequencies does the velocity approximate α or β, as defined above. Nevertheless, this does not detract from the usefulness of the ray method in many applications in seismology and seismic prospecting, particularly when considering problems involving the propagation of impulses which contain mainly higher frequencies.

We shall briefly mention some important conclusions which follow from the eikonal equation. This equation determines the *kinematic properties* of waves such as wave fronts, rays, and times of arrival. It should be noted that the derivation of the ray equations from the eikonal equation is a special case of the theory of characteristics of first-order partial differential equations (see Courant and Hilbert, 1962). Here we shall derive the ray equations and some other important relations directly, using only simple considerations.

We shall discuss here only the eikonal equation for the compressional wave, i.e. $(\nabla\tau)^2 = \alpha^{-2}$. All the results will apply mutatis mutandis to shear waves when we substitute β for α. In Cartesian coordinates, the eikonal equation can be written

$$\left(\frac{\partial\tau}{\partial x}\right)^2+\left(\frac{\partial\tau}{\partial y}\right)^2+\left(\frac{\partial\tau}{\partial z}\right)^2 = \frac{1}{\alpha^2}.$$

We now introduce s as the arc length along a ray (measured positively in the direction of propagation). As the rays are orthogonal trajectories to the wave fronts $t = \tau(x, y, z)$, we obtain from the eikonal equation the formula

$$\mathrm{d}\tau/\mathrm{d}s = 1/\alpha,$$

where $\mathrm{d}\tau/\mathrm{d}s$ is a directional derivative of τ along a ray. On integration, we have

$$\tau(s) = \tau(s_0)+\int_{s_0}^{s}\alpha^{-1}\,\mathrm{d}s, \tag{2.17}$$

where the integration path is along a ray. We can represent any ray by the following system of three equations:

$$x = x(s), \qquad y = y(s), \qquad z = z(s),$$

or, succinctly, by one vector equation

$$\mathbf{X} = \mathbf{X}(s),$$

where the vector $\mathbf{X}(s)$ has components $x(s)$, $y(s)$, $z(s)$. Since the ray is orthogonal to the wave front through which it passes, it must be parallel to $\nabla\tau$, i.e., $\mathrm{d}\mathbf{X}/\mathrm{d}s = \lambda\nabla\tau$ (where λ is some proportionality factor). We can determine λ by inserting $\nabla\tau = \lambda^{-1}(\mathrm{d}\mathbf{X}/\mathrm{d}s)$ into the eikonal equation. We obtain

$$\frac{1}{\lambda^2}\left\{\left(\frac{\mathrm{d}x}{\mathrm{d}s}\right)^2+\left(\frac{\mathrm{d}y}{\mathrm{d}s}\right)^2+\left(\frac{\mathrm{d}z}{\mathrm{d}s}\right)^2\right\} = \frac{1}{\alpha^2}.$$

As $\mathrm{d}s^2 = \mathrm{d}x^2+\mathrm{d}y^2+\mathrm{d}z^2$, we have finally $\lambda = \alpha$. Thus, we can write a differential equation for rays as follows:

$$\frac{\mathrm{d}\mathbf{X}}{\mathrm{d}s} = \alpha\nabla\tau.$$

Now we shall try to find other differential equations for rays, which are independent of the phase function τ. We divide the above equation for rays by α and differentiate with respect to s. As $\mathrm{d}(\nabla\tau)/\mathrm{d}s = \nabla(1/\alpha)$, we obtain

$$\frac{\mathrm{d}}{\mathrm{d}s}\left(\frac{1}{\alpha}\frac{\mathrm{d}\mathbf{X}}{\mathrm{d}s}\right) = \nabla\left(\frac{1}{\alpha}\right),$$

which represents three differential equations for $x(s)$, $y(s)$, $z(s)$:

$$\frac{\mathrm{d}}{\mathrm{d}s}\left(\frac{1}{\alpha}\frac{\mathrm{d}x}{\mathrm{d}s}\right) = \frac{\partial}{\partial x}\left(\frac{1}{\alpha}\right), \quad \frac{\mathrm{d}}{\mathrm{d}s}\left(\frac{1}{\alpha}\frac{\mathrm{d}y}{\mathrm{d}s}\right) = \frac{\partial}{\partial y}\left(\frac{1}{\alpha}\right), \quad \frac{\mathrm{d}}{\mathrm{d}s}\left(\frac{1}{\alpha}\frac{\mathrm{d}z}{\mathrm{d}s}\right) = \frac{\partial}{\partial z}\left(\frac{1}{\alpha}\right).$$

In seismology, it is common practice to represent the ray by the following two equations:

$$x = x(z), \qquad y = y(z).$$

Since $\mathrm{d}s^2 = \mathrm{d}x^2+\mathrm{d}y^2+\mathrm{d}z^2$, we have $\mathrm{d}s/\mathrm{d}z = [(x')^2+(y')^2+1]^{1/2}$, where the prime denotes a derivative with respect to z. Using this formula and the first two equations of the above system of differential equations for rays, we obtain

$$\frac{\mathrm{d}}{\mathrm{d}z}\left(\frac{x'}{\alpha[(x')^2+(y')^2+1]^{1/2}}\right) = [(x')^2+(y')^2+1]^{1/2}\frac{\partial}{\partial x}\left(\frac{1}{\alpha}\right),$$

$$\frac{\mathrm{d}}{\mathrm{d}z}\left(\frac{y'}{\alpha[(x')^2+(y')^2+1]^{1/2}}\right) = [(x')^2+(y')^2+1]^{1/2}\frac{\partial}{\partial y}\left(\frac{1}{\alpha}\right).$$

This is yet another system of differential equations for rays.

The above differential equations for rays are precisely the Euler equations for the *extremals* of the integral

$$\int_{M_0}^{M} \frac{ds}{\alpha} = \int_{M_0}^{M} \frac{1}{\alpha} [(x')^2 + (y')^2 + 1]^{1/2} \, dz$$

(see, e.g., Morse and Feshbach, 1953, pp. 276–80). It follows that the rays are extremals of these integrals. When we take into account the definition of extremals in the calculus of variations, we can conclude that *the actual path, between two points, taken by a signal is the one which renders the time of travel stationary* (Kline and Kay, 1965, p. 72). Less correctly, this means that *the rays represent the paths which require least time for a signal to travel from M_0 to M (Fermat's principle).*

We shall now present a simple example of the solution of the last system of differential equations for rays. We shall assume that the medium is *vertically inhomogeneous*, i.e. α depends only on the z-coordinate. The initial conditions for the ray under investigation are assumed to be

$$x(z_0) = x_0, \qquad y(z_0) = 0, \qquad x'(z_0) = \tan \theta(z_0), \qquad y'(z_0) = 0.$$

In other words, we shall investigate the ray which passes through the point

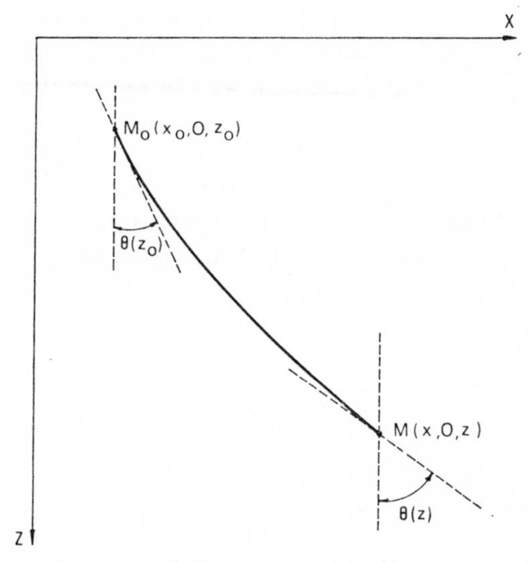

FIGURE 2.2 Element of a ray in a vertically inhomogeneous medium between the points $M_0(x_0, 0, z_0)$ and $M(x, 0, z)$.

$M_0(x_0, 0, z_0)$. The tangent to the ray at M_0 lies in the xz plane (as $(y'(z_0) = 0)$ and the acute angle between this tangent and the z-axis at M_0 is $\theta(z_0)$ (see figure 2.2). As

$$\frac{\partial}{\partial x}\left(\frac{1}{\alpha}\right) = \frac{\partial}{\partial y}\left(\frac{1}{\alpha}\right) = 0,$$

we obtain the system of differential equations for rays in the form

$$\frac{\mathrm{d}}{\mathrm{d}z}\left(\frac{x'}{\alpha[(x')^2+(y')^2+1]^{1/2}}\right) = 0, \qquad \frac{\mathrm{d}}{\mathrm{d}z}\left(\frac{y'}{\alpha[(x')^2+(y')^2+1]^{1/2}}\right) = 0.$$

From the second equation, it follows that $y'/\{\alpha[(x')^2+(y')^2+1]^{1/2}\} = C$, where C is a constant. However, C must be zero since $y'(z_0) = 0$. When we insert $C = 0$, we obtain $y'(z) = 0$ along the whole ray. Since $y(z_0) = 0$, then $y(z) = 0$ along the whole ray. Thus, the ray is a plane curve and lies in the plane (x, z). Inserting $y' = 0$ into the first equation and solving it, we obtain

$$x'/\{\alpha[(x')^2+1]^{1/2}\} = p,$$

where p = const. is the so-called *ray path parameter* which must remain constant along the whole ray. What does this mean geometrically? At the point M_0, we easily get, from the above formula and from the initial conditions, $p = \sin\theta(z_0)/\alpha(z_0)$. Since $x'/[(x')^2+1]^{1/2}$ can be interpreted as $\sin\theta(z)$, where $\theta(z)$ is the acute angle between the tangent to the ray and the z-axis at the depth z (see point M in figure 2.2), we can rewrite the above equation in the form

$$p = \sin\theta(z_0)/\alpha(z_0) = \sin\theta(z)/\alpha(z).$$

This gives *Snell's law* in its usual form. Now we shall solve the differential equation $x'/\{\alpha[(x')^2+1]^{1/2}\} = p$ to find $x = x(z)$. After some simple mathematics, we obtain

$$x'(z) = p\alpha(z)/[1-p^2\alpha^2(z)]^{1/2}.$$

The solution of this equation is:

$$x(z) = x(z_0)+\int_{z_0}^{z}\frac{p\alpha(z)}{[1-p^2\alpha^2(z)]^{1/2}}\mathrm{d}z,$$

which is the final formula for the ray in a vertically inhomogeneous medium.

From (2.17) we easily obtain an expression for τ for our present case of vertically inhomogeneous media. Since $\mathrm{d}s = [(x')^2+(y')^2+1]^{1/2}\mathrm{d}z$, we can write (2.17) in the form

$$\tau(M) = \tau(M_0)+\int_{M_0}^{M} \alpha^{-1}[(x')^2+(y')^2+1]^{1/2}\mathrm{d}z,$$

where the integration path is along the ray. We have found that along the ray $y' = 0$ and $x'/[(x')^2+1]^{1/2} = \alpha p$, i.e. $[(x')^2+1]^{1/2} = 1/[1-p^2\alpha^2]^{1/2}$. Inserting these expressions into the formula for $\tau(M)$, we get

$$\tau(z) = \tau(z_0)+\int_{z_0}^{z} \frac{1}{\alpha(z)[1-p^2\alpha^2(z)]^{1/2}}\mathrm{d}z.$$

The above formulae for $x(z)$ and $\tau(z)$ are very well known (see, e.g., Grant and West, 1965, p. 134). They are valid when $z > z_0$; otherwise, we must take the absolute values of the definite integrals, as in section 2.6.1.

It should be noted that for $\partial(1/\alpha)/\partial x \neq 0$ and $\partial(1/\alpha)/\partial y \neq 0$, solution of the differential equations for rays given above is not so simple, the best procedure being to use numerical methods. We shall present below a system of ordinary differential equations of first order for rays which can be solved simply by standard numerical techniques, such as the Runge-Kutta method. If we introduce three directional cosines at every point of the ray, viz. $\cos e_x = \mathrm{d}x/\mathrm{d}s$, $\cos e_y = \mathrm{d}y/\mathrm{d}s$, and $\cos e_z = \mathrm{d}z/\mathrm{d}s$, in the above differential equations (and the eikonal equation), we obtain, after some mathematical manipulation, the following system of six ordinary differential equations of first order:

$$\frac{\mathrm{d}x(\tau)}{\mathrm{d}\tau} = \alpha \cos e_x,$$

$$\frac{\mathrm{d}y(\tau)}{\mathrm{d}\tau} = \alpha \cos e_y,$$

$$\frac{\mathrm{d}z(\tau)}{\mathrm{d}\tau} = \alpha \cos e_z,$$

$$\frac{\partial e_x(\tau)}{\partial \tau} = \frac{\partial \alpha}{\partial x}\sin e_x-\frac{\partial \alpha}{\partial y}\cot e_x \cos e_y-\frac{\partial \alpha}{\partial z}\cot e_x \cos e_z,$$

$$\frac{\mathrm{d}e_y(\tau)}{\mathrm{d}\tau} = -\frac{\partial \alpha}{\partial x}\cos e_x \cot e_y+\frac{\partial \alpha}{\partial y}\sin e_y-\frac{\partial \alpha}{\partial z}\cot e_y \cos e_z,$$

$$\frac{\mathrm{d}e_z(\tau)}{\mathrm{d}\tau} = -\frac{\partial \alpha}{\partial x}\cos e_x \cot e_z-\frac{\partial \alpha}{\partial y}\cot e_z \cos e_y+\frac{\partial \alpha}{\partial z}\sin e_z,$$

where $\alpha = \alpha(x, y, z)$. In order to solve this system, we must know six initial conditions which determine the coordinates of the point on the ray at the initial time $\tau = \tau_0$ and also the direction of the ray at that point: $x(\tau_0) = x_0$, $y(\tau_0) = y_0$, $z(\tau_0) = z_0$, $e_x(\tau_0) = e_{x0}$, $e_y(\tau_0) = e_{y0}$, and $e_z(\tau_0) = e_{z0}$.

If the velocity α depends only on two coordinates, say x and z, and $y_0 = 0$, $e_{y0} = \pi/2$, the system becomes simpler:

$$\frac{dx(\tau)}{d\tau} = \alpha \cos e_x,$$

$$\frac{dz(\tau)}{d\tau} = \alpha \sin e_x,$$

$$\frac{de_x(\tau)}{d\tau} = \frac{\partial\alpha}{\partial x} \sin e_x - \frac{\partial\alpha}{\partial z} \cos e_x.$$

Note that $e_x = \pi/2 - e_z$ in this case. For more details see Yeliseyevnin (1964), where a numerical example is also given.

2.2.2 *Determination of Coefficients of the Ray Series*

Since the equations for τ for compressional and shear waves *are not coupled to each other*, we shall determine the coefficients $\mathbf{W}_k$ separately in the two cases. With respect to the ray in each case, it is convenient to introduce

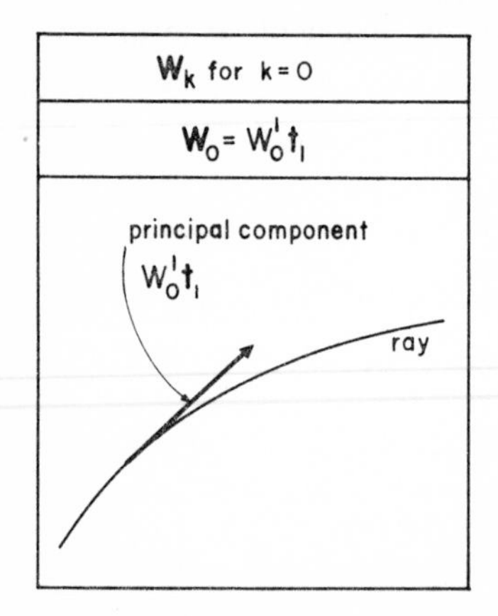

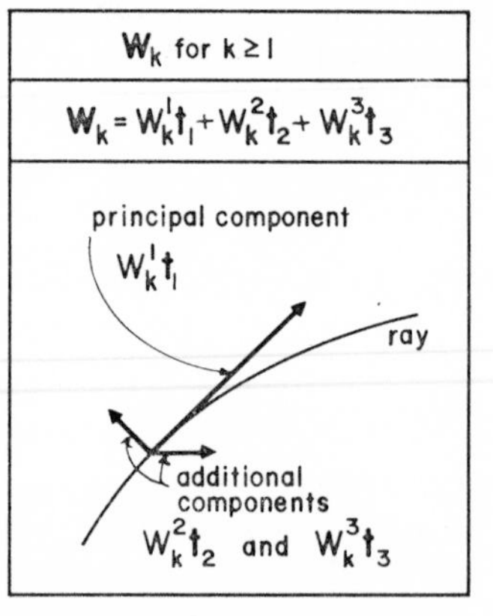

FIGURE 2.3 Principal and additional components of the coefficients of the ray series $\mathbf{W}_k$ for compressional waves. The principal components are in the direction of $\mathbf{t}_1$, and the additional ones are along $\mathbf{t}_2$ and $\mathbf{t}_3$ ($\mathbf{t}_1$, $\mathbf{t}_2$, and $\mathbf{t}_3$ are unit vectors tangent, principal normal, and binormal to the ray). For $k = 0$, the additional components vanish.

a unit tangent vector $\mathbf{t}_1$, a unit principal normal vector $\mathbf{t}_2$, and a unit binormal vector $\mathbf{t}_3$. (We use the symbols $\mathbf{t}_1$, $\mathbf{t}_2$, $\mathbf{t}_3$ rather than the usual symbols $\mathbf{t}$, $\mathbf{n}$, $\mathbf{b}$ since we shall need these for other purposes.) Then we can write

$$\mathbf{W}_k = W_k^1\,\mathbf{t}_1 + W_k^2\,\mathbf{t}_2 + W_k^3\,\mathbf{t}_3. \tag{2.18}$$

It was shown earlier that for a compressional wave $\mathbf{W}_0$ is tangent to the ray, i.e., $W_0^1 \neq 0$, $W_0^2 = W_0^3 = 0$. For a shear wave, $\mathbf{W}_0$ lies in the plane normal to the ray, i.e. $W_0^1 = 0$, $W_0^2 \neq 0$, and/or $W_0^3 \neq 0$. For $k \neq 0$, all three components may be different from zero, as was mentioned above. Hence the compressional wave is not purely longitudinal and the shear wave is not purely transverse. For compressional waves, we shall call the component $W_k^1\,\mathbf{t}_1$ (parallel to the ray) the *principal component*, and the other two components (perpendicular to the ray) the *additional components* (see figure 2.3). For shear waves, the principal components are those perpendicular to the ray ($W_k^2\,\mathbf{t}_2$ and $W_k^3\,\mathbf{t}_3$) and the additional component is parallel to it ($W_k^1\,\mathbf{t}_1$; see figure 2.5). We shall continue to discuss the two kinds of waves separately. As we shall use the basic system of partial differential equations in the form (2.12′), we must keep in mind that $\mathbf{W}_{-1} = \mathbf{W}_{-2} = 0$.

A Compressional Waves

For compressional waves, the phase function τ is a solution of the eikonal equation $(\nabla\tau)^2 = 1/\alpha^2$, where $\alpha = [(\lambda+2\mu)/\rho]^{1/2}$ (see (2.15′)). In this subsection we shall determine the coefficients of the ray series for the compressional wave $\mathbf{W}_k$ (see eq. (2.2)).

The determination of the additional components of $\mathbf{W}_k$ ($W_k^2\,\mathbf{t}_2$ and $W_k^3\,\mathbf{t}_3$; see figure 2.3) is simple. As $\mathbf{t}_2$ and $\mathbf{t}_3$ are perpendicular to $\nabla\tau$, we obtain from (2.11), (2.15′), and (2.18)

$$\mathbf{N}(\mathbf{W}_k)\cdot\mathbf{t}_2 = (-\rho+\mu\alpha^{-2})W_k^2, \qquad \mathbf{N}(\mathbf{W}_k)\cdot\mathbf{t}_3 = (-\rho+\mu\alpha^{-2})W_k^3. \tag{2.19}$$

Inserting the value for $\mathbf{N}(\mathbf{W}_k)$ from (2.12′) into these equations gives us

$$\begin{aligned} W_k^2 &= -[\alpha^2/(\lambda+\mu)][\mathbf{M}(\mathbf{W}_{k-1})-\mathbf{L}(\mathbf{W}_{k-2})]\cdot\mathbf{t}_2, \\ W_k^3 &= -[\alpha^2/(\lambda+\mu)][\mathbf{M}(\mathbf{W}_{k-1})-\mathbf{L}(\mathbf{W}_{k-2})]\cdot\mathbf{t}_3, \end{aligned} \tag{2.20}$$

from which it is seen that the additional components are obtained by differentiation of the vectors $\mathbf{W}_{k-1}$ and $\mathbf{W}_{k-2}$, without needing to know any initial conditions.

Equations (2.20) prove our earlier statement that the higher coefficients of the ray series for compressional waves have generally non-zero components perpendicular to the ray. The additional components vanish for the leading term of the ray series (see (2.20) for $k = 0$, $\mathbf{W}_{-1} = \mathbf{W}_{-2} = 0$).

To describe the additional components of $\mathbf{W}_k$, we shall also use the vector $\mathbf{W}_k^{\perp}$, which is given by the formula

$$\mathbf{W}_k^{\perp} = W_k^2 \mathbf{t}_2 + W_k^3 \mathbf{t}_3. \tag{2.21}$$

From (2.20) we obtain

$$\mathbf{W}_k^{\perp} = -[\alpha^2/(\lambda+\mu)][\mathbf{M}(\mathbf{W}_{k-1}) - \mathbf{L}(\mathbf{W}_{k-2})]. \tag{2.20'}$$

The determination of the principal component $W_k^1 \mathbf{t}_1$ is more complicated. We first find W_k^{τ}, where

$$W_k^{\tau}\, \nabla\tau = W_k^1 \mathbf{t}_1, \qquad \text{i.e. } W_k^{\tau} = \alpha W_k^1. \tag{2.22}$$

Taking the scalar product of (2.12′) with $\nabla\tau$, we obtain

$$\mathbf{M}(\mathbf{W}_k)\cdot \nabla\tau = \mathbf{L}(\mathbf{W}_{k-1})\cdot \nabla\tau, \tag{2.23}$$

as $\mathbf{N}(\mathbf{W}_k)\cdot \nabla\tau = 0$ for any k. Now we express $\mathbf{W}_k$ as a superposition of two vectors, $\mathbf{W}_k = W_k^{\tau}\, \nabla\tau + \mathbf{W}_k^{\perp}$ (see (2.18), (2.21), and (2.22)). Because $\mathbf{M}(\mathbf{W}_k)$ is a linear differential operator, we find from (2.23) that

$$\mathbf{M}(W_k^{\tau}\, \nabla\tau)\cdot \nabla\tau = [\mathbf{L}(\mathbf{W}_{k-1}) - \mathbf{M}(\mathbf{W}_k^{\perp})]\cdot \nabla\tau. \tag{2.24}$$

If the coefficients of the ray series $\mathbf{W}_j$ are known for $j = 0, 1, \ldots, k-1$, the right-hand side of this equation is known ($\mathbf{W}_k^{\perp}$ may be determined from (2.20′)), which enables us to find W_k^{τ}. First we must rewrite the left-hand side of the equation in a simpler form. Using some formulae from vector analysis (see Appendix) we find from the expression for $\mathbf{M}$ given in (2.11) the following formula:

$$\mathbf{M}(W_k^{\tau}\, \nabla\tau)\cdot \nabla\tau = 2\rho(\nabla\tau\cdot \nabla)W_k^{\tau} + W_k^{\tau}[\rho\nabla^2\tau + (\nabla\tau\cdot\nabla)\rho].$$

Since we can write

$$(\nabla\tau\cdot\nabla)W_k^{\tau} = \alpha^{-1}(\mathrm{d}W_k^{\tau}/\mathrm{d}s), \qquad (\nabla\tau\cdot\nabla)\rho = \alpha^{-1}(\mathrm{d}\rho/\mathrm{d}s),$$

we find that

$$\mathbf{M}(W_k^{\tau}\, \nabla\tau)\cdot \nabla\tau = 2\rho\alpha^{-1}(\mathrm{d}W_k^{\tau}/\mathrm{d}s) + W_k^{\tau}[\rho\nabla^2\tau + \alpha^{-1}(\mathrm{d}\rho/\mathrm{d}s)].$$

Inserting this expression into (2.24) we obtain

$$\frac{\mathrm{d}W_k^{\tau}}{\mathrm{d}s} + W_k^{\tau}\left(\frac{\alpha}{2}\nabla^2\tau + \frac{1}{2\rho}\frac{\mathrm{d}\rho}{\mathrm{d}s}\right) = \frac{\alpha}{2\rho}[\mathbf{L}(\mathbf{W}_{k-1}) - \mathbf{M}(\mathbf{W}_k^{\perp})]\cdot \nabla\tau. \tag{2.25}$$

This is a linear, ordinary, first-order differential equation for W_k^{τ}. It is generally inhomogeneous for $k \geqslant 1$, and becomes homogeneous only for

$k = 0$ since then $\mathbf{W}_{k-1} = \mathbf{W}_{-1} = 0$, $\mathbf{W}_k^\perp = \mathbf{W}_0^\perp = 0$. Thus we obtain for W_0^τ

$$\frac{\mathrm{d}W_0^\tau}{\mathrm{d}s} + W_0^\tau\left(\frac{\alpha}{2}\nabla^2\tau + \frac{1}{2\rho}\frac{\mathrm{d}\rho}{\mathrm{d}s}\right) = 0, \tag{2.25'}$$

which is one possible form of the so-called *transport equation*. Following Kline and Kay (1965), we shall call the inhomogeneous equation (2.25) the *higher transport equation* or simply the transport equation. The system of transport equations (2.25) for $k = 0, 1, 2, \ldots$ enables us to find all W_k^τ $(k = 0, 1, 2, \ldots)$. It should be noted that the system is recursive. We must first solve for $k = 0$ and find W_0^τ. Then we use W_0^τ for the determination of the right-hand side of (2.25) for $k = 1$, and solve this equation to find W_1^τ. This process can be repeated for any $k > 1$.

It is not difficult to find, from (2.25), an explicit formula for W_k^τ. The solution of (2.25) is (see, e.g., Schelkunoff, 1948, p. 182)

$$\begin{aligned} W_k^\tau(s) = {} & W_k^\tau(s_0)\exp\left\{-\frac{1}{2}\int_{s_0}^{s}\left(\alpha\nabla^2\tau + \frac{1}{\rho}\frac{\mathrm{d}\rho}{\mathrm{d}s}\right)\mathrm{d}s\right\} \\ & + \frac{1}{2}\int_{s_0}^{s}\frac{\alpha(s')}{\rho(s')}\{[\mathbf{L}(\mathbf{W}_{k-1}) - \mathbf{M}(\mathbf{W}_k^\perp)]\cdot\nabla\tau\} \\ & \times\exp\left\{-\frac{1}{2}\int_{s'}^{s}\left(\alpha\nabla^2\tau + \frac{1}{\rho}\frac{\mathrm{d}\rho}{\mathrm{d}s}\right)\mathrm{d}s\right\}\mathrm{d}s', \end{aligned} \tag{2.26}$$

where all the integrals are along the ray.

We shall now discuss (2.26) in detail. We can write

$$\exp\left\{-\frac{1}{2}\int_{s_0}^{s}\left(\alpha\nabla^2\tau + \frac{1}{\rho}\frac{\mathrm{d}\rho}{\mathrm{d}s}\right)\mathrm{d}s\right\} = \exp\left(-\frac{1}{2}\int_{s_0}^{s}\alpha\nabla^2\tau\,\mathrm{d}s\right)\exp\left(-\frac{1}{2}\int_{s_0}^{s}\frac{1}{\rho}\frac{\mathrm{d}\rho}{\mathrm{d}s}\mathrm{d}s\right).$$

For the second expression on the right-hand side, we obtain

$$\exp\left(-\frac{1}{2}\int_{s_0}^{s}\frac{1}{\rho}\frac{\mathrm{d}\rho}{\mathrm{d}s}\mathrm{d}s\right) = \left(\frac{\rho(s_0)}{\rho(s)}\right)^{1/2}.$$

The first expression also has a simple physical meaning. We introduce the so called *ray coordinates* (q_1, q_2, τ), where q_1 and q_2 characterize a ray and τ the position of a point on the ray. We shall call the coordinates q_1

and q_2 the *parameters of the ray*. They can be regarded as curvilinear coordinates on the wave front, at some specific moment of time $\tau(x, y, z) = t_0$; or, in the case of a point source, as angular coordinates with origin at the source. For any ray with parameters q_1 and q_2, we can write the parametric equation $\mathbf{X} = \mathbf{X}(\tau, q_1, q_2)$. For a fixed τ, as q_1 and q_2 vary, this equation becomes the equation of a wave front. By an *elementary ray tube* (or simply a ray tube), we shall understand the family of rays the parameters of which are within the limits $q_1, q_1+\mathrm{d}q_1$; $q_2, q_2+\mathrm{d}q_2$ (see figure 2.4). The cross-sectional area of this ray tube, $\mathrm{d}\sigma$, as given by the standard formula from the differential geometry of surfaces (see Schwartz *et al.*, 1960) is $\mathrm{d}\sigma = J\mathrm{d}q_1\,\mathrm{d}q_2$, where $J = |\mathbf{X}_{q_1} \times \mathbf{X}_{q_2}|$ and $\mathbf{X}_{q_i}$ denotes the derivative of $\mathbf{X}$ with respect to q_i. When the analytic expressions for the wave front are known, J can be calculated without difficulty.

Now we shall consider an infinitesimal curvilinear parallelepiped $(q_1, q_1+\mathrm{d}q_1; q_2, q_2+\mathrm{d}q_2; \tau, \tau+\mathrm{d}\tau)$ and calculate the flux of the vector $\nabla\tau$ through its surface. As a result we obtain for $\nabla^2\tau$

$$\nabla^2\tau = \nabla\cdot\nabla\tau = \frac{1}{J\alpha}\frac{\mathrm{d}}{\mathrm{d}\tau}\left(\frac{J}{\alpha}\right)$$

(see Babich and Alekseyev, 1958, for more details). Inserting this into our

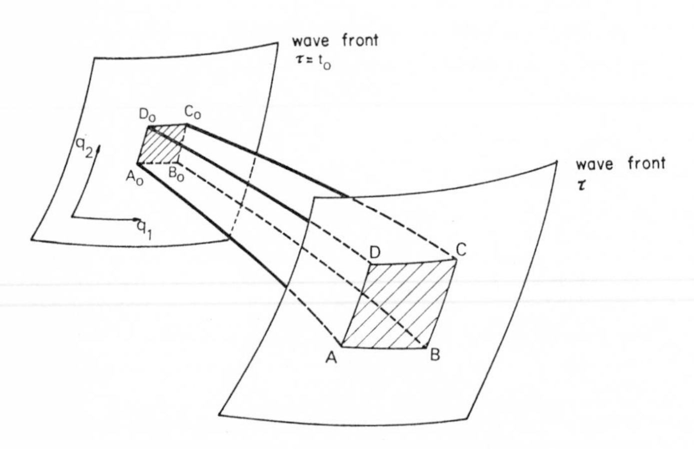

FIGURE 2.4 Elementary ray tube between wave fronts at $\tau = t_0$ and τ. The parameters of the rays passing through the points A_0, B_0, C_0, and D_0 are $(q_1; q_2)$, $(q_1+dq_1; q_2)$, $(q_1+dq_1; q_2+dq_2)$, and $(q_1; q_2+dq_2)$, respectively. The cross-sectional areas of the elementary ray tube are shaded.

integral above and taking into account that $d/d\tau = \alpha d/ds$ along a ray, we obtain

$$\exp\left(-\frac{1}{2}\int_{s_0}^{s} \alpha \nabla^2\tau \, ds\right) = \left(\frac{\alpha(s_0)\,J(s)}{\alpha(s)\,J(s_0)}\right)^{-1/2}.$$

As J can be calculated for any s, this may be accepted as the final result. The function $J(s)$, defined above, has been widely used in the ray theory by Russian scientists (Babich and Alekseyev, 1958; Alekseyev *et al.*, 1961; Yanovskaya, n.d.). Note that instead of $J(s)$ we can also use the so called *expansion coefficients* $K(s)$ (see, e.g., Kline and Kay, 1965; Vlaar, 1968), which is defined by the relation $K(s) = 1/J(s)$. Let us consider what these functions mean from a physical point of view. Denoting the cross-sectional area of the ray tube at s_0 by $d\sigma(s_0)$ and the area of the same tube at s by $d\sigma(s)$, we can write

$$d\sigma(s_0) = J(s_0)dq_1dq_2, \qquad d\sigma(s) = J(s)dq_1dq_2.$$

It follows that $J(s)/J(s_0) = d\sigma(s)/d\sigma(s_0)$. Therefore, $J(s)$ and $K(s)$ measure the expansion or contraction of the ray tube along a ray. From this we obtain (see also Luneburg, 1964)

$$\exp\left(-\frac{1}{2}\int_{s_0}^{s} \alpha \nabla^2\tau \, ds\right) = \left(\frac{\alpha(s)}{\alpha(s_0)}\frac{d\sigma(s_0)}{d\sigma(s)}\right)^{1/2}. \tag{2.27}$$

In the following, we shall use this formula consistently, as the quantities $d\sigma(s_0)$ and $d\sigma(s)$ have a simple and clear geometrical meaning. All the following formulae can be rewritten simply in terms of $K(s)$ and $J(s)$, using the relation

$$\frac{d\sigma(s_0)}{d\sigma(s)} = \frac{K(s)}{K(s_0)} = \frac{J(s_0)}{J(s)}, \tag{2.27'}$$

which can serve us as a precise definition of $d\sigma(s_0)/d\sigma(s)$, when we use $J(s) = |\mathbf{X}_{q_1} \times \mathbf{X}_{q_2}|$.

We shall now derive some useful formulae for $d\sigma(s_0)/d\sigma(s)$. We rewrite $\nabla^2\tau$ in the form

$$\nabla^2\tau = \nabla\cdot\nabla\tau = \nabla\cdot(\mathbf{t}_1/\alpha) = \frac{d}{ds}\left(\frac{1}{\alpha}\right) + \frac{1}{\alpha}(\nabla\cdot\mathbf{t}_1),$$

where $\mathbf{t}_1$ is a unit vector tangent to the ray. From differential geometry we know that

$$\nabla\cdot\mathbf{t}_1 = \operatorname{div}\mathbf{t}_1 = \frac{1}{r_1} + \frac{1}{r_2}$$

where r_1 and r_2 are the *principal radii of curvature* of the wave front (see detailed derivation in Kline and Kay, 1965, pp. 184–6). Inserting

$$\nabla^2\tau = \frac{d}{ds}\left(\frac{1}{\alpha}\right)+\frac{1}{\alpha}\left(\frac{1}{r_1}+\frac{1}{r_2}\right)$$

into (2.27), we obtain

$$\left(\frac{d\sigma(s_0)}{d\sigma(s)}\right)^{1/2} = \exp\left\{-\frac{1}{2}\int_{s_0}^{s}\left(\frac{1}{r_1(s)}+\frac{1}{r_2(s)}\right)ds\right\}, \tag{2.28}$$

where the integration path is along a ray. Thus, we have expressed $d\sigma(s_0)/d\sigma(s)$ in terms of the principal radii of curvature of the wave front.

For homogeneous media, we obtain from (2.28), taking into account that $r_1(s) = r_1(s_0)+(s-s_0)$ and $r_2(s) = r_2(s_0)+(s-s_0)$ in this case, the simple formula

$$\frac{d\sigma(s_0)}{d\sigma(s)} = \frac{r_1(s_0)r_2(s_0)}{r_1(s)r_2(s)} = \frac{r_1(s_0)r_2(s_0)}{[r_1(s_0)+l][r_2(s_0)+l]}, \tag{2.28'}$$

where $l = s-s_0$.

For the determination of $d\sigma(s_0)/d\sigma(s)$, we can also use (2.27′) and the definition of the function J. This yields

$$\frac{d\sigma(s_0)}{d\sigma(s)} = \frac{|\mathbf{X}_{q_1}\times\mathbf{X}_{q_2}|_{\tau=\tau(s_0)}}{|\mathbf{X}_{q_1}\times\mathbf{X}_{q_2}|_{\tau=\tau(s)}} \tag{2.28''}$$

The methods for evaluating $|\mathbf{X}_{q_1}\times\mathbf{X}_{q_2}|$ are known from differential geometry. Using the Laplace identity $|\mathbf{A}\times\mathbf{B}|^2 = (\mathbf{A}\cdot\mathbf{A})(\mathbf{B}\cdot\mathbf{B})-(\mathbf{A}\cdot\mathbf{B})^2$ (see Appendix), and putting

$$E = (\mathbf{X}_{q_1}\cdot\mathbf{X}_{q_1}) = \left(\frac{\partial x}{\partial q_1}\right)^2+\left(\frac{\partial y}{\partial q_1}\right)^2+\left(\frac{\partial z}{\partial q_1}\right)^2,$$

$$F = (\mathbf{X}_{q_2}\cdot\mathbf{X}_{q_2}) = \left(\frac{\partial x}{\partial q_2}\right)^2+\left(\frac{\partial y}{\partial q_2}\right)^2+\left(\frac{\partial z}{\partial q_2}\right)^2,$$

$$G = (\mathbf{X}_{q_1}\cdot\mathbf{X}_{q_2}) = \frac{\partial x}{\partial q_1}\frac{\partial x}{\partial q_2}+\frac{\partial y}{\partial q_1}\frac{\partial y}{\partial q_2}+\frac{\partial z}{\partial q_1}\frac{\partial z}{\partial q_2},$$

we can write

$$|\mathbf{X}_{q_1}\times\mathbf{X}_{q_2}| = (EF-G^2)^{1/2}.$$

Formula (2.28″) is convenient for calculation when the equations of the wave front,

$$x = x(\tau, q_1, q_2), \qquad y = y(\tau, q_1, q_2), \qquad z = z(\tau, q_1, q_2),$$

are known for fixed times $\tau = \tau(s_0)$ and $\tau = \tau(s)$.

Now, we revert to the formula (2.26). Inserting (2.27) and (2.22) into (2.26), we find the final expression for the principal component of the coefficient of the ray series for compressional waves:

$$W_k^1(s) = W_k^1(s_0)\left(\frac{\alpha(s_0)\rho(s_0)\mathrm{d}\sigma(s_0)}{\alpha(s)\rho(s)\mathrm{d}\sigma(s)}\right)^{1/2}$$

$$+\frac{1}{2}\left(\frac{1}{\alpha(s)\rho(s)}\right)^{1/2}\int_{s_0}^{s}\left(\frac{\alpha(s')}{\rho(s')}\right)^{1/2}\left(\frac{\mathrm{d}\sigma(s')}{\mathrm{d}\sigma(s)}\right)^{1/2}\{(\mathbf{L}(\mathbf{W}_{k-1}) - \mathbf{M}(\mathbf{W}_k^{\perp}))\cdot\nabla\tau\}\mathrm{d}s'. \tag{2.29}$$

From this equation we can determine the principal component W_k^1 at an arbitrary point s of the ray if we know its value at an earlier point s_0 of the ray.

Expressions (2.18), (2.20), and (2.29) enable us to determine successively all $\mathbf{W}_k$. For each k, however, we must know one initial value of W_k^1.

For the leading term of the ray series, $\mathbf{W}_0$, the formulae become considerably simpler. We have found from (2.20) that the additional components of the leading term, W_0^2 and W_0^3, vanish. For the principal component W_0^1, we get from (2.29)

$$W_0^1(s) = W_0^1(s_0)\left(\frac{\alpha(s_0)\rho(s_0)\mathrm{d}\sigma(s_0)}{\alpha(s)\rho(s)\mathrm{d}\sigma(s)}\right)^{1/2}. \tag{2.30}$$

The above formula for the zeroth approximation of the solution follows directly from some simple energy considerations also. We shall assume that energy propagates only *along the rays* and that no energy flows through the side walls of the ray tube. Then it follows from the law of conservation of energy that the energy flux through the cross-sectional area $\mathrm{d}\sigma(s_0)$ must be the same as through $\mathrm{d}\sigma(s)$ (see figure 2.4). Using the appropriate formulae for energy flux, we can write this condition in the form (see Alekseyev and Gel'chinskiy, 1959, p. 112)

$$\rho(s_0)\alpha(s_0)|\dot{\mathbf{W}}(s_0)|^2\mathrm{d}\sigma(s_0) = \rho(s)\alpha(s)|\dot{\mathbf{W}}(s)|^2\mathrm{d}\sigma(s),$$

where the dot denotes a derivative with respect to time. If we insert the zero-order approximations for $\mathbf{W}(s_0)$ and $\mathbf{W}(s)$, we obtain, after a few simple rearrangements, the formula

$$[\rho(s_0)\alpha(s_0)\mathrm{d}\sigma(s_0)]^{1/2}|W_0^1(s_0)| = [\rho(s)\alpha(s)\mathrm{d}\sigma(s)]^{1/2}|W_0^1(s)|,$$

which gives the same variation of $|W_0^1|$ with s as (2.30). (However, it does not determine the possible changes of the phase factor of W_0^1.) It should be emphasized again that the law of conservation of energy flux along the ray tube does not hold in general, but is applicable only in the zeroth approximation of the ray theory, i.e. for the geometrical optics approximation.

For the derivation of head waves, we need to know the values of the additional components of $\mathbf{W}_1$ in homogeneous media. From (2.20), we get

$$W_1^2 = -\alpha(\nabla W_0^1 \cdot \mathbf{t}_2), \qquad W_1^3 = -\alpha(\nabla W_0^1 \cdot \mathbf{t}_3). \tag{2.31}$$

B Shear Waves

For shear waves, the phase function τ is a solution of the eikonal equation $(\nabla\tau)^2 = \beta^{-2}$, where $\beta = (\mu/\rho)^{1/2}$ (see (2.16′)). We shall determine the coefficients $\mathbf{W}_k$ in the ray series for shear waves, but since the procedure is essentially the same as for compressional waves, we shall be brief. We shall first determine the additional component $W_k^1\,\mathbf{t}_1$, and then the principal components $W_k^2\,\mathbf{t}_2$ and $W_k^3\,\mathbf{t}_3$ (see figure 2.5).

From (2.11) and (2.16′), we get

$$\mathbf{N}(\mathbf{W}_k)\cdot\mathbf{t}_1 = (\lambda+\mu)\beta^{-2}W_k^1, \qquad \mathbf{N}(\mathbf{W}_k)\cdot\mathbf{t}_2 = \mathbf{N}(\mathbf{W}_k)\cdot\mathbf{t}_3 = 0. \tag{2.32}$$

Taking the scalar product of (2.12′) with $\mathbf{t}_1$ and inserting into the above equation yields

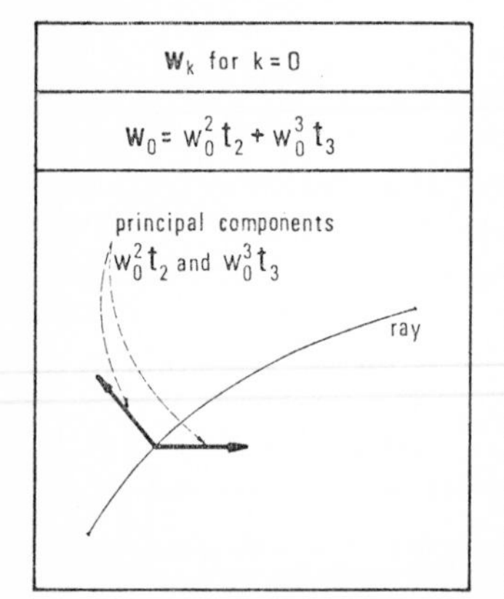

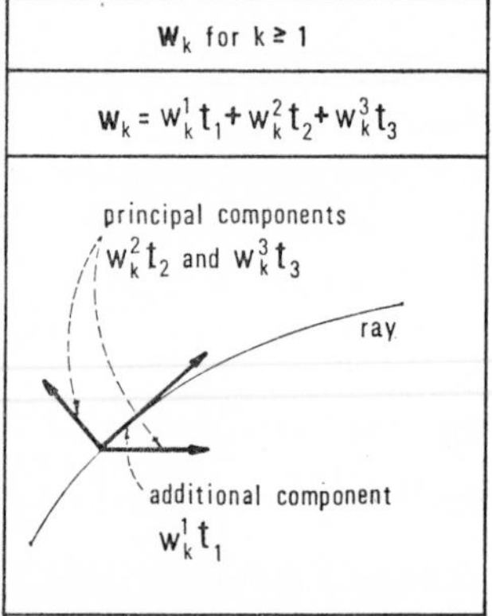

FIGURE 2.5 Principal and additional components of the coefficients of the ray series $\mathbf{W}_k$ for shear waves. The principal components are in the direction of $\mathbf{t}_2$ and $\mathbf{t}_3$, and the additional ones are along $\mathbf{t}_1$. For $k = 0$, the additional component vanishes. See figure 2.3 for comparison with compressional waves.

$$W_k^1 = [\beta^2/(\lambda+\mu)][\mathbf{M}(\mathbf{W}_{k-1})-\mathbf{L}(\mathbf{W}_{k-2})]\cdot\mathbf{t}_1. \tag{2.33}$$

We see from this expression that the additional component of $\mathbf{W}_k$, i.e. W_k^1, is obtained by simple differentiation of the vectors $\mathbf{W}_{k-1}$ and $\mathbf{W}_{k-2}$ (as in the case of compressional waves).

The principal components, W_k^2 and W_k^3, can be determined from the scalar product of (2.12′) with $\mathbf{t}_2$ and $\mathbf{t}_3$, successively, if we use (2.32). Then

$$\mathbf{M}(\mathbf{W}_k)\cdot\mathbf{t}_2 = \mathbf{L}(\mathbf{W}_{k-1})\cdot\mathbf{t}_2,$$
$$\mathbf{M}(\mathbf{W}_k)\cdot\mathbf{t}_3 = \mathbf{L}(\mathbf{W}_{k-1})\cdot\mathbf{t}_3.$$

Because $\mathbf{M}$ is a linear differential operator, we can write

$$\begin{aligned}\mathbf{M}(\mathbf{W}_k^\perp)\cdot\mathbf{t}_2 &= [\mathbf{L}(\mathbf{W}_{k-1})-\mathbf{M}(\mathbf{W}_k^1)]\cdot\mathbf{t}_2,\\ \mathbf{M}(\mathbf{W}_k^\perp)\cdot\mathbf{t}_3 &= [\mathbf{L}(\mathbf{W}_{k-1})-\mathbf{M}(\mathbf{W}_k^1)]\cdot\mathbf{t}_3,\end{aligned} \tag{2.34}$$

where $\mathbf{W}_k^1 = W_k^1\,\mathbf{t}_1$ and $\mathbf{W}_k^\perp$ is given by (2.21).

From the expression for $\mathbf{M}$ in (2.11), it follows that

$$\begin{aligned}\mathbf{M}(\mathbf{W}_k^\perp)\cdot\mathbf{t}_2 &= 2\mu\beta^{-1}\left(\mathbf{t}_2\cdot\frac{\mathrm{d}\mathbf{W}_k^\perp}{\mathrm{d}s}\right)+W_k^2[\mu\nabla^2\tau+(\nabla\tau\cdot\nabla)\mu],\\ \mathbf{M}(\mathbf{W}_k^\perp)\cdot\mathbf{t}_3 &= 2\mu\beta^{-1}\left(\mathbf{t}_3\cdot\frac{\mathrm{d}\mathbf{W}_k^\perp}{\mathrm{d}s}\right)+W_k^3[\mu\nabla^2\tau+(\nabla\tau\cdot\nabla)\mu].\end{aligned} \tag{2.35}$$

Upon differentiating (2.21) with respect to s and using Frenet's formulae for the derivatives of $\mathbf{t}_2$ and $\mathbf{t}_3$, we obtain (Schwartz *et al.*, 1960, p. 246)

$$\frac{\mathrm{d}\mathbf{W}_k^\perp}{\mathrm{d}s} = \frac{\mathrm{d}W_k^2}{\mathrm{d}s}\mathbf{t}_2+\frac{\mathrm{d}W_k^3}{\mathrm{d}s}\mathbf{t}_3+W_k^2\left(\frac{\mathbf{t}_3}{T}-\frac{\mathbf{t}_1}{K}\right)+W_k^3\left(-\frac{\mathbf{t}_2}{T}\right), \tag{2.36}$$

where T and K are the radii of torsion and curvature of the ray.

Inserting (2.36) and (2.35) into (2.34) yields

$$\begin{aligned}\frac{\mathrm{d}W_k^2}{\mathrm{d}s}-\frac{1}{T}W_k^3+\frac{\beta}{2\mu}W_k^2(\nabla\cdot\mu\nabla\tau) &= \frac{\beta}{2\mu}[\mathbf{L}(\mathbf{W}_{k-1})-\mathbf{M}(\mathbf{W}_k^1)]\cdot\mathbf{t}_2,\\ \frac{\mathrm{d}W_k^3}{\mathrm{d}s}+\frac{1}{T}W_k^2+\frac{\beta}{2\mu}W_k^3(\nabla\cdot\mu\nabla\tau) &= \frac{\beta}{2\mu}[\mathbf{L}(\mathbf{W}_{k-1})-\mathbf{M}(\mathbf{W}_k^1)]\cdot\mathbf{t}_3.\end{aligned} \tag{2.37}$$

Equations (2.37) are the higher transport equations for shear waves, corresponding to (2.25) for compressional waves. For $k = 0, 1, 2, \ldots$ they enable us to find successively all W_k^2 and W_k^3 ($k = 0, 1, 2, \ldots$), starting with W_0^2 and W_0^3, when two initial conditions for each k are given. However, the transport equations for shear waves are not so simple as for compressional waves. The two principal components, W_k^2 and W_k^3, are coupled to each other, except when the *torsion of the ray* vanishes along the whole ray (i.e. $1/T = 0$), i.e. when the ray is a plane curve. Equations (2.37) can be transformed into a single equation when we introduce a complex quantity $V_k = W_k^2 + iW_k^3$. Multiplying the second equation by i (where $i = (-1)^{1/2}$) and adding to the first, we obtain the following differential equation for V_k:

$$\frac{\mathrm{d}V_k}{\mathrm{d}s} + V_k\left(\frac{i}{T} + \frac{\beta}{2\mu}(\nabla\cdot\mu\nabla\tau)\right) = \frac{\beta}{2\mu}[\mathbf{L}(\mathbf{W}_{k-1}) - \mathbf{M}(\mathbf{W}_k^1)]\cdot\mathbf{l}, \tag{2.38}$$

where $\mathbf{l} = \mathbf{t}_2 + i\mathbf{t}_3$. This is a first-order ordinary differential equation, the solution of which is

$$V_k(s) = V_k(s_0)\exp\left\{-\int_{s_0}^{s}\left(\frac{i}{T} + \frac{\beta}{2\mu}(\nabla\cdot\mu\nabla\tau)\right)\mathrm{d}s\right\}$$
$$+ \int_{s_0}^{s}\left\{\frac{\beta}{2\mu}[\mathbf{L}(\mathbf{W}_{k-1}) - \mathbf{M}(\mathbf{W}_k^1)]\cdot\mathbf{l}\right\}\exp\left\{-\int_{s'}^{s}\left(\frac{i}{T} + \frac{\beta}{2\mu}(\nabla\cdot\mu\nabla\tau)\right)\mathrm{d}s\right\}\mathrm{d}s'. \tag{2.39}$$

This equation can be simplified by defining an angle $\Theta_T(s, s_0)$ by

$$\Theta_T(s, s_0) = \int_{s_0}^{s} T^{-1}\mathrm{d}s, \tag{2.40}$$

and taking into consideration that

$$\exp\left(-\int_{s_0}^{s}\frac{\beta}{2\mu}(\nabla\cdot\mu\nabla\tau)\mathrm{d}s\right) = \exp\left(-\frac{1}{2}\int_{s_0}^{s}\beta\nabla^2\tau\mathrm{d}s\right)\exp\left(-\frac{1}{2}\int_{s_0}^{s}\frac{1}{\mu}\frac{\mathrm{d}\mu}{\mathrm{d}s}\mathrm{d}s\right)$$
$$= \left(\frac{\mu(s_0)\beta(s)\mathrm{d}\sigma(s_0)}{\mu(s)\beta(s_0)\mathrm{d}\sigma(s)}\right)^{1/2} = \left(\frac{\rho(s_0)\beta(s_0)\mathrm{d}\sigma(s_0)}{\rho(s)\beta(s)\mathrm{d}\sigma(s)}\right)^{1/2},$$

to give

$$V_k(s) = V_k(s_0)\exp[-i\Theta_T(s, s_0)]\left(\frac{\rho(s_0)\beta(s_0)\mathrm{d}\sigma(s_0)}{\rho(s)\beta(s)\mathrm{d}\sigma(s)}\right)^{1/2}$$
$$+ \frac{1}{2}\left(\frac{1}{\rho(s)\beta(s)}\right)^{1/2}\int_{s_0}^{s}\{[\mathbf{L}(\mathbf{W}_{k-1}) - \mathbf{M}(\mathbf{W}_k^1)]\cdot\mathbf{l}\}$$
$$\times\exp[-i\Theta_T(s, s')]\left(\frac{\mathrm{d}\sigma(s')}{\mathrm{d}\sigma(s)}\right)^{1/2}\left(\frac{1}{\rho(s')\beta(s')}\right)^{1/2}\mathrm{d}s'. \tag{2.41}$$

Separating the real and imaginary parts yields

$$W_k^2(s) = \left(\frac{\rho(s_0)\beta(s_0)\mathrm{d}\sigma(s_0)}{\rho(s)\beta(s)\mathrm{d}\sigma(s)}\right)^{1/2} \{W_k^2(s_0)\cos\Theta_T(s, s_0) + W_k^3(s_0)\sin\Theta_T(s, s_0)\}$$

$$+\frac{1}{2}\frac{1}{[\rho(s)\beta(s)]^{1/2}}\int_{s_0}^{s}\frac{L_k^{(2)}\cos\Theta_T(s, s') + L_k^{(3)}\sin\Theta_T(s, s')}{[\rho(s')\beta(s')]^{1/2}}\left(\frac{\mathrm{d}\sigma(s')}{\mathrm{d}\sigma(s)}\right)^{1/2}\mathrm{d}s', \tag{2.42a}$$

$$W_k^3(s) = \left(\frac{\rho(s_0)\beta(s_0)\mathrm{d}\sigma(s_0)}{\rho(s)\beta(s)\mathrm{d}\sigma(s)}\right)^{1/2} \{W_k^3(s_0)\cos\Theta_T(s, s_0) - W_k^2(s_0)\sin\Theta_T(s, s_0)\}$$

$$+\frac{1}{2}\frac{1}{[\rho(s)\beta(s)]^{1/2}}\int_{s_0}^{s}\frac{L_k^{(3)}\cos\Theta_T(s, s') - L_k^{(2)}\sin\Theta_T(s, s')}{[\rho(s')\beta(s')]^{1/2}}\left(\frac{\mathrm{d}\sigma(s')}{\mathrm{d}\sigma(s)}\right)^{1/2}\mathrm{d}s', \tag{2.42b}$$

where

$$L_k^{(i)} = [\mathbf{L}(\mathbf{W}_{k-1}) - \mathbf{M}(\mathbf{W}_k^1)]\cdot\mathbf{t}_i \qquad (i = 2, 3).$$

From (2.42) we can see that for the determination of the principal components of $\mathbf{W}_k$ in the case of shear waves, we must know two initial values, viz., $W_k^2(s_0)$ and $W_k^3(s_0)$, for each k. The expressions (2.33) and (2.42) then enable us to determine successively all $\mathbf{W}_k$.

For $k = 0$ we have from (2.33) and (2.42)

$$\begin{aligned} W_0^1(s) &= 0, \\ W_0^2(s) &= \left(\frac{\rho(s_0)\beta(s_0)\mathrm{d}\sigma(s_0)}{\rho(s)\beta(s)\mathrm{d}\sigma(s)}\right)^{1/2}\{W_0^2(s_0)\cos\Theta_T(s, s_0) \\ &\qquad\qquad + W_0^3(s_0)\sin\Theta_T(s, s_0)\}, \\ W_0^3(s) &= \left(\frac{\rho(s_0)\beta(s_0)\mathrm{d}\sigma(s_0)}{\rho(s)\beta(s)\mathrm{d}\sigma(s)}\right)^{1/2}\{W_0^3(s_0)\cos\Theta_T(s, s_0) \\ &\qquad\qquad - W_0^2(s_0)\sin\Theta_T(s, s_0)\}, \end{aligned} \tag{2.43}$$

from which it follows that even for $k = 0$ the components W_0^2 and W_0^3 are coupled to each other when $1/T \neq 0$ (i.e., when the ray is not a plane curve). Therefore, the formulae for the variation of W_0^2 and W_0^3 with the arc length s are not so simple as for W_0^1 for compressional waves (see formula (2.30)). The standard energy considerations do not apply to the components W_0^2 and W_0^3 separately in this case.

We shall now briefly investigate the vector $\mathbf{W}_0$ for shear waves

($\mathbf{W}_0 = W_0^2\mathbf{t}_2 + W_0^3\mathbf{t}_3$). It follows from (2.43) that $\mathbf{W}_0$ lies in the plane normal to the ray and rotates around the ray as the wave progresses. In many cases, we are not interested in the direction of $\mathbf{W}_0$, but only in $|\mathbf{W}_0|$. From (2.43), we obtain simply (see also Karal and Keller, 1959, p. 703)

$$|\mathbf{W}_0(s)| = \left(\frac{\rho(s_0)\beta(s_0)\mathrm{d}\sigma(s_0)}{\rho(s)\beta(s)\mathrm{d}\sigma(s)}\right)^{1/2} |\mathbf{W}_0(s_0)|. \tag{2.43'}$$

We see that $|\mathbf{W}_0(s)|$ varies with s precisely in accordance with the principle of conservation of energy within a narrow tube of rays (as in the case of compressional waves). Formula (2.43′) applies to inhomogeneous media also when the ray may not be a plane curve; an example is a medium which is inhomogeneous in three dimensions.

If the ray is a plane curve, as in homogeneous or vertically inhomogeneous media, then $1/T = 0$ and $\Theta_T = 0$, and the expressions for $W_0^2(s)$ and $W_0^3(s)$ become simpler:

$$\begin{aligned} W_0^2(s) &= W_0^2(s_0)\left(\frac{\rho(s_0)\beta(s_0)\mathrm{d}\sigma(s_0)}{\rho(s)\beta(s)\mathrm{d}\sigma(s)}\right)^{1/2}, \\ W_0^3(s) &= W_0^3(s_0)\left(\frac{\rho(s_0)\beta(s_0)\mathrm{d}\sigma(s_0)}{\rho(s)\beta(s)\mathrm{d}\sigma(s)}\right)^{1/2}. \end{aligned} \tag{2.44}$$

In this case, the principal components W_0^2 and W_0^3 are not coupled to each other and propagate as if they were independent waves; energy considerations can be applied to the two components separately. The formulae (2.44) are the same as (2.30) for the principal component of compressional waves, except that the shear wave velocity β replaces the compressional wave velocity α.

As in the case of compressional waves, we shall need to know the value of the additional component of $\mathbf{W}_1$ in homogeneous media for the derivation of head waves. From (2.33) we easily find that

$$W_1^1 = \beta(\nabla \cdot \mathbf{W}_0). \tag{2.45}$$

2.2.3 *Recapitulation of the General Procedure*

Let us summarize briefly the results of this section. If we write the solution of the equation of motion (2.1) in the form of a ray series (2.2), the phase function τ satisfies one of the eikonal equations (2.15′) and (2.16′). We call the wave a compressional wave or a shear wave according as it satisfies (2.15′) or (2.16′). The eikonal equation may be solved by the usual methods of geometrical optics. The surfaces $\tau(x, y, z) = \text{const.}$ represent wave fronts and the orthogonal trajectories of these surfaces are the rays. When the eikonal equation has been solved, the coefficients of the ray series $\mathbf{W}_k$ may be determined successively. The most important coefficient is $\mathbf{W}_0$, which has a simple physical interpretation. Its magnitude varies in accordance with the law of conservation of energy within a narrow ray tube. For the compressional wave $\mathbf{W}_0$ is tangent to the ray, whereas for the shear wave it lies in the plane normal to the ray.

The higher coefficients $\mathbf{W}_k$ ($k \geqslant 1$), however, have both components, tangent to the ray and perpendicular to it. Hence the compressional wave is not purely longitudinal and the shear wave is not purely transverse. The additional components can be determined by simple differentiation of $\mathbf{W}_{k-1}$ and $\mathbf{W}_{k-2}$, and the principal components by solving first-order ordinary differential equations. For the solution of these differential equations, we must know some initial data (two for shear waves and one for compressional waves for each k). It was shown above that construction of the ray series requires successive differentiation of the coefficients of the series. As the zeroth term of the ray series depends on elastic parameters, the kth term $\mathbf{W}_k$ depends on the kth derivatives of these parameters.

2.3 BEHAVIOUR OF WAVES AT AN INTERFACE

2.3.1 *Discussion of Interface Conditions*

The formulae derived in the previous section can be applied only when λ, μ, and ρ and all their derivatives are continuous along the ray. We mentioned earlier that the coefficients $\mathbf{W}_k$ in the ray series (2.2) depend on the kth derivatives of the elastic parameters. Therefore, when the ray strikes a surface on which the kth derivatives of the elastic parameters are discontinuous, $\mathbf{W}_k$ also becomes discontinuous at this surface, and it is not possible to determine the subsequent terms in the ray series. To "compensate" for the discontinuity in $\mathbf{W}_k$ we must introduce new waves, viz. *reflected* and *refracted* waves. In all the problems which follow, it will be assumed that the elastic media are in *welded contact* with each other across the surface at which the elastic parameters, or their derivatives, are discontinuous. This implies continuity of displacement and stress across the interface. The six conditions expressing this continuity enable us to find properties of reflected and refracted waves.

We shall call the surfaces at which the nth derivatives of the elastic parameters are discontinuous *interfaces of* $(n+1)$*th order*; at the *interface of first order*, the elastic parameters themselves are discontinuous. For simplicity, the interfaces of nth order for $n \geqslant 2$ will be called *weak interfaces* also. In other words, weak interfaces are surfaces across which the elastic parameters themselves are continuous but some of their derivatives are discontinuous.

Reflection and refraction of elastic waves from an interface is a basic problem in seismology and seismic prospecting. However, only relatively simpler situations have been considered in seismic literature. The main attention has been devoted to interfaces of the first order. Weak interfaces have been investigated only exceptionally; see, e.g., Chekin (1959), Tsepelev (1959), Gupta (1966a, 1966b), for plane interfaces of second order. Usually several other serious restrictions are introduced in the treatment of even first-order interfaces, some of the usual ones being as follows: (1) the interfaces are plane, (2) the wave front of the incident wave is plane (or spherical), (3) the media which are in contact at an interface are homogeneous. In reality, however, the interfaces are likely to be curved, the media are likely to be heterogeneous, and the incident waves come in many shapes.

To deal with curved interfaces and arbitrary types of incident waves, the *principle of isolated element* (also called the *principle of local field character*) has often been used. It states that an incident wave with an arbitrary wave front is reflected (or refracted) from any point of a curved interface Σ in the same way that a plane wave would be from a plane interface at the *same point*. Both planes are assumed to be tangent to the corresponding surfaces at the point of reflection (or refraction). This principle was used for the theoretical investigation of some seismological problems by Skuridin (1957), where other references are also given. It should be noted that in seismic measurements, particularly in seismic prospecting, seismic waves reflected from an irregular boundary are interpreted essentially on the basis of this principle. We shall discuss the limitations of this principle later.

In this section, we shall derive (from the interface conditions) a general set of linear equations which will enable us to find successively the coefficients of the ray series for reflected and refracted waves in a very general case of a *curved interface*, of *any order*, between two *inhomogeneous media* for an *arbitrarily shaped incident wave front*. However, we shall also impose a restriction, viz., that the interface is *smooth* (see more details later). We must also remember that our method is not exact, but only asymptotic. We feel that the set of linear equations obtained can be the basis for the solution of many problems of seismological interest, although in this book they are used for the investigation of head waves only.

We shall first consider an *interface of first-order* (across which at least one elastic parameter is discontinuous), denoted by Σ. We shall assume that Σ is analytic; in other words, that the principle curvatures of Σ, $1/R_1$ and $1/R_2$, are analytic functions. Suppose that the displacement vector of the incident wave $\mathbf{W}^0$ is given by

$$\mathbf{W}^0 = \exp[i\omega(t-\tau_0)] \sum_{k=0}^{\infty} (i\omega)^{-k}\mathbf{W}_k^0, \tag{2.46}$$

where τ and $\mathbf{W}_k^0$ are known. We shall consider one arbitrary ray of this wave, and denote the point at which it strikes the interface Σ by O. We use λ_1, μ_1, and ρ_1 to designate the elastic parameters on one side of the interface and λ_2, μ_2, and ρ_2 on the other side. In addition, we define

$$\alpha_i = \left(\frac{\lambda_i+2\mu_i}{\rho_i}\right)^{1/2}, \qquad \beta_i = \left(\frac{\mu_i}{\rho_i}\right)^{1/2}, \qquad \gamma_i = \frac{\beta_i}{\alpha_i} = \left(\frac{\mu_i}{\lambda_i+2\mu_i}\right)^{1/2}.$$

We shall call the medium with the parameters λ_1, μ_1, and ρ_1 the *first* medium, and the other the *second* medium. Note that the elastic parameters need not be constant throughout each medium, but may change continuously.

We shall define the *plane of incidence* ($E_{\parallel}$) as a plane determined by the normal to the interface and the tangent to the ray at the point O (see figure 2.6). If the ray is a space curve (i.e. a curve of double curvature, with some

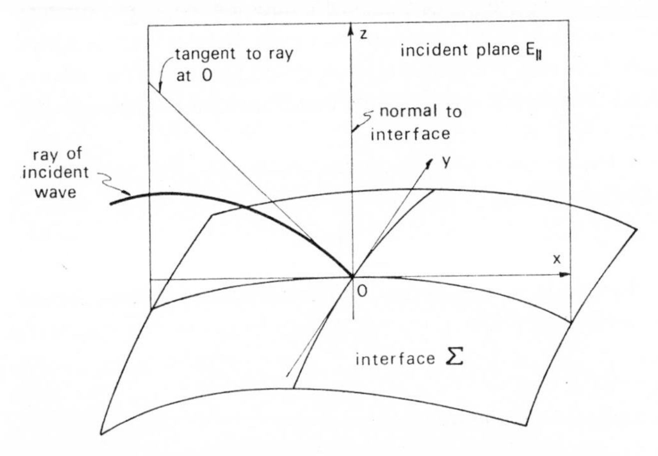

FIGURE 2.6 Geometry of incidence. The incident ray may be a curve of double curvature when the medium is inhomogeneous and, therefore, need not lie in the plane of incidence.

torsion), the entire ray does not lie in the plane of incidence. We introduce a Cartesian coordinate system x, y, z with its origin at the point O. The z-axis is normal to Σ with the positive direction in the first medium. The x-axis lies along the intersection of $E_{\|}$ with the plane tangent to the interface (at the point O), with the positive direction in the direction of propagation of the incident wave. The positive direction of the y-axis is determined so that the system is right-handed. Now we introduce three unit vectors $\mathbf{i}, \mathbf{j}, \mathbf{k}$ in the directions of the x, y, and z axes, respectively. We do not assume that the wave is incident on the interface from the first medium; it may reach the interface from either side.

There are six conditions expressing continuity of displacement and stress across the boundary. It is easy to see that the incident wave alone cannot satisfy these six conditions simultaneously. It is necessary to consider other waves arising at the interface, viz., reflected and refracted compressional and shear waves. We shall designate the type of waves generated by an index ν, which may assume values 1, 2, 3, and 4. $\nu = 1$ and 2 refer to compressional and shear waves in the first medium and $\nu = 3$ and 4 to compressional and shear waves in the second medium, respectively. We shall assume that for each of these waves a ray series expansion similar to (2.2) can be written:

$$\mathbf{W}^{\nu} = \exp[i\omega(t-\tau_{\nu})] \sum_{k=0}^{\infty} (i\omega)^{-k} \mathbf{W}_k^{\nu}. \tag{2.47}$$

The purpose of this section is to find τ_{ν} and $\mathbf{W}_k^{\nu}$ (for $\nu = 1, 2, 3, 4$ and $k = 0, 1, 2, \ldots$) from a knowledge of the elastic and geometric parameters of the media and properties of the incident wave (i.e. from a knowledge of τ_0 and $\mathbf{W}_k^0$, $k = 0, 1, 2, \ldots$). We shall first determine the phase function τ_{ν} for each wave, and then apply the interface conditions and determine $\mathbf{W}_k^{\nu}$ successively for $k = 0, 1, 2, \ldots$

We use V_{ν} to denote the velocity corresponding to the velocity of propagation of the wave front of the νth wave, i.e.

$$V_1 = \alpha_1, \qquad V_2 = \beta_1, \qquad V_3 = \alpha_2, \qquad V_4 = \beta_2. \tag{2.48}$$

Then the phase functions τ_{ν} are given by the solutions of the eikonal equations (2.15′) and (2.16′), i.e.

$$(\nabla\tau_{\nu})^2 = V_{\nu}^{-2} \qquad (\nu = 1, 2, 3, 4), \tag{2.49}$$

with the initial conditions at the interface

$$\tau_{\nu} = \tau_0 \qquad (\nu = 1, 2, 3, 4). \tag{2.50}$$

From the above equation, it follows that

$$\frac{\partial \tau_\nu}{\partial x} = \frac{\partial \tau_0}{\partial x} = \frac{\sin \theta_\nu}{V_\nu}, \qquad \frac{\partial \tau_\nu}{\partial y} = \frac{\partial \tau_0}{\partial y} = 0, \tag{2.51}$$

for $\nu = 1, 2, 3, 4$ at the point O, where θ_ν is the acute angle between the z-axis and the ray of the νth wave at the point O.

We can see from the second formula in (2.51) that the tangents to the rays of reflected and refracted waves at O lie in the incident plane $E_{\parallel}$. If we express $\partial\tau_0/\partial x$ in terms of the angle of incidence and the velocity of propagation of the incident wave at O, we obtain, from the first formula in (2.51), the well-known *Snell's law*.

The conditions (2.50) and (2.51) do not guarantee the uniqueness of the solution of (2.49). We must take into account the direction of propagation of the wave front of the νth wave and the radiation condition. In other words, we must determine the sign of the derivative $\partial\tau_\nu/\partial z$ at the point O. It follows from (2.49) and (2.51) that

$$\frac{\partial \tau_\nu}{\partial z} = \pm\left[V_\nu^{-2} - \left(\frac{\partial \tau_\nu}{\partial x}\right)^2\right]^{1/2} = \pm\left[V_\nu^{-2} - \left(\frac{\partial \tau_0}{\partial x}\right)^2\right]^{1/2}.$$

As the waves $\mathbf{W}^1$ and $\mathbf{W}^2$ propagate from the interface into the first medium, $\partial\tau_\nu/\partial z > 0$ for $\nu = 1, 2$ at O. Similarly, $\partial\tau_\nu/\partial z < 0$ for $\nu = 3, 4$ at O. Then we can write, at the point O,

$$\frac{\partial \tau_\nu}{\partial z} = (-1)^{\varepsilon_\nu + 1} \frac{\cos \theta_\nu}{V_\nu}, \tag{2.51$'$}$$

where ε_ν assumes the value 1 or 2, depending on whether the νth wave propagates in the first or the second medium, i.e.

$$\begin{aligned} \varepsilon_\nu &= 1 \quad \text{for } \nu = 1, 2, \\ \varepsilon_\nu &= 2 \quad \text{for } \nu = 3, 4. \end{aligned} \tag{2.52}$$

If $\partial\tau_0/\partial x > V_\nu^{-1}$, then $\sin\theta_\nu > 1$ and $\cos\theta_\nu = \pm i(\sin^2\theta_\nu - 1)^{1/2}$. The radiation condition dictates that we must choose

$$\cos\theta_\nu = -i(\sin^2\theta_\nu - 1)^{1/2}, \tag{2.53}$$

since the other choice leads to an exponential increase in the amplitude of the wave $\mathbf{W}^\nu$ with distance from the interface, a physically untenable situation.

The conditions (2.50), (2.51), (2.51′), and (2.53) are sufficient initial conditions for a unique solution of the eikonal equations (2.49). The phase functions τ_ν, the rays, and the wave fronts of reflected and refracted waves can then be determined using the methods described in section 2.2.1.

Now we shall determine the coefficients $\mathbf{W}_k^\nu$ from the interface conditions. At the point O, we define three perpendicular unit vectors $\mathbf{n}_\mathrm{P}^\nu$, $\mathbf{n}_\mathrm{SV}^\nu$, and $\mathbf{n}_\mathrm{SH}^\nu$ for every ν connected with the rays of the waves under consideration. $\mathbf{n}_\mathrm{P}^\nu$ is the unit vector tangent to the ray, pointing in the direction of propagation; it corresponds to the vector $\mathbf{t}_1$ in the previous section. $\mathbf{n}_\mathrm{SV}^\nu$ is perpendicular to the ray and lies in the plane $E_\parallel$ (chosen such that its projection on the x-axis is positive). $\mathbf{n}_\mathrm{SH}^\nu$ corresponds directly to the unit vector $\mathbf{j}$ (see figure 2.7). The unit vectors $\mathbf{n}_\mathrm{SV}^\nu$ and $\mathbf{n}_\mathrm{SH}^\nu$ lie in the same plane (normal to the ray) as the unit vectors $\mathbf{t}_2$ and $\mathbf{t}_3$ defined in the previous section. If we know the projections of an arbitrary vector on the directions

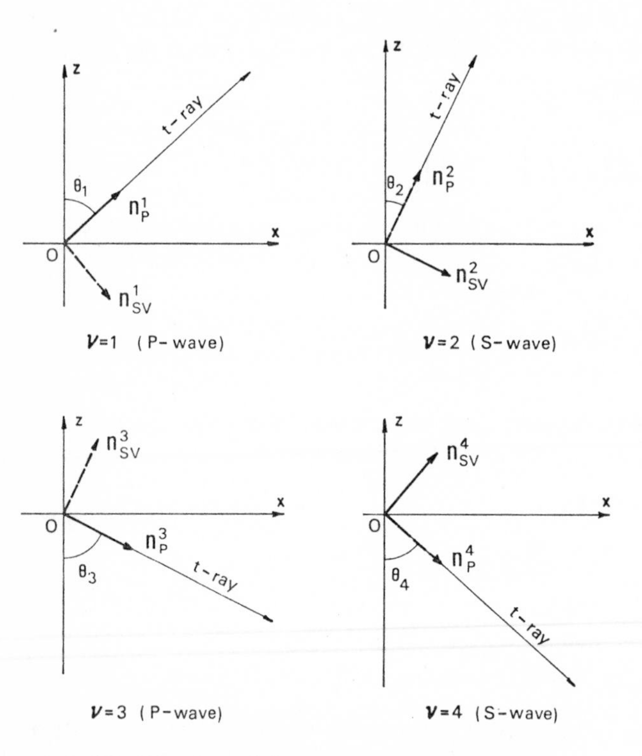

FIGURE 2.7 Directions of the unit vectors $\mathbf{n}_\mathrm{P}^\nu$ and $\mathbf{n}_\mathrm{SV}^\nu$ for reflected and refracted waves ($\nu = 1, 2, 3, 4$). $\mathbf{n}_\mathrm{SH}^\nu$ is not shown here, as it is always perpendicular to the xz plane. The straight line "t-ray" denotes the tangent to the ray of the wave under consideration at O. The unit vectors corresponding to the principal components are given by solid lines, and those corresponding to additional components by dashed lines.

of $\mathbf{t}_1$, $\mathbf{t}_2$, and $\mathbf{t}_3$, we can easily determine its projections on the directions $\mathbf{n}_P^\nu$, $\mathbf{n}_{SV}^\nu$, and $\mathbf{n}_{SH}^\nu$, and vice versa, from simple geometrical considerations.

We can write for $\mathbf{W}_k^\nu$

$$\mathbf{W}_k^\nu = W_{k,P}^\nu \,\mathbf{n}_P^\nu + W_{k,SV}^\nu \,\mathbf{n}_{SV}^\nu + W_{k,SH}^\nu \,\mathbf{n}_{SH}^\nu. \tag{2.54}$$

Corresponding to the four generated waves, we have twelve components, $W_{k,P}^\nu$, $W_{k,SV}^\nu$, and $W_{k,SH}^\nu$ ($\nu = 1, 2, 3, 4$), for each k. Six of these are principal components ($W_{k,P}^1$, $W_{k,P}^3$, $W_{k,SV}^2$, $W_{k,SV}^4$, $W_{k,SH}^2$, and $W_{k,SH}^4$), and the other six are additional components ($W_{k,P}^2$, $W_{k,P}^4$, $W_{k,SV}^1$, $W_{k,SV}^3$, $W_{k,SH}^1$, and $W_{k,SH}^3$). For the determination of the additional components, we do not need to know any initial data, since they can be determined by differentiation of the preceding terms in the ray series, $\mathbf{W}_{k-1}^\nu$ and $\mathbf{W}_{k-2}^\nu$ (see (2.20) and (2.33)). However, for the determination of the principal components at any point of the ray we *must* know their values at some previous point of the ray (see (2.29) and (2.42)). Therefore, when we want to find the expression for the coefficients $\mathbf{W}_k^\nu$ for reflected and refracted waves at any point of their rays, we must first determine the values of the principal components of $\mathbf{W}_k^\nu$ at the interface and these components will serve as initial data in (2.29) and (2.42). It follows that the only unknown quantities, for a given k, are the principal components $W_{k,P}^1$, $W_{k,P}^3$, $W_{k,SV}^2$, $W_{k,SV}^4$, $W_{k,SH}^2$, $W_{k,SH}^4$.

We denote the x, y, and z components of $\mathbf{W}^\nu$ and $\mathbf{W}_k^\nu$ by W_x^ν, W_y^ν, W_z^ν and W_{kx}^ν, W_{ky}^ν, W_{kz}^ν, respectively. We shall use a similar notation for the incident wave ($\nu = 0$) also. The components of $\mathbf{W}_k^\nu$ ($\nu = 1, 2, 3, 4$) are given by the expressions (see figure 2.7)

$$\begin{aligned} W_{kx}^\nu &= W_{k,P}^\nu \sin\theta_\nu + W_{k,SV}^\nu \cos\theta_\nu, \\ W_{ky}^\nu &= W_{k,SH}^\nu, \\ W_{kz}^\nu &= [W_{k,P}^\nu \cos\theta_\nu - W_{k,SV}^\nu \sin\theta_\nu](-1)^{\varepsilon_\nu+1}, \end{aligned} \tag{2.55}$$

where ε_ν is given by (2.52).

We have assumed that solid elastic media are welded together at the surface of contact. Therefore, all stress and displacement components must be continuous across the boundary. More specifically, we shall require continuity of x, y, and z components of the displacement vectors and continuity of η_{zz}, η_{xz}, and η_{yz} stress components. The stress components can be expressed by means of displacements, in the following manner (see Ewing *et al.*, 1957, p. 6):

$$\eta_{xz} = \mu\left(\frac{\partial W_x}{\partial z} + \frac{\partial W_z}{\partial x}\right),$$

$$\eta_{yz} = \mu\left(\frac{\partial W_y}{\partial z}+\frac{\partial W_z}{\partial y}\right),$$

$$\eta_{zz} = \lambda(\nabla\cdot\mathbf{W})+2\mu\frac{\partial W_z}{\partial z}.$$

Taking into account the fact that the waves $\mathbf{W}^1$ and $\mathbf{W}^2$ exist in the first medium, the waves $\mathbf{W}^3$ and $\mathbf{W}^4$ in the second medium, and the incident wave $\mathbf{W}^0$ may strike the interface from any side, we can write the interface conditions in the following form:

$$\begin{aligned}
&(\text{I}) && \sum_{\nu=1}^{4}(-1)^{\varepsilon_\nu}W_x^\nu &&= (-1)^{\varepsilon_0+1}W_x^0,\\
&(\text{II}) && \sum_{\nu=1}^{4}(-1)^{\varepsilon_\nu}W_z^\nu &&= (-1)^{\varepsilon_0+1}W_z^0,\\
&(\text{III}) && \sum_{\nu=1}^{4}(-1)^{\varepsilon_\nu}\eta_{zz}(\mathbf{W}^\nu) &&= (-1)^{\varepsilon_0+1}\eta_{zz}(\mathbf{W}^0),\\
&(\text{IV}) && \sum_{\nu=1}^{4}(-1)^{\varepsilon_\nu}\eta_{xz}(\mathbf{W}^\nu) &&= (-1)^{\varepsilon_0+1}\eta_{xz}(\mathbf{W}^0),\\
&(\text{V}) && \sum_{\nu=1}^{4}(-1)^{\varepsilon_\nu}W_y^\nu &&= (-1)^{\varepsilon_0+1}W_y^0,\\
&(\text{VI}) && \sum_{\nu=1}^{4}(-1)^{\varepsilon_\nu}\eta_{yz}(\mathbf{W}^\nu) &&= (-1)^{\varepsilon_0+1}\eta_{yz}(\mathbf{W}^0),
\end{aligned} \tag{2.56}$$

where ε_ν ($\nu = 1, 2, 3, 4$) is given by (2.52) and ε_0 assumes the value 1 or 2, depending on whether the incident wave is in the first or the second medium.

To express the interface conditions (2.56) in terms of the principal and additional components of the coefficients of the ray series for reflected and refracted waves (R-waves), we must use (2.47), (2.51), (2.51′), (2.54), and (2.55). As an example, we shall investigate condition IV in detail. From (2.47) it follows that

$$W_x^\nu = \exp[i\omega(t-\tau_\nu)]\sum_{k=0}^{\infty}(i\omega)^{-k}W_{kx}^\nu,$$

$$W_z^\nu = \exp[i\omega(t-\tau_\nu)]\sum_{k=0}^{\infty}(i\omega)^{-k}W_{kz}^\nu.$$

Inserting this into the expression for η_{xz}, we obtain

$$\eta_{xz}(\mathbf{W}^\nu) = \mu_{\varepsilon_\nu}\sum_{k=0}^{\infty}(i\omega)^{-k}\left\{\frac{\partial}{\partial z}\{\exp[i\omega(t-\tau_\nu)]W^\nu_{kx}\}+\frac{\partial}{\partial x}\{\exp[i\omega(t-\tau_\nu)]W^\nu_{kz}\}\right\}$$

$$= \mu_{\varepsilon_\nu}\exp[i\omega(t-\tau_\nu)]\sum_{k=0}^{\infty}(i\omega)^{-k}\left\{\left(\frac{\partial W^\nu_{kx}}{\partial z}+\frac{\partial W^\nu_{kz}}{\partial x}\right)-i\omega\left(W^\nu_{kx}\frac{\partial\tau_\nu}{\partial z}+W^\nu_{kz}\frac{\partial\tau_\nu}{\partial x}\right)\right\}$$

$$= -\mu_{\varepsilon_\nu}\exp[i\omega(t-\tau_\nu)]\sum_{k=0}^{\infty}(i\omega)^{-k+1}\left\{W^\nu_{kx}\frac{\partial\tau_\nu}{\partial z}+W^\nu_{kz}\frac{\partial\tau_\nu}{\partial x}-\frac{\partial W^\nu_{k-1,x}}{\partial z}-\frac{\partial W^\nu_{k-1,z}}{\partial x}\right\},$$

where $W^\nu_{-1,x} = W^\nu_{-1,z} = 0$ (as $\mathbf{W}^\nu_{-1} = 0$). The same expression is applicable for $\nu = 0$ (incident wave). Inserting the above into condition IV and taking into account that $\tau_\nu = \tau_0$ (see (2.50)), we obtain

$$\sum_{k=0}^{\infty}(i\omega)^{-k+1}\left\{\sum_{\nu=1}^{4}(-1)^{\varepsilon_\nu}\mu_{\varepsilon_\nu}\left(W^\nu_{kx}\frac{\partial\tau_\nu}{\partial z}+W^\nu_{kz}\frac{\partial\tau_\nu}{\partial x}-\frac{\partial W^\nu_{k-1,x}}{\partial z}-\frac{\partial W^\nu_{k-1,z}}{\partial x}\right)\right.$$

$$\left.-(-1)^{\varepsilon_0+1}\mu_{\varepsilon_0}\left(W^0_{kx}\frac{\partial\tau_0}{\partial z}+W^0_{kz}\frac{\partial\tau_0}{\partial x}-\frac{\partial W^0_{k-1,x}}{\partial z}-\frac{\partial W^0_{k-1,z}}{\partial x}\right)\right\} = 0.$$

As this series must vanish for any ω, the coefficient of each power of ω must be zero. Therefore

$$\sum_{\nu=1}^{4}(-1)^{\varepsilon_\nu+1}\mu_{\varepsilon_\nu}\left(W^\nu_{kx}\frac{\partial\tau_\nu}{\partial z}+W^\nu_{kz}\frac{\partial\tau_\nu}{\partial x}\right) = a_1+a_2+a_3,$$

where

$$a_1 = (-1)^{\varepsilon_0}\mu_{\varepsilon_0}\left(W^0_{kx}\frac{\partial\tau_0}{\partial z}+W^0_{kz}\frac{\partial\tau_0}{\partial x}\right),$$

$$a_2 = (-1)^{\varepsilon_0+1}\mu_{\varepsilon_0}\left(\frac{\partial W^0_{k-1,x}}{\partial z}+\frac{\partial W^0_{k-1,z}}{\partial x}\right),$$

$$a_3 = \sum_{\nu=1}^{4}(-1)^{\varepsilon_\nu+1}\mu_{\varepsilon_\nu}\left(\frac{\partial W^\nu_{k-1,x}}{\partial z}+\frac{\partial W^\nu_{k-1,z}}{\partial x}\right).$$

Now we insert the expressions for W^ν_{kx} and W^ν_{kz}, given by (2.55), and the expressions for $\partial\tau_\nu/\partial x$ and $\partial\tau_\nu/\partial z$, given by (2.51) and (2.51'). Then we can write the equation in the form

$$\sum_{\nu=1}^{4}(-1)^{\varepsilon_\nu+1}\mu_{\varepsilon_\nu}\left((W^\nu_{k,\mathrm{P}}\sin\theta_\nu+W^\nu_{k,\mathrm{SV}}\cos\theta_\nu)(-1)^{\varepsilon_\nu+1}\frac{\cos\theta_\nu}{V_\nu}\right.$$

$$\left.+(W^\nu_{k,\mathrm{P}}\cos\theta_\nu-W^\nu_{k,\mathrm{SV}}\sin\theta_\nu)(-1)^{\varepsilon_\nu+1}\frac{\sin\theta_\nu}{V_\nu}\right) = a_1+a_2+a_3,$$

and from this obtain

$$\sum_{\nu=1}^{4}\left(\frac{\mu_{\varepsilon}}{V_\nu}\sin 2\theta_\nu\, W^{\nu}_{k,\mathrm{P}}+\frac{\mu_{\varepsilon_\nu}}{V_\nu}\cos 2\theta_\nu\, W^{\nu}_{k,\mathrm{SV}}\right) = a_1+a_2+a_3.$$

In the above equation, we have four principal components ($W^1_{k,\mathrm{P}}$, $W^3_{k,\mathrm{P}}$, $W^2_{k,\mathrm{SV}}$, and $W^4_{k,\mathrm{SV}}$) and four additional components ($W^2_{k,\mathrm{P}}$, $W^4_{k,\mathrm{P}}$, $W^1_{k,\mathrm{SV}}$, $W^3_{k,\mathrm{SV}}$). If we transfer the additional components to the right-hand side, insert (2.48) for V_ν ($\nu = 1, 2, 3, 4$), and introduce

$$\begin{aligned} a_4 = & -W^1_{k,\mathrm{SV}}\,\rho_1\,\beta_1\,\gamma_1\cos 2\theta_1 - W^2_{k,\mathrm{P}}\,\rho_1\,\beta_1\sin 2\theta_2 \\ & - W^3_{k,\mathrm{SV}}\,\rho_2\,\beta_2\,\gamma_2\cos 2\theta_3 - W^4_{k,\mathrm{P}}\,\rho_2\,\beta_2\sin 2\theta_4, \end{aligned}$$

we obtain finally

$$\begin{aligned} & W^1_{k,\mathrm{P}}\,\rho_1\,\gamma_1\,\beta_1\sin 2\theta_1 + W^2_{k,\mathrm{SV}}\,\rho_1\,\beta_1\cos 2\theta_2 + W^3_{k,\mathrm{P}}\,\rho_2\,\beta_2\,\gamma_2\sin 2\theta_3 \\ & \qquad + W^4_{k,\mathrm{SV}}\,\rho_2\,\beta_2\cos 2\theta_4 = a_1+a_2+a_3+a_4, \end{aligned}$$

where $\gamma_1 = \beta_1/\alpha_1$, $\gamma_2 = \beta_2/\alpha_2$. This is the final form of condition IV. It holds for each k ($k = 0, 1, 2, \ldots$). Using the same procedure for other interface conditions, we obtain, for each k,

$$\begin{aligned} &\text{(I)} \quad W^1_{k,\mathrm{P}}\sin\theta_1 + W^2_{k,\mathrm{SV}}\cos\theta_2 - W^3_{k,\mathrm{P}}\sin\theta_3 - W^4_{k,\mathrm{SV}}\cos\theta_4 = F_1, \\ &\text{(II)} \quad W^1_{k,\mathrm{P}}\cos\theta_1 - W^2_{k,\mathrm{SV}}\sin\theta_2 + W^3_{k,\mathrm{P}}\cos\theta_3 - W^4_{k,\mathrm{SV}}\sin\theta_4 = F_2, \\ &\text{(III)} \quad W^1_{k,\mathrm{P}}\,\rho_1\,\alpha_1\cos 2\theta_2 - W^2_{k,\mathrm{SV}}\,\rho_1\,\beta_1\sin 2\theta_2 \\ &\qquad - W^3_{k,\mathrm{P}}\,\rho_2\,\alpha_2\cos 2\theta_4 + W^4_{k,\mathrm{SV}}\,\rho_2\,\beta_2\sin 2\theta_4 = F_3, \\ &\text{(IV)} \quad W^1_{k,\mathrm{P}}\,\rho_1\,\gamma_1\,\beta_1\sin 2\theta_1 + W^2_{k,\mathrm{SV}}\,\rho_1\,\beta_1\cos 2\theta_2 \\ &\qquad + W^3_{k,\mathrm{P}}\,\rho_2\,\beta_2\,\gamma_2\sin 2\theta_3 + W^4_{k,\mathrm{SV}}\,\rho_2\,\beta_2\cos 2\theta_4 = F_4, \\ &\text{(V)} \quad W^2_{k,\mathrm{SH}} - W^4_{k,\mathrm{SH}} = F_5, \\ &\text{(VI)} \quad W^2_{k,\mathrm{SH}}\,\rho_1\,\beta_1\cos\theta_2 + W^4_{k,\mathrm{SH}}\,\rho_2\,\beta_2\cos\theta_4 = F_6, \end{aligned} \tag{2.57}$$

where the functions F_i ($i = 1, 2, \ldots, 6$) are given by the expressions

$$F_i = F_{i1}+F_{i2}+F_{i3}+F_{i4} \qquad (i = 1, 2, \ldots, 6), \tag{2.57'}$$

with

$$\left.\begin{aligned} F_{11} &= (-1)^{\varepsilon_0} W^0_{kx}, \\ F_{21} &= (-1)^{\varepsilon_0} W^0_{kz}, \\ F_{31} &= (-1)^{\varepsilon_0}\left(\lambda_{\varepsilon_0}\frac{\partial\tau_0}{\partial x}\,W^0_{kx} + (\lambda_{\varepsilon_0}+2\mu_{\varepsilon_0})W^0_{kz}\,\frac{\partial\tau_0}{\partial z}\right), \end{aligned}\right\} \tag{2.58}$$

$$\left.\begin{aligned}
F_{41} &= (-1)^{\varepsilon_0}\mu_{\varepsilon_0}\left(\frac{\partial\tau_0}{\partial z}W^0_{kx}+\frac{\partial\tau_0}{\partial x}W^0_{kz}\right),\\
F_{51} &= (-1)^{\varepsilon_0}W^0_{ky},\\
F_{61} &= (-1)^{\varepsilon_0}\mu_{\varepsilon_0}\frac{\partial\tau_0}{\partial z}W^0_{ky},
\end{aligned}\right\}\quad (2.58)\ (cont.)$$

$$\left.\begin{aligned}
F_{12} &= F_{22} = F_{52} = 0,\\
F_{32} &= (-1)^{\varepsilon_0+1}\Bigg(\lambda_{\varepsilon_0}\frac{\partial W^0_{k-1,x}}{\partial x}+\lambda_{\varepsilon_0}\frac{\partial W^0_{k-1,y}}{\partial y}\\
&\qquad\qquad +(\lambda_{\varepsilon_0}+2\mu_{\varepsilon_0})\frac{\partial W^0_{k-1,z}}{\partial z}\Bigg),\\
F_{42} &= (-1)^{\varepsilon_0+1}\mu_{\varepsilon_0}\left(\frac{\partial W^0_{k-1,x}}{\partial z}+\frac{\partial W^0_{k-1,z}}{\partial x}\right),\\
F_{62} &= (-1)^{\varepsilon_0+1}\mu_{\varepsilon_0}\left(\frac{\partial W^0_{k-1,y}}{\partial z}+\frac{\partial W^0_{k-1,z}}{\partial y}\right),
\end{aligned}\right\}\quad (2.58')$$

$$\left.\begin{aligned}
F_{13} &= -W^1_{k,\mathrm{SV}}\cos\theta_1-W^2_{k,\mathrm{P}}\sin\theta_2+W^3_{k,\mathrm{SV}}\cos\theta_3+W^4_{k,\mathrm{P}}\sin\theta_4,\\
F_{23} &= W^1_{k,\mathrm{SV}}\sin\theta_1-W^2_{k,\mathrm{P}}\cos\theta_2+W^3_{k,\mathrm{SV}}\sin\theta_3-W^4_{k,\mathrm{P}}\cos\theta_4,\\
F_{33} &= W^1_{k,\mathrm{SV}}\rho_1\beta_1\gamma_1\sin 2\theta_1-W^2_{k,\mathrm{P}}\rho_1\beta_1{}^{-1}(\alpha_1{}^2-2\beta_1{}^2\sin^2\theta_2)\\
&\quad -W^3_{k,\mathrm{SV}}\rho_2\beta_2\gamma_2\sin 2\theta_3+W^4_{k,\mathrm{P}}\rho_2\beta_2{}^{-1}(\alpha_2{}^2-2\beta_2{}^2\sin^2\theta_4),\\
F_{43} &= -W^1_{k,\mathrm{SV}}\rho_1\beta_1\gamma_1\cos 2\theta_1-W^2_{k,\mathrm{P}}\rho_1\beta_1\sin 2\theta_2\\
&\qquad -W^3_{k,\mathrm{SV}}\rho_2\beta_2\gamma_2\cos 2\theta_3-W^4_{k,\mathrm{P}}\rho_2\beta_2\sin 2\theta_4,\\
F_{53} &= -W^1_{k,\mathrm{SH}}+W^3_{k,\mathrm{SH}},\\
F_{63} &= -W^1_{k,\mathrm{SH}}\rho_1\beta_1\gamma_1\cos\theta_1-W^3_{k,\mathrm{SH}}\rho_2\beta_2\gamma_2\cos\theta_3,
\end{aligned}\right\}\quad (2.58'')$$

$$\left.\begin{aligned}
F_{14} &= F_{24} = F_{54} = 0,\\
F_{34} &= \sum_{\nu=1}^{4}(-1)^{\varepsilon_\nu+1}\Bigg(\lambda_{\varepsilon_\nu}\frac{\partial W^\nu_{k-1,x}}{\partial x}+\lambda_{\varepsilon_\nu}\frac{\partial W^\nu_{k-1,y}}{\partial y}\\
&\qquad\qquad +(\lambda_{\varepsilon_\nu}+2\mu_{\varepsilon_\nu})\frac{\partial W^\nu_{k-1,z}}{\partial z}\Bigg),\\
F_{44} &= \sum_{\nu=1}^{4}(-1)^{\varepsilon_\nu+1}\mu_{\varepsilon_\nu}\left(\frac{\partial W^\nu_{k-1,x}}{\partial z}+\frac{\partial W^\nu_{k-1,z}}{\partial x}\right),\\
F_{64} &= \sum_{\nu=1}^{4}(-1)^{\varepsilon_\nu+1}\mu_{\varepsilon_\nu}\left(\frac{\partial W^\nu_{k-1,y}}{\partial z}+\frac{\partial W^\nu_{k-1,z}}{\partial y}\right).
\end{aligned}\right\}\quad (2.58''')$$

C

(2.57) is a recursive set of six linear equations for six unknown components of reflected and refracted waves at the point O, viz., $W^1_{k,\text{P}}(O)$, $W^3_{k,\text{P}}(O)$, $W^2_{k,\text{SV}}(O)$, $W^4_{k,\text{SV}}(O)$, $W^2_{k,\text{SH}}(O)$, and $W^4_{k,\text{SH}}(O)$. All other quantities in (2.57) are also taken at the point O. A similar system of equations was derived by Alekseyev *et al.* (1961).

We must first solve the set of equations (2.57) for $k = 0$, use this solution to determine functions F_i on the right-hand side, and then solve (2.57) for $k = 1$. Generally, when we want the solution for any $k = n$, we must know the solutions of the set for all $k < n$. It should be mentioned that the right-hand sides of (2.57) contain derivatives (including z-derivatives) of the components of the displacement vector $\mathbf{W}^{\nu}_{k-1}$ (see expressions for F_{i4}, (2.58‴)), and for their evaluation we must know values of $\mathbf{W}^{\nu}_{k-1}$ not only just at the interface, but also in its neighbourhood. Therefore, for every k, the coefficients of the ray series of R-waves $\mathbf{W}^{\nu}_k$ must be determined in the neighbourhood of the interface also, using the methods described in the previous section (see (2.29) and (2.42); see also subsection 2.3.2). The principal components of $\mathbf{W}^{\nu}_k$ determined by the kth solution of (2.57) serve as initial conditions for this purpose.

When the functions F_i are known, (2.57) can be solved without any difficulty. The matrix of coefficients of the principal components of R-waves, on the left-hand side of (2.57) does not depend on the subscript k; therefore, the same procedure can be used for the solution for any k. (This matrix depends only on the elastic parameters at O and on the angle of incidence.)

The determination of F_{i1} and F_{i2} (see (2.58) and (2.58′)) is simple. They can be evaluated from the given expressions for the incident wave. F_{i1} contains components of $\mathbf{W}^0_k$, whereas F_{i2} contains derivatives of the components of $\mathbf{W}^0_{k-1}$. We know from the preceding section that the additional components of $\mathbf{W}^{\nu}_k$ can be found by differentiating $\mathbf{W}^{\nu}_{k-1}$ and $\mathbf{W}^{\nu}_{k-2}$ (see (2.20) and (2.33)). It follows that the determination of F_{i3} involves similar problems to the determination of F_{i4} (see above).

The set of equations (2.57) can be divided into two independent systems. The first system, consisting of four equations, I, II, III, and IV, involves only the unknown principal components $W^1_{k,\text{P}}$, $W^3_{k,\text{P}}$, $W^2_{k,\text{SV}}$, and $W^4_{k,\text{SV}}$, which lie in the plane of incidence (P and SV components). The second system, consisting of two equations, V and VI, involves the unknown principal components $W^2_{k,\text{SH}}$ and $W^4_{k,\text{SH}}$, normal to the plane of incidence (SH-components). Both systems can be solved independently. However, to find the right-hand sides of the equations for either of these systems for $k = n$, we must know the solution of *both* systems for $k < n$. Indeed, we can see from (2.58′) and (2.58‴) that the right-hand side of the first system (for

P and SV components) also contains the y-components of $\mathbf{W}^0_{k-1}$ and $\mathbf{W}^\nu_{k-1}$, and similarly the right-hand side of the second system (for SH-components) contains the z-components of $\mathbf{W}^\nu_{k-1}$ and $\mathbf{W}^0_{k-1}$. Thus, when an SH-wave is incident on a curved interface, between two inhomogeneous media, *higher-order* P and SV R-waves can be generated at the interface, and vice versa. (By higher-order R-waves, we understand the waves for which at least the zeroth term in the ray series (2.47) vanishes.)

Now we shall consider the system (2.57) for $k = 0$ (zeroth approximation of ray theory). We can easily see that $F_{i2} = F_{i3} = F_{i4} = 0$, as $\mathbf{W}^0_{-1} = \mathbf{W}^\nu_{-1} = 0$ and the additional components of $\mathbf{W}^\nu_0$ also vanish. As $\partial\tau_0/\partial x$ and $\partial\tau_0/\partial z$ in the expression for F_{i1} can be expressed by means of the angle of incidence and velocities at O, the zeroth terms in the ray series for R-waves depend explicitly only on $\mathbf{W}^0_O(O)$, the elastic parameters at O, and the angle of incidence. Moreover, the right-hand sides of the first system of equations, I–IV, contain only the components of $\mathbf{W}^1_O$ lying in the plane of incidence (P and SV components) and, similarly, the second system of equations, V–VI, contains only the SH-components of $\mathbf{W}^0_O$. From this follow some important conclusions valid for the *zeroth approximation* of the ray theory for reflected and refracted waves from an interface of first-order:

(1) When a P or SV wave is incident on an interface of first order, only P and SV reflected and refracted waves can appear.

(2) When an SH wave is incident on an interface of first order, only SH R-waves can appear.

(3) The ratio of any principal component of a reflected (or refracted) wave to any principal component of the incident wave at the point O does not depend on the curvature of the interface, on the curvature of the wave front of the incident wave, or on any derivatives of the elastic parameters. It depends only on the angle of incidence and on the elastic parameters at O.

Note that all these conclusions are valid only in the zeroth approximation of the ray theory, and generally do not hold for higher terms. We can also see that the *principle of isolated element*, mentioned earlier, follows directly from the third conclusion. This principle applies only to the zeroth approximation of the ray theory and cannot be used generally.

It also follows from the above conclusions that the ratios of principal components, as mentioned there, must be given by the standard coefficients of reflection and refraction of plane waves on the plane interface between two homogeneous media (with elastic parameters $\lambda_1(O)$, $\mu_1(O)$, $\rho_1(O)$ and $\lambda_2(O)$, $\mu_2(O)$, $\rho_2(O)$, respectively). Indeed, the system (2.57) for $k = 0$ is the same as is well known from the theory of reflection and refraction of plane waves on a plane interface. More detailed discussion pertaining to this will be given in the following section.

The situation involving *higher terms* of the ray series of R-waves, $\mathbf{W}_k^\nu$ ($k > 0$), is more complicated. As was shown earlier, the right-hand side of (2.57) contains derivatives of the components $\mathbf{W}_{k-1}^0$, $\mathbf{W}_{k-2}^0$, . . ., and of $\mathbf{W}_{k-1}^\nu$, $\mathbf{W}_{k-2}^\nu$, . . ., at the point O. It follows from (2.30), (2.43), and (2.28) or (2.28′) that the right-hand side of (2.57) will contain the principal radii of curvature of the wave fronts of the incident and R waves at O, the elastic parameters at O, and also some derivatives of these quantities. We shall show at the end of this section that the principal radii of curvature of the wave front of R-waves at O depend on the principal radii of curvature (R_1 and R_2) of the interface Σ at O. Hence we can finally conclude that the *higher terms* of the ray series for R-waves at O depend, among other things, on the following quantities and some of their derivatives: (a) on the principal radii of curvature of the incident wave at O, (b) on the principal radii of curvature of the interface Σ at O, (c) on the elastic parameters at O. The higher the value of k, the higher are the derivatives of the above quantities which must be known. If the kth derivatives of R_1 (or R_2) are discontinuous, $\mathbf{W}_k^\nu$ cannot be determined. In this case diffracted waves also arise (see Alekseyev *et al.*, 1961, for further details).

Now we shall consider *weak interfaces.* First we introduce some notation: we shall call the refracted wave which is of the same type as the incident wave the *non-transformed* refracted wave. (If the incident wave is of P or S type, the non-transformed refracted wave is also of P or S type, respectively.) We shall use $\bar{\nu}$ to denote the index ν corresponding to the non-transformed refracted wave. It should be noted that if the interface vanishes, reflected waves and the transformed refracted wave do not exist and the incident wave formally changes into the non-transformed refracted wave at O. In this case, all the formulae of the preceding section can be used in the vicinity of O.

We shall assume that the interface Σ is of $(n+1)$th order, i.e. the elastic parameters and their derivatives of order less than n are continuous across Σ but the nth derivatives are discontinuous. The coefficients of the ray series for the incident wave $\mathbf{W}_k^0$ ($k < n$) are then continuous across the interface and can be determined by the methods described in the previous section. Strictly speaking, this means that for $k < n$, $\mathbf{W}_k^{\bar{\nu}}(O) = \mathbf{W}_k^0(O)$ and $\mathbf{W}_k^\nu(O) = 0$ for $\nu \neq \bar{\nu}$. We can also use, formally, the set of equations (2.57) successively for $k = 0, 1, \ldots, n-1$ and obtain the same result: the coefficients of the ray series for reflected and transformed refracted waves vanish, but the coefficients for the non-transformed refracted wave are the same as those for the incident wave for $k < n$. The situation changes for

$k = n$, since $\mathbf{W}_n^0$ is then discontinuous at O, and it is necessary to introduce reflected and refracted waves to "compensate" for the discontinuity. The corresponding $\mathbf{W}_n^\nu(O)$ can be determined from (2.57) for $k = n$. They are generally non-zero for all ν, $\nu = 1, 2, 3, 4$. The set of equations (2.57) can then be used for all subsequent $k > n$.

Thus, we can finally say that the reflected and transformed refracted waves from an interface of $(n+1)$th order are of nth order. (The leading term of their ray series will contain a factor $(1/i\omega)^n$ as contrasted with the factor 1 in the leading term of the incident wave.) The only exception is the non-transformed refracted wave, which is of the same order as the incident wave.

2.3.2 *Waves Moving away from an Interface*

Equations (2.57) enable us to find the coefficients of the ray series for R-waves just at the interface. We shall now consider the problem of determining these coefficients in the *neighbourhood of the interface*. Let us, for simplicity, look at the zero-order approximation and a homogeneous medium. We know that in this case we can use formulae (2.30) and (2.44) where the coefficients determined at the interface serve as initial values. As the values of ρ and α (or β) are supposed to be known, the only difficulty which can arise in determining the coefficients in the neighbourhood of the interface is connected with the calculation of $d\sigma(s_0)/d\sigma(s)$ (where $d\sigma(s_0)$ denotes the cross-sectional area of an elementary ray tube of a reflected (refracted) wave at the point O on the side of the interface where the reflected (refracted) wave is generated, and $d\sigma(s)$ is the cross-sectional area at any point of the ray). The value of $d\sigma(s_0)/d\sigma(s)$ for reflected (refracted) waves can often be calculated by simple geometrical considerations (e.g., in vertically inhomogeneous media, see section 2.6.1), or by using (2.28″). For curved and dipping interfaces, it is more convenient to express $d\sigma(s_0)/d\sigma(s)$ in terms of the principal radii of curvature of the wave front of reflected (refracted) waves along Σ. (A similar situation occurs not only for the zeroth approximation of the ray theory but also for higher-order terms.) In this connection, there arises a very important problem of ray theory, viz., the determination of the principal radii of curvature of the wave front of reflected (refracted) waves at O, which was studied by Gel'chinskiy (1961) for the general case of an interface between two inhomogeneous media. We shall not repeat his derivation here, but we shall present some results.

It would be helpful to mention, first, some useful concepts from the differential geometry of surfaces. Let us take a surface Σ and construct the normal to this surface at the point O. Any intersection of the surface and a plane containing the normal is called a *normal section of the surface*. Using standard formulae, we can calculate R_s, the radius of curvature of this normal section at O. There exist two *principal normal sections* associated with the smallest and the largest value of the radius of curvature. The planes of the principal normal sections are perpendicular to each other. The radii of curvature of the principal normal sections at O, R_1 and R_2, are called the *principal radii of curvature* of the surface at O. The radius of curvature R_s is connected with R_1 and R_2 by *Euler's theorem*, as follows:

$$\frac{1}{R_s} = \frac{1}{R_1}\cos^2\phi + \frac{1}{R_2}\sin^2\phi,$$

where ϕ is an oblique angle between the plane of the normal section under consideration (associated with R_s) and the first principal normal section (associated with R_1).

Now we shall return to the problem of determining the principal radii of curvature of the wave front of reflected (refracted) waves. We shall denote the wave front of the incident wave by S_0, the wave front of the reflected (or refracted) wave by S_ν, the interface by Σ, the plane of incidence by $E_\parallel$, and the plane perpendicular to $E_\parallel$ by $E_\perp$. (All these surfaces are understood to contain the point O.) As before, θ_0 and θ_ν are the angles of incidence and reflection (refraction), and V_0 and V_ν are the velocities of propagation of the incident and reflected (refracted) waves.

We assume that all parameters connected with S_0 and Σ are known, and we want to determine the parameters connected with S_ν. More specifically, the known values are: r_1 and r_2, the principal radii of curvature of S_0 at O; R_1 and R_2, the principal radii of curvature of Σ at O; and φ and ϕ, the angles between $E_\parallel$ and the first principal normal sections of S_0 and Σ, respectively.

In addition to r_1, r_2 and R_1, R_2, we shall also use other radii of curvature, $r_\parallel$, $r_\perp$ and $R_\parallel$, $R_\perp$. $r_\parallel$ and $r_\perp$ are the radii of curvature associated with the normal sections of S_0 determined by $E_\parallel$ and by the plane perpendicular to it, respectively; $R_\parallel$ and $R_\perp$ are the radii of curvature associated with the normal sections of Σ by $E_\parallel$ and $E_\perp$, respectively. (These radii are positive or negative depending on whether the interface is convex or concave at O in the corresponding normal sections for an observer situated on the side of the incident wave.) They are connected with the corresponding principal radii by Euler's theorem as follows:

$$\frac{1}{r_\parallel} = \frac{\cos^2\varphi}{r_1} + \frac{\sin^2\varphi}{r_2}, \qquad \frac{1}{r_\perp} = \frac{\sin^2\varphi}{r_1} + \frac{\cos^2\varphi}{r_2},$$

with similar expressions connecting radii of curvature of Σ (R_1, R_2, $R_{\parallel}$, and $R_{\perp}$) when we insert ϕ for φ.

The unknown values are: r_1', r_2', the principal radii of curvature of S_v at O; and φ', the angle between $E_{\parallel}$ and the first principal normal section of S_v. Instead of r_1' and r_2', we can use $r_{\parallel}'$ and $r_{\perp}'$, which are defined in the same way as $r_{\parallel}$ and $r_{\perp}$. (Euler's theorem is again applicable when we insert φ' for φ.)

The formulae connecting r_1', r_2', and φ' with known parameters of S_0 and Σ, derived by Gel'chinskiy (1961), are as follows:

$$\frac{1}{r'_{1,2}} = \frac{1}{2}\left((A+B) \pm \frac{[(A-B)^2 \cos^2\theta_v + 4C^2]^{1/2}}{\cos\theta_v}\right),$$
$$\tan 2\varphi' = \frac{2C}{(A-B)\cos\theta_v}, \tag{2.59}$$

where the upper sign holds for r_1' and the lower for r_2'. A, B, and C are given by

$$A = \frac{1}{\cos^2\theta_v}\left(\frac{n\cos^2\theta_0}{r_{\parallel}} + \frac{n\cos\theta_0 \pm \cos\theta_v}{R_{\parallel}}\right)$$
$$+ \frac{1}{\cos\theta_v}\left[\frac{\sin\theta_v}{V_v}\frac{\partial V_v}{\partial l_{\parallel}'} + \frac{1}{\cos\theta_v}\left(\sin\theta_0\frac{\partial n}{\partial x} - \frac{n\sin 2\theta_0}{2V_0}\cdot\frac{\partial V_0}{\partial l_{\parallel}}\right)\right],$$
$$B = \frac{n}{r_{\perp}} + \frac{n\cos\theta_0 \pm \cos\theta_v}{R_{\perp}}, \tag{2.60}$$
$$C = -\frac{n\cos\theta_0 \sin 2\varphi}{2}\left(\frac{1}{r_1} - \frac{1}{r_2}\right)$$
$$-\frac{1}{2}\sin 2\phi\left(\frac{1}{R_1} - \frac{1}{R_2}\right)(n\cos\theta_0 \pm \cos\theta_v) + \sin\theta_0\frac{\partial n}{\partial y},$$

where the upper sign applies to reflected waves and the lower to refracted waves. $\partial/\partial y$, $\partial/\partial l_{\parallel}$, and $\partial/\partial l_{\parallel}'$ denote derivatives along the interface in the direction perpendicular to $E_{\parallel}$, along the intersection of $E_{\parallel}$ with S_0, and along the intersection of $E_{\parallel}$ with S_v, respectively, all at the point O. n is the refractive index, $n = V_v/V_0$.

The expressions (2.59) are very cumbersome, since they hold under very general conditions, but for simpler situations they can be simplified considerably. (For some special situations, we obtain, from (2.59), formulae derived for electromagnetic waves by Riblet and Barker (1948) and Fock (1950).) For example, for homogeneous media and $\varphi = \phi = 0$, we have $r_1 = r_{\parallel}$, $r_2 = r_{\perp}$, $R_1 = R_{\parallel}$, $R_2 = R_{\perp}$, and then (2.59) yields

$$r_1' = r_{\|}' = r_{\|}/\Delta_{\|}, \qquad \varphi' = 0, \tag{2.61}$$
$$r_2' = r_{\perp}' = r_{\perp}/\Delta_{\perp},$$

where

$$\Delta_{\|} = \frac{V_v \cos^2 \theta_0}{V_0 \cos^2 \theta_v} + \frac{r_{\|}}{R_{\|} \cos^2 \theta_v}\left(\frac{V_v}{V_0} \cos \theta_0 \pm \cos \theta_v\right),$$
$$\Delta_{\perp} = \frac{V_v}{V_0} + \frac{r_{\perp}}{R_{\perp}}\left(\frac{V_v}{V_0} \cos \theta_0 \pm \cos \theta_v\right). \tag{2.61'}$$

Formulae (2.61) and (2.61′) will be used very often in the following sections.

We shall now show how to use (2.59) for the calculation of displacement vectors of reflected and refracted waves from a curved interface of arbitrary shape between two homogeneous media. We shall choose the ray of the R-wave starting at O. The expressions for the coefficients of the ray series of reflected (refracted) waves will always contain a factor $(\mathrm{d}\sigma'(O)/\mathrm{d}\sigma(M))^{1/2}$, where $\mathrm{d}\sigma'(O)$ is the cross-sectional area of the ray tube of the reflected (refracted) wave at O and $\mathrm{d}\sigma(M)$ is the area of the same ray tube at M. Using (2.28′), we can write

$$\left(\frac{\mathrm{d}\sigma'(O)}{\mathrm{d}\sigma(M)}\right)^{1/2} = \left(\frac{r_1' \, r_2'}{(r_1'+l)(r_2'+l)}\right)^{1/2},$$

where l is the distance between M and O. Inserting values for r_1' and r_2', we obtain

$$\left(\frac{\mathrm{d}\sigma'(O)}{\mathrm{d}\sigma(M)}\right)^{1/2} = \frac{\cos \theta_v}{[\cos^2 \theta_v + (A+B)l \cos^2 \theta_v + (AB \cos^2 \theta_v - C^2)l^2]^{1/2}}. \tag{2.62}$$

This equation contains only the angle of reflection (or refraction), the distance l, and known parameters of S_0 and Σ. For $\varphi = \phi = 0$, it reduces to the form

$$\left(\frac{\mathrm{d}\sigma'(O)}{\mathrm{d}\sigma(M)}\right)^{1/2} = \left\{\left(1+\frac{l}{r_{\|}}\Delta_{\|}\right)\left(1+\frac{l}{r_{\perp}}\Delta_{\perp}\right)\right\}^{-1/2}. \tag{2.62'}$$

When the ray of the wave reflected (refracted) at Σ strikes another interface, say Σ_2, at the point O_2, the procedure described in this section can be repeated. Our reflected (refracted) wave becomes an incident wave at the point O_2 with the parameters $r_1(O_2)$, $r_2(O_2)$, and $\varphi(O_2)$. When the medium between O and O_2 is homogeneous, these parameters can be determined from the values of $r_1'(O)$, $r_2'(O)$, and $\varphi'(O)$ very simply. The principal radii of the incident wave at O_2 are given by the obvious relations

$r_1(O_2) = r_1'(O)+l$, $r_2(O_2) = r_2'(O)+l$, where l is now the distance between O and O_2. The angle $\varphi(O_2)$ can also be determined without difficulty from $\varphi'(O)$, as it does not change along the ray in a homogeneous medium. When the incident planes at O and O_2 coincide, $\varphi(O_2) = \varphi'(O)$. When they differ, $\varphi(O_2)$ can be determined from $\varphi'(O)$ by simple geometrical considerations. From the values of the parameters $r_1(O_2)$, $r_2(O_2)$, and $\varphi(O_2)$, we can again determine the parameters of other reflected (refracted) waves at O_2, i.e. $r_1'(O_2)$, $r_2'(O_2)$, and $\varphi'(O_2)$, using formulae (2.59) and (2.60). This procedure can be repeated for any number of interfaces; some examples will be given in section 2.6.4.

2.4
FORMULAE FOR THE ZEROTH APPROXIMATION

For the derivation of displacement vectors of head waves, we shall need to know the properties of the basic body waves in the zeroth approximation of the ray theory. In this section, we shall therefore specify the general results of the preceding sections for the particular case of $k = 0$. These results will be used for an investigation of head waves in the following chapters. However, the results are very important in themselves if we are interested in the propagation of multiply reflected and refracted waves in some complicated media.

We shall not consider here the general problem of ray paths which are curves of double curvature. The reason is that, for the general problem, it is very difficult to find the final expression for the displacement vector in a *compact form*. The best method is to solve the problem successively, step by step, following the ray from one interface to another and applying the formulae of the preceding sections at each step. All the formulae necessary for the solution of this general problem have been derived in the preceding sections. Here we shall consider only situations in which the ray is a *plane curve*. In other words, we assume that the torsion of the ray is zero along the ray and that the normals to all interfaces (at the points where the ray strikes them) lie in the same plane.

We introduce a rectangular coordinate system x, y, z and suppose that the whole ray lies in a vertical plane xz. Then, as in the previous section, we can introduce three perpendicular unit vectors $\mathbf{n}_\mathrm{P}$, $\mathbf{n}_\mathrm{SV}$, and $\mathbf{n}_\mathrm{SH}$ at every point of the ray: $\mathbf{n}_\mathrm{P}$ is a unit vector tangent to the ray, pointing in the direction of propagation; $\mathbf{n}_\mathrm{SV}$ is perpendicular to the ray and lies in the vertical plane (such that its projection on the x-axis is positive); and $\mathbf{n}_\mathrm{SH}$ is a unit vector in the direction of the y-axis. As the torsion of the ray is assumed to be zero, the SV and SH components of shear waves are not coupled to each other (see (2.44)). We may, therefore, consider them as two separate waves.

In the following, for simplicity, we shall use the notation

$$\mathbf{U} = \mathbf{W}_0. \tag{2.63}$$

Then we can write (see (2.3))

$$\mathbf{W} = \mathbf{U}\exp[i\omega(t-\tau)], \tag{2.63'}$$

or

$$\mathbf{W} = U\exp[i\omega(t-\tau)]\mathbf{n}, \tag{2.63''}$$

where $\mathbf{n}$ may be $\mathbf{n}_P$ (for P-waves), $\mathbf{n}_{SV}$ (for SV-waves), or $\mathbf{n}_{SH}$ (for SH-waves).

2.4.1 *Continuation of the Displacement Vector along the Ray*

Suppose we know $\mathbf{W}$ at the point M_0 which lies on the ray:

$$\mathbf{W}(M_0) = U(M_0)\exp\{i\omega[t-\tau(M_0)]\}\mathbf{n}(M_0).$$

According to formulae (2.17), (2.30), and (2.44), the displacement vector at another point, M, on the ray is given by

$$\mathbf{W}(M) = U(M)\exp\{i\omega[t-\tau(M)]\}\mathbf{n}(M), \tag{2.64}$$

where

$$U(M) = U(M_0)\left(\frac{v(M_0)\rho(M_0)}{v(M)\rho(M)}\right)^{1/2}\left(\frac{\mathrm{d}\sigma(M_0)}{\mathrm{d}\sigma(M)}\right)^{1/2},$$

$$\tau(M) = \tau(M_0)+\int_{M_0}^{M} v^{-1}\mathrm{d}s, \tag{2.64'}$$

where v denotes the velocity of propagation, which may be either α (in the case of P-waves) or β (in the case of SV or SH waves). The unit vector $\mathbf{n}(M)$ is constructed by the usual methods. (For SV-waves, an additional multiplier -1 must be included in the expression for $U(M)$ if the ray is descending at M_0 and ascending at M, or vice versa. See more details in 2.4.4.)

The value of $[\mathrm{d}\sigma(M_0)/\mathrm{d}\sigma(M)]^{1/2}$ in (2.64′) can often be calculated from simple geometrical considerations. In more complicated cases, we can use (2.28) or (2.28″). In *homogeneous media*, expressions for $U(M)$ and $\tau(M)$ become simpler because $v(M_0)\rho(M_0)/v(M)\rho(M) = 1$ and $\int_{M_0}^{M} v^{-1}\mathrm{d}s = l/v$, where l is the length of the ray between the points M and M_0. Further, using (2.28′) for $\mathrm{d}\sigma(M_0)/\mathrm{d}\sigma(M)$, we obtain

$$U(M) = U(M_0)\left(\frac{r_1(M_0)r_2(M_0)}{[r_1(M_0)+l][r_2(M_0)+l]}\right)^{1/2}, \qquad (2.65)$$
$$\tau(M) = \tau(M_0)+l/v,$$

where $r_1(M_0)$ and $r_2(M_0)$ are the principal radii of curvature of the wave front at the point M_0.

The expressions (2.64) and (2.64′) are applicable if the elastic parameters are continuous functions of coordinates. If there exists an interface where these parameters undergo a discontinuous jump, it is necessary to take into account the interface conditions discussed in section 2.3.

2.4.2 *Influence of an Interface*

We shall assume that the wave (2.64) is incident on a smooth interface of first order. The point at which a particular ray strikes the interface will be denoted by O, as in section 2.3, and we again introduce the local rectangular coordinate system with the origin at the point O as described there. Similarly, the notation for the parameters of the media remains the same. It was shown in section 2.3 that the set of linear equations (2.57) can be divided into two independent groups, the first group for P and SV components and the second for SH components. We shall first study the reflection and refraction of P and SV waves, and later that of SH waves.

A P and SV Waves

We shall indicate the type of incident wave by an index m, where m may be 1, 2, 3, or 4. $m = 1$ denotes a compressional wave (P) incident from the first medium; $m = 2$, a shear wave (SV) incident from the first medium; and $m = 3$ and 4 characterize compressional and shear waves incident from the second medium. The unit vector $\mathbf{n}_m$ may be either $\mathbf{n}_P$ (for $m = 1$ or 3) or $\mathbf{n}_{SV}$ (for $m = 2$ or 4). We denote the angle of incidence by θ_m (i.e. θ_m is the acute angle between the normal to the interface and the ray at the point O). Similarly, we use V_m to denote the velocity of propagation of the wave (i.e. $V_1 = \alpha_1$, $V_2 = \beta_1$, $V_3 = \alpha_2$, $V_4 = \beta_2$).

The displacement vector of the incident wave at the point O is given by

$$\mathbf{W}^m(O) = U^m(O)\exp\{i\omega[t-\tau_m(O)]\}\mathbf{n}_m(O), \qquad (2.66)$$

where $U^m(O)$ and $\tau_m(O)$ are given by (2.64′) when we have substituted O for M.

The x and z components of $\mathbf{W}^m$ in the local coordinate system are given by

$$\begin{bmatrix} W_x^m(O) \\ W_z^m(O) \end{bmatrix} = U^m(O)\exp\{i\omega[t-\tau_m(O)]\}\begin{bmatrix} n_{mx}(O) \\ n_{mz}(O) \end{bmatrix}, \qquad (2.66')$$

TABLE 2.1

Projections of $\mathbf{n}_m(O)$ on the x and z axes

	$m = 1$	$m = 2$	$m = 3$	$m = 4$
$n_{mx}(O)$	$\sin\theta_1$	$\cos\theta_2$	$\sin\theta_3$	$\cos\theta_4$
$n_{mz}(O)$	$-\cos\theta_1$	$\sin\theta_2$	$\cos\theta_3$	$-\sin\theta_4$

where n_{mx} and n_{mz} are projections of the unit vector $\mathbf{n}_m(O)$ on the x and z axes of the local coordinate system, respectively; they are given in table 2.1.

Four waves are produced when the wave $\mathbf{W}^m$ strikes the interface. Their displacement vectors will be given by

$$\mathbf{W}^{mn}(O) = U^{mn}(O)\exp\{i\omega[t-\tau_{mn}(O)]\}\mathbf{n}_n(O), \tag{2.67}$$

where the index n indicates the type of wave generated (it corresponds to the index ν in section 2.3). Again we have $n = 1, 2, 3$, or 4, with $n = 1$ and 2 referring to compressional and shear waves in the first medium and $n = 3$ and 4 to compressional and shear waves in the second medium, respectively. All possible types of reflected and refracted waves are shown schematically in figure 2.8 and the corresponding unit vectors $\mathbf{n}_n$ are given in figure 2.9. The unit vector $\mathbf{n}_n$ may be $\mathbf{n}_{\mathrm{P}}$ (for $n = 1$ and 3) or $\mathbf{n}_{\mathrm{SV}}$ (for $n = 2$ and 4). θ_n are the angles of reflection, and V_n the velocity of propaga-

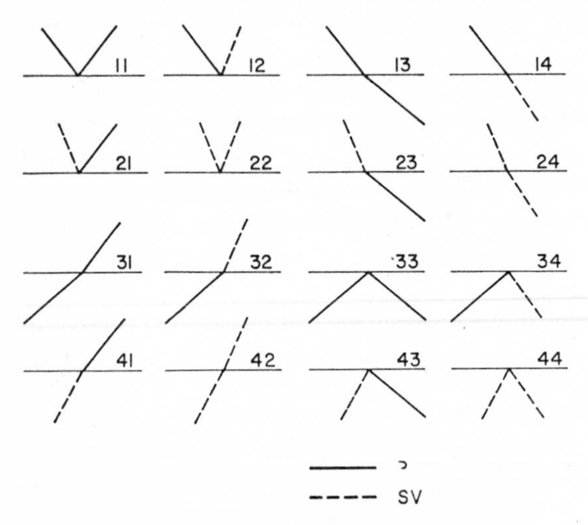

FIGURE 2.8 Schematic ray representation of reflected and refracted waves for a single interface. Each type of wave is characterized by two numbers, *mn*.

tion of the nth wave. The phase function τ_{mn} is a solution of the eikonal equation $(\nabla\tau_{mn})^2 = 1/V_n^2$, with the initial condition at the surface $\tau_{mn}(O) = \tau_m(O)$ (see (2.49) and (2.50)). The x and z components of $\mathbf{W}^{mn}$ in the local coordinate system are given by

$$\begin{bmatrix} W_x^{mn}(O) \\ W_z^{mn}(O) \end{bmatrix} = U^{mn}(O)\exp\{i\omega[t-\tau_{mn}(O)]\}\begin{bmatrix} n_{nx}(O) \\ n_{nz}(O) \end{bmatrix}, \tag{2.67'}$$

where n_{nx} and n_{nz} are projections of the unit vector $\mathbf{n}_n$ on the x and z axes of this system, respectively; they are given in table 2.2.

Note that tables 2.2 and 2.1 are essentially the same except for the

TABLE 2.2
Projections of $\mathbf{n}_n(O)$ on the x and z axes

	$n = 1$	$n = 2$	$n = 3$	$n = 4$
$n_{nx}(O)$	$\sin\theta_1$	$\cos\theta_2$	$\sin\theta_3$	$\cos\theta_4$
$n_{nz}(O)$	$\cos\theta_1$	$-\sin\theta_2$	$-\cos\theta_3$	$\sin\theta_4$

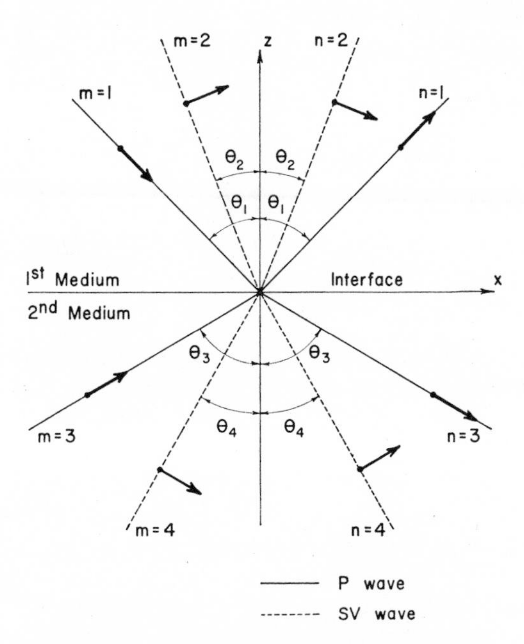

FIGURE 2.9 Unit vectors $\mathbf{n}_m$ and $\mathbf{n}_n$ (shown by the arrows) of the incident wave ($m = 1, 2, 3, 4$) and the reflected and refracted waves ($n = 1, 2, 3, 4$).

difference in the signs of the projections on the z-axis. This difference arises because of the change in the direction of propagation of the wave. In the remainder of this section $\mathbf{W}^m$, $\mathbf{W}^{mn}$, τ^{mn}, etc. are all to be understood at the point O.

The first equation of (2.51) gives Snell's law, viz.,

$$\frac{\partial \tau_m}{\partial x} = \frac{\partial \tau_{mn}}{\partial x} = \frac{\sin \theta_m}{V_m} = \frac{\sin \theta_n}{V_n} = \Theta, \tag{2.68}$$

where Θ is a constant, independent of the indices m and n.

Now we shall use the equations I–IV from the system of equations (2.57) to determine U^{mn}. Note that in the notation of this section $W^1_{k,\mathrm{P}} = U^{m1}$, $W^2_{k,\mathrm{SV}} = U^{m2}$, $W^3_{k,\mathrm{P}} = U^{m3}$, $W^4_{k,\mathrm{SV}} = U^{m4}$.

For $k = 0$, $F_{i2} = F_{i3} = F_{i4} = 0$ and we have $F_i = F_{i1}$. Inserting the expression for the incident wave into (2.58), we obtain from (2.57) a set of four linear equations for U^{mn}:

$$\sum_{n=1}^{4} a_{in}\, U^{mn} = b_{mi}\, U^m \qquad (i = 1, 2, 3, 4), \tag{2.69}$$

where the coefficients a_{in} are given by the matrix

$$A = \begin{bmatrix} \sin\theta_1 & \cos\theta_2 & -\sin\theta_3 & -\cos\theta_4 \\ \cos\theta_1 & -\sin\theta_2 & \cos\theta_3 & -\sin\theta_4 \\ \rho_1\alpha_1\cos 2\theta_2 & -\rho_1\alpha_1\gamma_1\sin 2\theta_2 & -\rho_2\alpha_2\cos 2\theta_4 & \rho_2\alpha_2\gamma_2\sin 2\theta_4 \\ \rho_1\alpha_1\gamma_1{}^2\sin 2\theta_1 & \rho_1\alpha_1\gamma_1\cos 2\theta_2 & \rho_2\alpha_2\gamma_2{}^2\sin 2\theta_3 & \rho_2\alpha_2\gamma_2\cos 2\theta_4 \end{bmatrix} \tag{2.70}$$

and

$$b_{mi} = (-1)^i a_{im} \tag{2.71}$$

(where we have introduced $\gamma_i = \beta_i/\alpha_i = [\mu_i/(\lambda_i+2\mu_i)]^{1/2}$). All the angles θ_i in the matrix A are connected with the angle of incidence θ_m by the relation (2.68).

Solving (2.69), we obtain

$$U^{mn} = R_{mn}\, U^m, \tag{2.72}$$

where R_{mn} is the solution of a set of four linear equations

$$\sum_{n=1}^{4} a_{in}\, R_{mn} = b_{mi} \qquad (i = 1, 2, 3, 4), \tag{2.73}$$

and represents the coefficients of reflection and refraction (R-coefficients), which are the same as for plane waves incident on a plane interface. We see, therefore, that locally the wave $\mathbf{W}^m$ behaves as if it were a plane wave and the interface were plane. It should be noted that R_{mn} are the coefficients of reflection and refraction in terms of *displacements.* These coefficients are often calculated on the basis of *displacement potentials.* If we denote the latter by $\tilde{R}_{mn}$, we obtain

$$R_{mn} = (V_m/V_n)\tilde{R}_{mn}. \tag{2.73'}$$

We have sixteen different R-coefficients, corresponding to sixteen different R-waves (see figure 2.8). It is convenient to calculate R_{mn} directly from the set of equations (2.73). (We must not forget that a_{in} may be complex.) However, it is useful to give explicit expressions for the coefficients as follows:

$$\begin{aligned}
R_{11} &= -1+2P_1D^{-1}(\alpha_2\beta_2P_2X^2+\beta_1\alpha_2\rho_1\rho_2P_4+q^2\Theta^2P_2P_3P_4),\\
R_{12} &= -2\alpha_1\Theta P_1D^{-1}(qP_3P_4Y+\alpha_2\beta_2XZ),\\
R_{13} &= 2\alpha_1\rho_1P_1D^{-1}(\beta_2P_2X+\beta_1P_4Y),\\
R_{14} &= -2\alpha_1\rho_1\Theta P_1D^{-1}(qP_2P_3-\beta_1\alpha_2Z),\\
R_{21} &= -2\beta_1\Theta P_2D^{-1}(qP_3P_4Y+\alpha_2\beta_2XZ),\\
R_{22} &= 1-2P_2D^{-1}(\alpha_2\beta_2P_1X^2+\alpha_1\beta_2\rho_1\rho_2P_3+q^2\Theta^2P_1P_3P_4)\\
R_{23} &= 2\beta_1\rho_1\Theta P_2D^{-1}(qP_1P_4-\alpha_1\beta_2Z),\\
R_{24} &= 2\beta_1\rho_1P_2D^{-1}(\alpha_1P_3Y+\alpha_2P_1X),\\
R_{31} &= 2\alpha_2\rho_2P_3D^{-1}(\beta_1P_4Y+\beta_2P_2X),\\
R_{32} &= 2\alpha_2\rho_2\Theta P_3D^{-1}(qP_1P_4-\alpha_1\beta_2Z),\\
R_{33} &= -1+2P_3D^{-1}(\alpha_1\beta_1P_4Y^2+\alpha_1\beta_2\rho_1\rho_2P_2+q^2\Theta^2P_1P_2P_4),\\
R_{34} &= 2\alpha_2\Theta P_3D^{-1}(qP_1P_2X+\alpha_1\beta_1YZ),\\
R_{41} &= -2\beta_2\rho_2\Theta P_4D^{-1}(qP_2P_3-\beta_1\alpha_2Z),\\
R_{42} &= 2\beta_2\rho_2P_4D^{-1}(\alpha_2P_1X+\alpha_1P_3Y),\\
R_{43} &= 2\beta_2\Theta P_4D^{-1}(qP_1P_2X+\alpha_1\beta_1YZ),\\
R_{44} &= 1-2P_4D^{-1}(\alpha_1\beta_1P_3Y^2+\beta_1\alpha_2\rho_1\rho_2P_1+q^2\Theta^2P_1P_2P_3),
\end{aligned} \tag{2.74}$$

where

$$\begin{aligned}
D = {} & \alpha_1\alpha_2\beta_1\beta_2\Theta^2Z^2+\alpha_2\beta_2P_1P_2X^2+\alpha_1\beta_1P_3P_4Y^2\\
& +\rho_1\rho_2(\beta_1\alpha_2P_1P_4+\alpha_1\beta_2P_2P_3)+q^2\Theta^2P_1P_2P_3P_4,
\end{aligned} \tag{2.74'}$$

and

$$\begin{aligned} q &= 2(\rho_2\beta_2{}^2 - \rho_1\beta_1{}^2), \\ X &= \rho_2 - q\Theta^2, \qquad Y = \rho_1 + q\Theta^2, \qquad Z = \rho_2 - \rho_1 - q\Theta^2, \\ \Theta &= \sin\theta_i/V_i, \quad V_1 = \alpha_1, \quad V_2 = \beta_1, \quad V_3 = \alpha_2, \quad V_4 = \beta_2, \\ P_i &= (1 - V_i{}^2\Theta^2)^{1/2} \qquad (i = 1, 2, 3, 4). \end{aligned} \tag{2.74''}$$

All velocities and densities are to be taken at the point O. For $\Theta > 1/V_i$, we must take $P_i = -i(V_i{}^2\Theta^2 - 1)^{1/2}$ (see (2.53)).

We can write two simple relations, connecting different types of R-coefficients, which will be useful later for finding corresponding relations between head wave coefficients. We can see from (2.74) that the R-coefficients R_{mn} and R_{nm} are connected by the formula

$$P_n V_n \rho_{\varepsilon_n} R_{mn} = P_m V_m \rho_{\varepsilon_m} R_{nm},$$

where $\varepsilon_i = 1$ for $i = 1, 2$, and $\varepsilon_i = 2$ for $i = 3, 4$. The other formula connecting R-coefficients is quite clear from the physical properties of reflected waves. The sixteen possible R-waves can be divided into two groups (see figure 2.8). The first group (eight waves) corresponds to incidence from the first medium and the second group corresponds to incidence from the second medium. If we interchange the parameters of the media (i.e. take the first medium as the second, and the second as the first), the expressions for the R-coefficients of the two groups of reflected waves must interchange, too. If we define $\bar{m}$ such that $\bar{m} = 3, 4, 1,$ or 2 when $m = 1, 2, 3,$ or 4, respectively (and $\bar{n}$ similarly), we can write

$$R_{mn}(\alpha_1, \beta_1, \rho_1, \alpha_2, \beta_2, \rho_2) = R_{\bar{m}\bar{n}}(\alpha_2, \beta_2, \rho_2, \alpha_1, \beta_1, \rho_1).$$

From (2.67) and (2.72), we obtain the following expressions for the displacement vector of a generated wave at the point O:

$$\mathbf{W}^{mn}(O) = R_{mn}\, U^m(O) \exp\{i\omega[t - \tau_{mn}(O)]\}\mathbf{n}_n(O), \tag{2.75}$$

where R_{mn} are the R-coefficients at the point O, $\tau_{mn}(O) = \tau_m(O)$, and $\mathbf{n}_n$ may be either $\mathbf{n}_\mathrm{P}$ (for $n = 1$ and 3) or $\mathbf{n}_\mathrm{SV}$ (for $n = 2$ and 4).

From the expression (2.72) we can see that the ratio $U^{mn}(O)/U^m(O)$ does not depend on the radii of curvature of the interface, or on the radii of curvature of the wave front of the incident wave, or on the derivatives of the elastic parameters.

B SH Waves

The displacement vector of the incident wave at the point O is given by

$$\mathbf{W}^m(O) = U^m(O) \exp\{i\omega[t - \tau_m(O)]\}\mathbf{n}_\mathrm{SH}, \tag{2.76}$$

where $\mathbf{n}_{SH}$ is the unit vector in the direction of the y-axis. The displacement vector $\mathbf{W}^m(O)$ has only one non-zero component, which can be obtained directly from (2.76). The index m may be 2 or 4, depending on whether the wave is incident from the first or the second medium. The angle of incidence is again denoted by θ_m.

Two waves are produced when the wave $\mathbf{W}^m$ is incident on the interface, viz., reflected and refracted SH waves. Their displacement vectors are given by

$$\mathbf{W}^{mn}(O) = U^{mn}(O)\exp\{i\omega[t-\tau_{mn}(O)]\}\mathbf{n}_{SH}, \tag{2.77}$$

where the index n equals 2 for the wave propagating in the first medium and 4 for the wave propagating in the second medium. We denote the angles of reflection and refraction by θ_n. Again, the y-component is the only non-vanishing component of the displacement vector. As in the case of P and SV waves, τ_{mn} is a solution of the eikonal equation $(\nabla\tau_{mn})^2 = 1/V_n^{\,2}$, with the initial condition at the interface $\tau_{mn}(O) = \tau_m(O)$. For the determination of U^{mn} we use the equations v and vi from the set (2.57). We obtain

$$U^{mn}(O) = R^{SH}_{mn}\, U^m(O), \tag{2.78}$$

where R^{SH}_{mn} is the solution of a set of two linear equations

$$\begin{aligned} &R^{SH}_{m2} \qquad\qquad - R^{SH}_{m4} \qquad\qquad = (-1)^{m/2}, \\ &R^{SH}_{m2}\, \rho_1\, \beta_1 \cos\theta_2 + R^{SH}_{m4}\, \rho_2\, \beta_2 \cos\theta_4 = \rho_{m/2}\, \beta_{m/2} \cos\theta_m. \end{aligned} \tag{2.79}$$

The angles of reflection and refraction θ_n are again connected with the angle of incidence θ_m by the relation $\sin\theta_n = (V_n/V_m)\sin\theta_m$, where $V_2 = \beta_1$, $V_4 = \beta_2$. The explicit form of the R-coefficients for SH waves is as follows:

$$\begin{aligned} R^{SH}_{22} &= (\rho_1\beta_1\cos\theta_2 - \rho_2\beta_2\cos\theta_4)/D, \\ R^{SH}_{24} &= 2\rho_1\beta_1\cos\theta_2/D, \\ R^{SH}_{42} &= 2\rho_2\beta_2\cos\theta_4/D, \\ R^{SH}_{44} &= (\rho_2\beta_2\cos\theta_4 - \rho_1\beta_1\cos\theta_2)/D, \end{aligned} \tag{2.80}$$

where

$$D = \rho_1\beta_1\cos\theta_2 + \rho_2\beta_2\cos\theta_4. \tag{2.81}$$

From (2.77) and (2.78) we obtain the displacement vector of either of the waves generated at the point O, viz.,

$$\mathbf{W}^{mn}(O) = R^{SH}_{mn}\, U^m(O)\exp\{i\omega[t-\tau_{mn}(O)]\}\mathbf{n}_{SH}, \tag{2.82}$$

where $\tau_{mn}(O) = \tau_m(O)$; R_{mn}^{SH} are the R-coefficients at the point O, given by (2.80), and $U^m(O)$ is given by (2.64) (where we substitute O for M).

C Free Surface

The formulae (2.74) for R-coefficients can also be used in the case of a *free surface* (the surface of the earth). If the first medium is a vacuum, the R-coefficients R_{33}, R_{34}, R_{43}, and R_{44} are given by (see (2.74))

$$\begin{aligned} R_{33} &= [-(1-2\beta_2{}^2\Theta^2)^2+4\Theta^2P_3P_4\gamma_2\beta_2{}^2]/D, \\ R_{34} &= 4P_3\Theta\beta_2(1-2\beta_2{}^2\Theta^2)/D, \\ R_{43} &= 4\gamma_2\beta_2\Theta P_4(1-2\beta_2{}^2\Theta^2)/D, \\ R_{44} &= [(1-2\beta_2{}^2\Theta^2)^2-4P_3P_4\Theta^2\gamma_2\beta_2{}^2]/D, \end{aligned} \tag{2.83}$$

where

$$D = (1-2\beta_2{}^2\Theta^2)^2+4P_3P_4\Theta^2\gamma_2\beta_2{}^2, \tag{2.84}$$

and the other symbols are the same as in (2.74). D is known as the *Rayleigh function.*

If the receiver is located on the earth's surface, which is a very common situation in practical applications, it registers, in addition to the other waves, the waves reflected from the earth's surface. We shall assume that the wave (2.66) is incident on the earth's surface (m must, of course, be 3 or 4). Then two new waves are generated on the interface, viz., $\mathbf{W}^{m3}$ and $\mathbf{W}^{m4}$. For the displacement vector of the resulting wave, $\mathbf{W}^{m\Sigma}$, we obtain

$$\mathbf{W}^{m\Sigma}(O) = \mathbf{W}^m(O)+\mathbf{W}^{m3}(O)+\mathbf{W}^{m4}(O). \tag{2.85}$$

As $\tau_{m3}(O) = \tau_{m4}(O) = \tau_m(O)$, we can write

$$\mathbf{W}^{m\Sigma}(O) = U^m(O)\exp\{i\omega[t-\tau_m(O)]\}\left(\mathbf{n}_m(O)+\sum_{n=3}^{4} R_{mn}\,\mathbf{n}_n(O)\right). \tag{2.86}$$

We introduce a vector $\mathbf{q}_m$ given by the relation

$$\mathbf{q}_m(O) = \mathbf{n}_m(O)+\sum_{n=3}^{4} R_{mn}\,\mathbf{n}_n(O). \tag{2.87}$$

Inserting this into (2.86) yields

$$\mathbf{W}^{m\Sigma}(O) = U^m(O)\exp\{i\omega[t-\tau_m(O)]\}\mathbf{q}_m(O). \tag{2.88}$$

Comparing this equation with (2.66), we can see that the expression for the displacement vector at the point O on the earth's surface is just the same as the expression for the displacement vector of the incident wave;

only the unit vector $\mathbf{n}_m(O)$ is replaced by the vector $\mathbf{q}_m(O)$. We shall call the vector $\mathbf{q}_m(O)$ the *conversion vector*. The x and z components of the conversion vector, q_{mx} and q_{mz}, are called *conversion coefficients*. Inserting $\mathbf{n}_m$, $\mathbf{n}_n$, and R_{mn} into (2.87), we find the following values for them:

(a) Incident P-wave ($m = 3$):

$$q_{3x} = 4P_3P_4\Theta\beta_2/D,\quad q_{3z} = 2P_3(1-2\beta_2{}^2\Theta^2)/D. \tag{2.89}$$

(b) Incident SV-wave ($m = 4$):

$$q_{4x} = 2P_4(1-2\beta_2{}^2\Theta^2)/D,\quad q_{4z} = -4P_3P_4\beta_2\gamma_2\Theta/D, \tag{2.90}$$

where D is given by (2.84).

The displacement components of the wave registered on the earth's surface are then given by

$$\begin{bmatrix} W_x^{m\Sigma} \\ W_z^{m\Sigma} \end{bmatrix} = U^m(O)\exp\{i\omega[t-\tau_m(O)]\}\begin{bmatrix} q_{mx} \\ q_{mz} \end{bmatrix}, \tag{2.91}$$

from which an important conclusion follows. If the receiver does not lie *inside* the medium but is located on the earth's surface, the expressions for the displacement components remain the same, but the components of the unit vector $\mathbf{n}_m$, n_{mx} and n_{mz}, are to be replaced by the conversion coefficients q_{mx} and q_{mz}. In the following sections we shall, therefore, usually assume that the receiver lies inside the medium; if the receiver is on the earth's surface, the solution can be easily adapted.

Note that the direction of the conversion vector $\mathbf{q}_m$ is generally different from the direction of the unit vector $\mathbf{n}_m$, and the length of the conversion vector is different from unity. For normal incidence, $q_{3z} = q_{4x} = 2$ and $q_{3x} = q_{4z} = 0$, and for grazing incidence, $q_{3x} = q_{3z} = q_{4x} = q_{4z} = 0$.

For SH-waves, the directions of the unit vector $\mathbf{n}_{SH}$ and the conversion vector are the same; however, the length of the conversion vector is 2.

Figures 2.10a and 2.10b show the relation between the conversion coefficients q_{3x} and q_{3z} (incident P-wave) and the angle of incidence of the P-wave (θ_3). Note that the angle of incidence θ_3 is connected with Θ by the relation $\sin\theta_3 = \alpha_2\Theta$.

D Some Remarks on the R-coefficients

The R-coefficients are very important in the investigation of the propagation of seismic body waves in layered media. Knott (1899) and Zöppritz (1919) were the first to derive general equations for these coefficients.

References to the relevant literature concerning these equations, and analytic expressions for R-coefficients, can be found in any textbook of seismology or seismic wave propagation. These analytic expressions are very cumbersome for routine calculations, but in some cases it is possible to use simpler approximations without sacrificing too much accuracy (see Bortfeld, 1961). Many useful relations between different types of R-coefficients are given in Pod"yapol'skiy (1959b), and tables and graphs of several types are given in a number of books and papers (e.g., Muskat and Meres, 1940; Gutenberg, 1944; Petrashen, 1957b; Nafe, 1957; Vasil'yev, 1959; Steinhart and Meyer, 1961; Richards, 1961; McCamy *et al.*, 1962; Koefoed, 1962; Costain *et al.*, 1963, 1965; Červený *et al.*. 1964; Tooley

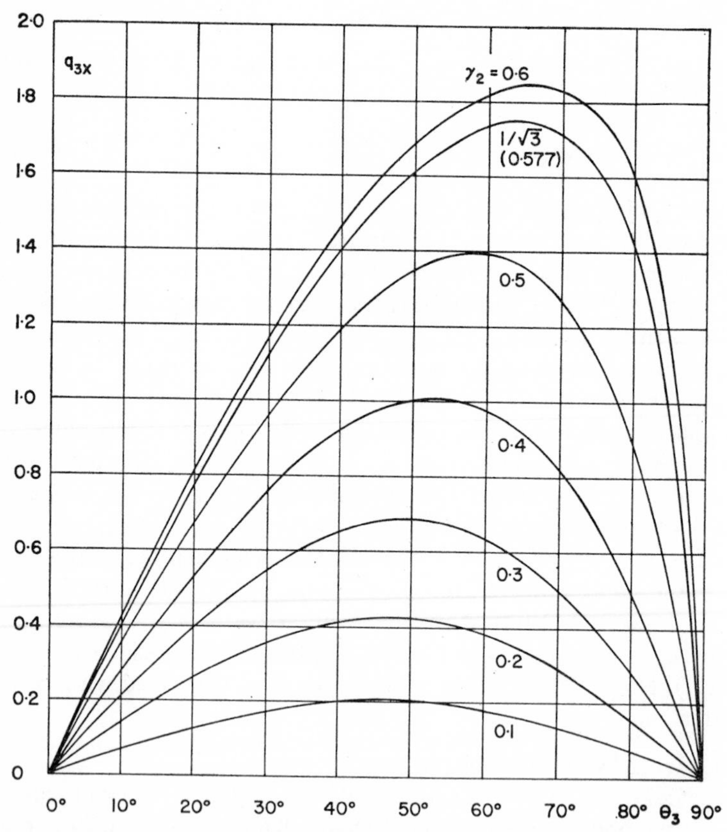

FIGURE 2.10a The x conversion coefficient q_{3x} as a function of the angle of incidence θ_3 (for different values of γ_2).

et al., 1963, 1965; Grant and West, 1965; Yepinat'yeva and Červený, 1965; Berzon *et al.*, 1966). By far the most comprehensive tables are those of Petrashen (1957b), which contain numerical values and graphs of R-coefficients of all types (including SH-waves) for a very broad system of parameters of an interface, as well as many other useful values such as conversion coefficients and R-coefficients for a free interface. All the above-mentioned references for R-coefficients must be used very carefully, however, as some authors use definitions of R-coefficients which are different (displacement potential coefficients, energy coefficients) from those given in this section. Also, unfortunately, many misprints and other errors have crept into several of these works.

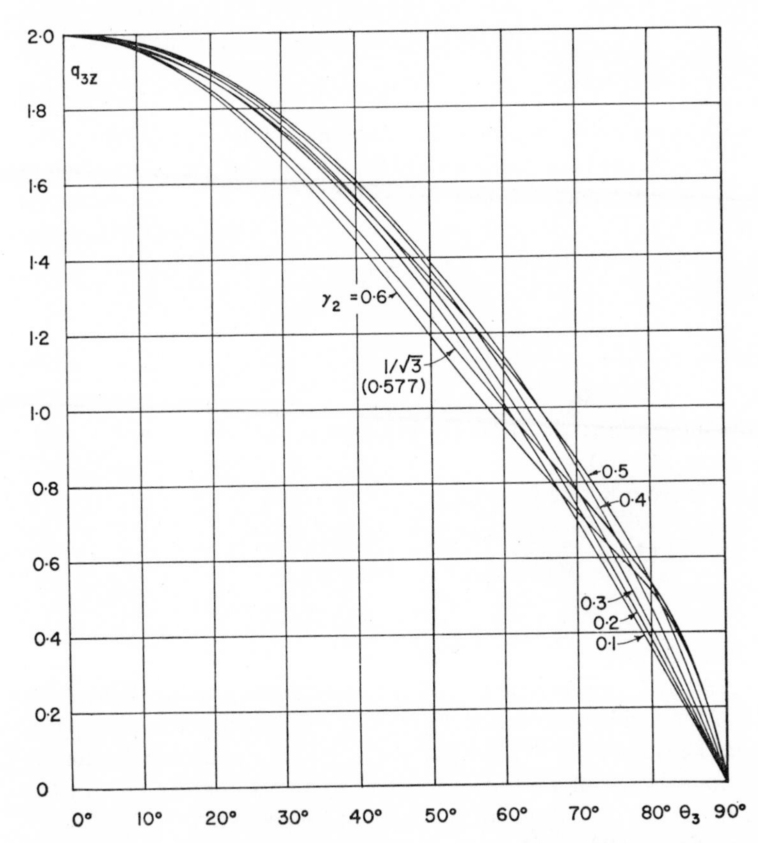

FIGURE 2.10b The z conversion coefficient q_{3z} as a function of the angle of incidence θ_3 (for different values of γ_2).

2.4.3 *Displacement Vector of Reflected and Refracted Waves at an Arbitrary Point of the Ray*

In the previous sections we have determined displacement vectors of the reflected and refracted waves (R-waves) at the point O lying on the interface. We obtained, in general,

$$\mathbf{W}^{mn}(O) = R_{mn}\, U^m(O)\exp\{i\omega[t-\tau_{mn}(O)]\}\mathbf{n}_n(O), \tag{2.92}$$

where $\mathbf{n}_n(O)$ may be $\mathbf{n}_{\mathrm{P}}$, $\mathbf{n}_{\mathrm{SV}}$, or $\mathbf{n}_{\mathrm{SH}}$, depending on the type of wave, and $\tau_{mn}(O)$, $U^m(O)$, and R_{mn} were determined above. (In the case of SH-waves, we must write R_{mn}^{SH} instead of R_{mn}.)

Using the formula (2.64′) for the continuation of the displacement vector along a ray and substituting the expression for the incident wave, we obtain $\mathbf{W}^{mn}$ at an arbitrary point, M, lying on the ray of the R-wave, i.e.

$$\mathbf{W}^{mn}(M) = U^{mn}(M)\exp\{i\omega[t-\tau_{mn}(M)]\}\mathbf{n}_n(M), \tag{2.93}$$

where $\mathbf{n}_n(M)$ may be $\mathbf{n}_{\mathrm{P}}$, $\mathbf{n}_{\mathrm{SV}}$, or $\mathbf{n}_{\mathrm{SH}}$, depending on the type of wave. For P and SV waves, the projections of $\mathbf{n}_n(M)$ on the x and z axes are given in table 2.2 (where the angles θ_n must be understood as the angles measured at the point M). An additional multiplier -1 may appear in (2.93) in the case of SV-waves; see details in 2.4.4. $\mathbf{n}_{\mathrm{SH}}$ is always $\mathbf{j}$. For $\tau_{mn}(M)$ we get, from (2.64′),

$$\tau_{mn}(M) = \tau(M_0)+\int_{M_0}^{O}\frac{\mathrm{d}s}{V_m}+\int_{O}^{M}\frac{\mathrm{d}s}{V_n}, \tag{2.94}$$

where the integrations are along the ray path (of course, V_m and V_n may change along the ray).

$U^{mn}(M)$ is given by the formula

$$U^{mn}(M) = U(M_0)\left(\frac{V_m(M_0)\rho(M_0)}{V_n(M)\rho(M)}\right)^{1/2}\left(\frac{\mathrm{d}\sigma(M_0)}{\mathrm{d}\sigma(M)}\right)^{1/2} \times\left(\frac{V_n(O)\rho'(O)}{V_m(O)\rho(O)}\right)^{1/2}\left(\frac{\mathrm{d}\sigma'(O)}{\mathrm{d}\sigma(O)}\right)^{1/2}R_{mn}, \tag{2.95}$$

where $V_m(O)$ and $\rho(O)$ denote the velocity and density at the point O on the side of the incident wave and $V_n(O)$ and $\rho'(O)$ denote the same parameters on the side of the R-wave. Similarly, $\mathrm{d}\sigma(O)$ and $\mathrm{d}\sigma'(O)$ denote the cross-sectional areas of the ray tube at the point O on the side of the incident wave and on the side of the R-wave, respectively.

The equations (2.93) to (2.95) give us the necessary general expressions for determining the displacement vector of the R-wave, if the displacement vector of the incident wave is known at an arbitrary point M_0.

2.4.4 *Arbitrary Number of Interfaces*

We can easily generalize the procedure of the preceding subsection to an arbitrary number of interfaces (see figure 2.11). We shall call the layer in which the point M_0 lies the 1st layer, and the other layers successively the 2nd layer, 3rd layer, . . ., $(k+1)$th layer, following the ray of the wave under consideration. Note that the same layer may be encountered several times depending on the number of times the wave under consideration passes through it. We shall denote the points of reflection or refraction (lying on the interfaces) by $O_1, O_2, \ldots, O_k$ and the interface on which the point O_j lies by Σ_j. (The same interface may again have different symbols Σ_j along different segments of the ray if the wave is reflected or refracted more than once at it.) The interfaces need not be continuous along the whole profile, but they must be smooth in the neighbourhood of the points O_j $(j = 1, 2, \ldots, k)$. More specifically, the principal radii of curvature of the interfaces must be continuous in the neighbourhood of O_j $(j = 1, 2, \ldots, k)$. We shall also assume that the principal radii of curvature and the thickness of individual layers are large in comparison with the wavelength, and that gradients of velocity in individual layers are small in comparison with the frequency. We use $v(O_j)$, $\rho(O_j)$, and $d\sigma(O_j)$ to denote the velocity.

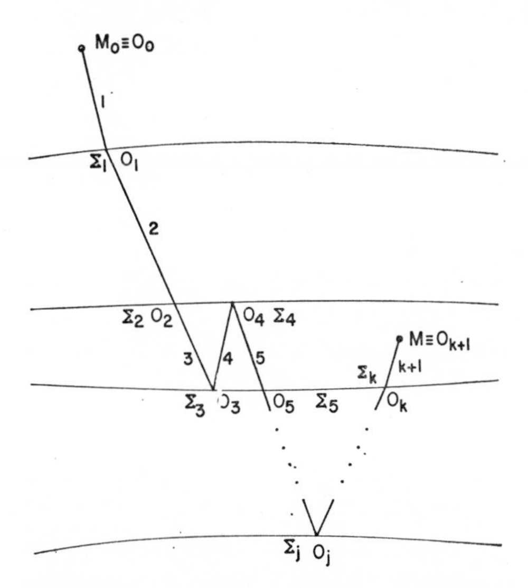

FIGURE 2.11 Explanation of various symbols for a ray in a medium with an arbitrary number of interfaces which may not be plane. The ray paths may be curved.

density, and cross-sectional area of an elementary ray tube at the point O_j on the side from which the wave is incident, and $v'(O_j)$, $\rho'(O_j)$, and $d\sigma'(O_j)$ for the same parameters on the side of the generated wave. $v(O_j)$ and $v'(O_j)$ may be compressional or shear wave velocities depending on the type of wave we are considering. (Note the difference in the definition of $v(O_j)$ and $v'(O_j)$ from the definition of V_i ($i = 1, 2, 3, 4$) given by (2.48).) The elements of the ray between M_0 and O_1, O_1 and $O_2, \ldots, O_k$ and M will be designated as the 1st element, 2nd element, ..., $(k+1)$th element of the ray, the length of the jth element of the ray being l_j. The point M_0 is also denoted by O_0 and the point M by O_{k+1}.

We shall assume that the whole ray lies in a vertical plane, which we shall call the ray plane or the plane $E_{\|}$. In this case, of course, the normals to the interfaces Σ_j at the points O_j also lie in $E_{\|}$ for each j ($j = 1, 2, \ldots, k$).

We now introduce a basic Cartesian coordinate system (x, y, z) defined thus: the z-axis is vertical (positive upwards), the x-axis lies in the $E_{\|}$ plane and is perpendicular to the z-axis (positive in the direction of propagation of the wave), and the y-axis is perpendicular to the $E_{\|}$ plane (positive direction determined so that the coordinate system is right-handed). The point M_0 lies on the z-axis. At every point O_j, we introduce a local rectangular system (x_j, y_j, z_j). The z_j-axis coincides with the normal to the Σ_j interface at the point O_j and the x_j-axis lies in the $E_{\|}$ plane and is tangent to the Σ_j interface at the same point, the direction of both being positive if they form an acute angle with the corresponding z and x axes of the basic coordinate system. The y_j-axis is parallel to the y-axis of the basic coordinate system. The local coordinate systems at the points M_0 and M are defined so that the axes are parallel to those of the basic coordinate system. We shall assume that the wave does not propagate in any layer in the direction opposite to the x-axis of the basic system.

The unit vector $\mathbf{n}_P$ is always measured along the ray, pointing in the direction of propagation (its x-component is always positive). The unit vector $\mathbf{n}_{SH}$ always lies along the y-axis. Some difficulties may arise with the unit vector $\mathbf{n}_{SV}$ when we pass from one local system $(x_{j-1}, y_{j-1}, z_{j-1})$ to another (x_j, y_j, z_j). This vector lies in the $E_{\|}$ plane and is perpendicular to the ray, such that its projection on the x_j-axis is positive (in the local coordinate system (x_j, y_j, z_j)). The positive direction of $\mathbf{n}_{SV}$ in one local coordinate system $(x_{j-1}, y_{j-1}, z_{j-1})$ may be negative in another (x_j, y_j, z_j), so that when we pass from the $(j-1)$th coordinate system to the jth we must multiply the displacement vector by -1. If we want to generalize our one-interface results, we must find the number, ϵ, of elements of the ray where $\mathbf{n}_{SV}$ changes sign. We introduce the number ϵ_j, corresponding to the jth element of the ray. If the wave propagates along the jth element as a P or SH wave, $\epsilon_j = 0$. If it propagates as an SV-wave, the value of ϵ_j is given by the following rule: $\epsilon_j = 0$ if the projection of $\mathbf{n}_{SV}(O_{j-1})$ on the z_{j-1}-axis has the same sign as the projection of $\mathbf{n}_{SV}(O_j)$ on the z_j-axis; $\epsilon_j = 1$ if the signs of the two projections are different.

Note that the vectors $\mathbf{n}_{SV}$ (O_{j-1}) and $\mathbf{n}_{SV}(O_j)$ correspond to the jth element of the ray, i.e. the projection of $\mathbf{n}_{SV}(O_{j-1})$ is to be measured at the point O_{j-1} on the side of the generated wave and the projection of $\mathbf{n}_{SV}(O_j)$ at the point O_j on the side from which the wave is incident. The number ϵ is then given by the expression

$$\epsilon = \sum_{j=1}^{k+1} \epsilon_j. \tag{2.96}$$

For SH-waves and for waves which travel as P-waves along the entire ray, $\epsilon = 0$. For a wave which propagates as an SV-wave along any element of the ray, ϵ may be different from zero and we must multiply the results obtained for the displacement vector by $(-1)^\epsilon$, where ϵ is determined by (2.96). Examples of the calculation of ϵ for different R-waves are given in figure 2.12.

Using the formulae (2.93) to (2.95), we easily obtain

$$\mathbf{W}(M) = U(M)\exp\{i\omega[t-\tau(M)]\}\mathbf{n}(M), \tag{2.97}$$

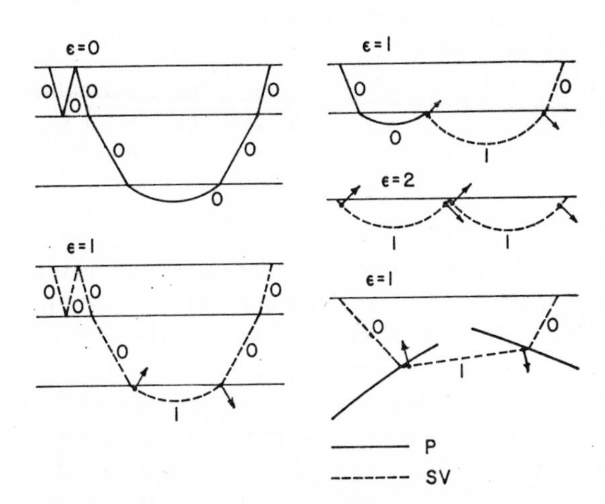

FIGURE 2.12 Examples of the calculation of ϵ for different R-waves. The arrows indicate the directions of $\mathbf{n}_{SV}$ in local coordinate systems. The number given for each element of the ray is the value of ϵ_j. The value of ϵ is the sum of these from $j = 1$ to $j = k+1$.

where $\mathbf{n}(M)$ may be $\mathbf{n}_P(M)$, $\mathbf{n}_{SV}(M)$, or $\mathbf{n}_{SH}(M)$ depending on the type of wave and $\tau(M)$ is given by

$$\tau(M) = \tau(M_0) + \sum_{j=1}^{k+1} \int_{O_{j-1}}^{O_j} \frac{ds}{v}, \tag{2.98}$$

where the integration paths are along the ray and v denotes the velocity, which may not be a constant, and may be α or β, depending on the type of wave along a given part of the ray. $U(M)$ is given by

$$U(M) = U(M_0)\frac{(-1)^\epsilon}{L(M, M_0)}\left(\frac{v(M_0)\rho(M_0)}{v(M)\rho(M)}\right)^{1/2} \prod_{j=1}^{k} \left(\frac{v'(O_j)\rho'(O_j)}{v(O_j)\rho(O_j)}\right)^{1/2} R_j, \tag{2.99}$$

where R_j is the appropriate R-coefficient at the point O_j and $L(M, M_0)$ is given by the relation

$$L(M, M_0) = \left(\frac{d\sigma(M)}{d\sigma(M_0)}\right)^{1/2} \prod_{j=1}^{k} \left(\frac{d\sigma(O_j)}{d\sigma'(O_j)}\right)^{1/2}. \tag{2.100}$$

Formulae (2.97) to (2.100) determine the displacement vector at the point M from knowledge of the displacement vector at M_0 lying on the same ray.

If the point M lies on the earth's surface, the unit vector $\mathbf{n}(M)$ in (2.97) is to be replaced by the conversion vector $\mathbf{q}(M)$ (see (2.87)).

It is possible to determine easily all the parameters in (2.97) to (2.100) except for the function $L(M, M_0)$. We shall give some suitable formulae for computing this function when a point source is located at the point M_0.

2.5
POINT SOURCE

Sources radiating seismic waves have been studied theoretically by many seismologists. In general, it is very difficult to determine the wave field in the close proximity of a source. If the receiver lies at large distances from the source (in comparison with the wavelength), the problem usually becomes simpler. In this book we shall not investigate the so-called source problem of seismology. We shall consider only the simplest possible situation in which the source is assumed to be a *point source* (at M_0) radiating any or all of the three types of seismic waves, viz., P, SV, SH waves, as long as we can treat these waves separately. Moreover, we shall be concerned with the wave field only at large distances from the source. Even then it is very difficult to determine the displacement vector of a wave emanating from this source if the material around it is inhomogeneous.

Therefore, we shall assume that it is possible to neglect any velocity changes in the neighbourhood of the source.

We denote the density in this volume by ρ_0 and the velocity by v_0, which may be α_0 or β_0 depending on whether the source radiates P or S waves. The displacement vector **W** at an arbitrary point N in this volume is assumed to be given by

$$\mathbf{W}(N) = U(N)\exp[i\omega(t - R_0/v_0)]\mathbf{n}(N), \tag{2.101}$$

where R_0 is the distance of the point N from the source. To determine $U(N)$ we must know values of U on some surface enclosing the source. Assume that we know U on a unit sphere with its centre at the source. U will naturally be a function of the spherical coordinates ϑ_0 and φ_0. We denote the function characterizing U on a unit sphere by $g(\vartheta_0, \varphi_0)$ and call it the *directional characteristic* of the source. The cross-sectional area of an elementary ray tube of the wave emanating from the source, on the unit sphere, will be

$$d\sigma = \sin\vartheta_0 \, d\vartheta_0 \, d\varphi_0, \tag{2.102}$$

and at a point N (at a distance R_0 from the source and still within the uniform medium)

$$d\sigma(N) = R_0^2 \sin\vartheta_0 \, d\vartheta_0 \, d\varphi_0. \tag{2.103}$$

From (2.64), (2.64′), and the above equations, we obtain

$$U(N) = g(\vartheta_0, \varphi_0)/R_0 \tag{2.104}$$

and

$$\mathbf{W}(N) = [g(\vartheta_0, \varphi_0)/R_0]\exp[i\omega(t - R_0/v_0)]\mathbf{n}(N). \tag{2.105}$$

The expression (2.105) will be used generally for characterizing the wave originating from the source. According to whether $\mathbf{n}(N)$ is $\mathbf{n}_P$, $\mathbf{n}_{SV}$, or $\mathbf{n}_{SH}$, we shall call the source a source of compressional P, shear SV, or shear SH waves, respectively.

If the wave (2.105) propagates from the source, we obtain for the displacement vector $\mathbf{W}(M)$, at an arbitrary point M (using formulae (2.97) to (2.100)),

$$\mathbf{W}(M) = U(M)\exp\{i\omega[t - \tau(M)]\}\mathbf{n}(M), \tag{2.106}$$

where

$$U(M) = \frac{g(\vartheta_0, \varphi_0)}{L}(-1)^{\epsilon}\left(\frac{v_0\,\rho_0}{v(M)\,\rho(M)}\right)^{1/2}\prod_{j=1}^{k}\left(\frac{v'(O_j)\rho'(O_j)}{v(O_j)\rho(O_j)}\right)^{1/2} R_j \tag{2.107}$$

and L is the so-called *spreading function* (or simply *spreading*), given by the formula

$$L = \left(\frac{\mathrm{d}\sigma(M)}{\sin\vartheta_0\,\mathrm{d}\vartheta_0\,\mathrm{d}\varphi_0}\right)^{1/2}\prod_{j=1}^{k}\left(\frac{\mathrm{d}\sigma(O_j)}{\mathrm{d}\sigma'(O_j)}\right)^{1/2}. \tag{2.108}$$

All parameters in (2.106) to (2.108) are the same as defined in the previous section; $\tau(M)$ is given by (see (2.98) when $\tau(M_0) = 0$)

$$\tau(M) = \sum_{j=1}^{k+1}\int_{O_{j-1}}^{O_j}\frac{\mathrm{d}s}{v} = \int_{M_0}^{O_1}\frac{\mathrm{d}s}{v}+\sum_{j=1}^{k-1}\int_{O_j}^{O_{j+1}}\frac{\mathrm{d}s}{v}+\int_{O_k}^{M}\frac{\mathrm{d}s}{v}. \tag{2.109}$$

The spreading function L will be calculated in the next section for some types of media.

As mentioned in chapter 1, ray formulae cannot be used in the immediate vicinity of a caustic. At the caustics, the ray tube shrinks to zero and $L = 0$ (see (2.108)). Therefore, the amplitudes of the R-waves (in the ray approximation) reach infinite values there. The zero-order formulae are applicable at sufficient distances on either side of a caustic; however, as we cross the caustic, an additional multiplier, $e^{\pm i\pi/2}$, appears in the formulae for **W**. When the time factor in the expression for **W** is given by $\exp[i\omega(t-\tau)]$ (as in our case), the multiplier is $e^{+i\pi/2}$; for the complex conjugate time factor $\exp[-i\omega(t-\tau)]$, the multiplier would be $e^{-i\pi/2}$. If the wave (2.106) passes s times through the caustic, we must use the multiplier $e^{+is\pi/2}$. This can be explained formally when we consider the orientation of the elementary ray tube $\mathrm{d}\sigma$ in the expression for L. In crossing a caustic the orientation of $\mathrm{d}\sigma$ changes and the factor $\sqrt{(-1)}$ appears in the expression for L. If the R-wave under consideration passes s times through the caustic, $L = |L|\exp(i\arg L)$, where $\arg L = -s\pi/2$. See more details in Alekseyev and Gel'chinskiy (1959) and Yanovskaya (n.d.).

2.6
FINAL EXPRESSION FOR R-WAVES FOR A POINT SOURCE FOR SOME TYPES OF MEDIA

In this section the final formulae for the displacement vector of an arbitrary R-wave propagating in some important types of media will be given. The source, lying at the point M_0, is assumed to be symmetric. We shall take $g(\vartheta_0, \varphi_0) = 1$ for simplicity. The displacement vector $\mathbf{W}^0$ of the wave originating at the source will be given, in the neighbourhood of the source, by (see (2.105))

$$\mathbf{W}^0 = (1/R_0)\exp[i\omega(t-R_0/v_0)]\mathbf{n}. \tag{2.110}$$

For the calculation of the displacement vector of the R-wave, **W**, at an arbitrary point M, we may use equations (2.106) through (2.109). These formulae are, however, not yet suitable for calculation, as the expressions for the spreading function L and for τ were not specified in detail. They may be calculated simply for some types of media. (Note that all results which follow remain valid for non-symmetric sources also, except for the additional multiplier $g(\vartheta_0, \varphi_0)$, where ϑ_0 and φ_0 correspond to the ray under consideration.)

2.6.1 *Velocity a Function of Depth*

If the velocity depends only on one coordinate, say z, we obtain rather simple formulae for L and τ. The dependence of velocity on z may be arbitrary, but the velocity gradients must be small in comparison with the frequency. If the velocity changes rapidly within a short distance (comparable with the wavelength), the formulae will be less precise, or inapplicable.

We shall assume that the source M_0 lies on the z-axis. As the

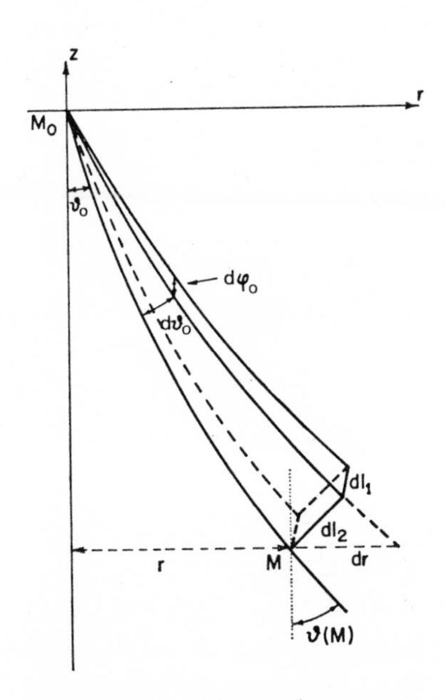

FIGURE 2.13 Cross-section at the point M of an elementary ray tube originating at the point M_0.

direction of the z-axis is upwards, we introduce a new parameter d such that

$$d = -z. \tag{2.111}$$

The parameter d will sometimes be used instead of z in the following sections and will be used to denote *depth*. By r we denote the horizontal distance of the point M (receiver) from the z-axis; r will be called the *epicentral distance*. We shall first calculate the spreading function L with the help of equation (2.108). To start with, we find a general expression for $(\mathrm{d}\sigma(M)/\sin\vartheta_0\,\mathrm{d}\vartheta_0\,\mathrm{d}\varphi_0)^{1/2}$ which depends only on the position of the source and the receiver. From simple geometrical considerations, we obtain for $\mathrm{d}\sigma(M)$ (see figure 2.13)

$$\mathrm{d}\sigma(M) = \mathrm{d}l_1\,\mathrm{d}l_2, \tag{2.112}$$

where

$$\begin{aligned} \mathrm{d}l_1 &= r\,\mathrm{d}\varphi_0, \\ \mathrm{d}l_2 &= \mathrm{d}r\cos\vartheta(M) = \frac{\partial r}{\partial\vartheta_0}\cos\vartheta(M)\,\mathrm{d}\vartheta_0. \end{aligned} \tag{2.113}$$

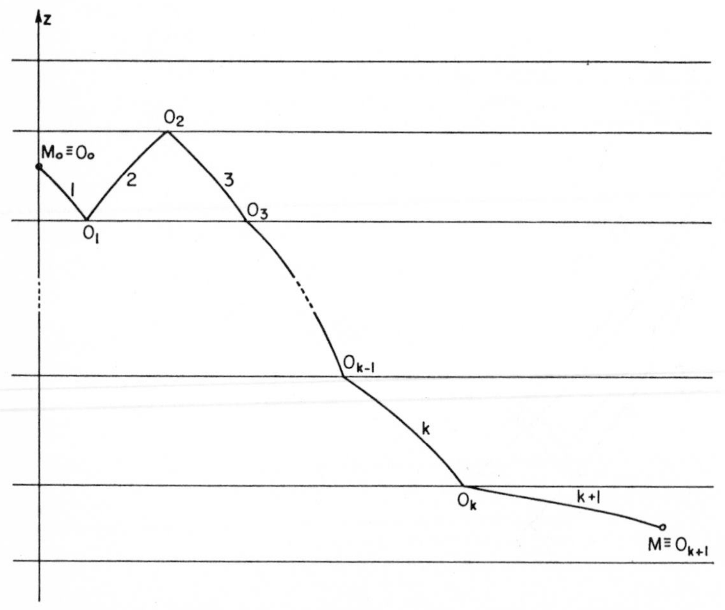

FIGURE 2.14 Explanation of various symbols for a ray in a vertically inhomogeneous medium with an arbitrary number of plane, horizontal interfaces.

The above leads to

$$\left(\frac{\mathrm{d}\sigma(M)}{\sin\vartheta_0\,\mathrm{d}\varphi_0\,\mathrm{d}\vartheta_0}\right)^{1/2} = \left(r\frac{\partial r}{\partial\vartheta_0}\frac{\cos\vartheta(M)}{\sin\vartheta_0}\right)^{1/2}. \tag{2.114}$$

This formula is valid generally, even if the wave under consideration is reflected or refracted at any interfaces between M_0 and M.

We introduce an arbitrary number of interfaces between M_0 and M. This is a special case of section 2.4.4 with interfaces all horizontal, and is illustrated in figures 2.14 and 2.15 using the same notation as before. The vertical projection of the jth element of the ray is h_j. Except for the first and the last elements, h_j is the thickness of the layer in which the jth element lies. The velocity of propagation in the jth element will be denoted by $v_j(z)$ (it may, of course, be compressional or shear velocity). It is obvious that

$$v_j(z(O_{j-1})) = v'(O_{j-1}), \qquad v_j(z(O_j)) = v(O_j).$$

We shall denote the angles of incidence and refraction (or reflection) at O_j by $\vartheta(O_j)$ and $\vartheta'(O_j)$, respectively.

Now we can determine the remainder of the terms in the formula for L, (2.108). For the point O_j, lying on the interface, we see, from figure 2.16, that

$$\left(\frac{\mathrm{d}\sigma(O_j)}{\mathrm{d}\sigma'(O_j)}\right)^{1/2} = \left(\frac{\cos\vartheta(O_j)}{\cos\vartheta'(O_j)}\right)^{1/2}. \tag{2.115}$$

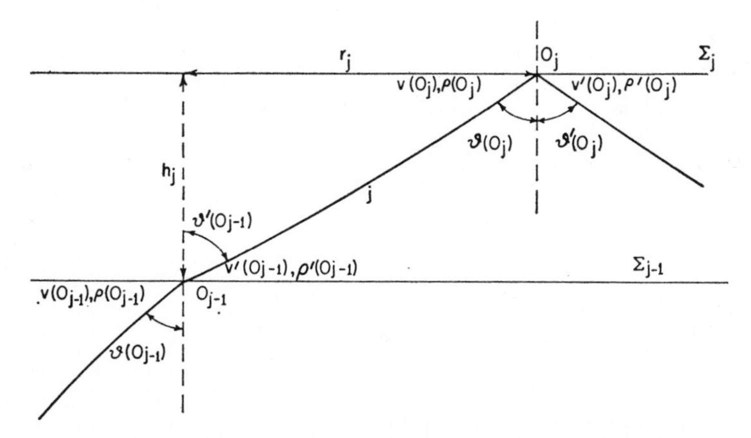

FIGURE 2.15 Explanation of the various symbols for the jth element of a ray in a vertically inhomogeneous medium; see figure 2.14.

Inserting (2.114) and (2.115) into (2.108) yields

$$L = \left(r\frac{\partial r}{\partial \vartheta_0}\frac{\cos\vartheta(M)}{\sin\vartheta_0}\right)^{1/2}\prod_{j=1}^{k}\left(\frac{\cos\vartheta(O_j)}{\cos\vartheta'(O_j)}\right)^{1/2}. \tag{2.116}$$

This is the general formula for the spreading function for an arbitrary dependence of the wave velocity on depth.

To determine L we must know r and $\partial r/\partial\vartheta_0$. The epicentral distance of the point M is given by

$$r = \sum_{j=1}^{k+1} r_j, \tag{2.117}$$

where r_j is the length of the horizontal projection of the jth element of the ray (see figure 2.15). From simple geometrical considerations we can write for r_j

$$r_j = \left|\int_{z(O_{j-1})}^{z(O_j)} \tan\vartheta_j(z)\,dz\right|, \tag{2.118}$$

where $\vartheta_j(z)$ is the acute angle between the z-axis and the jth element of

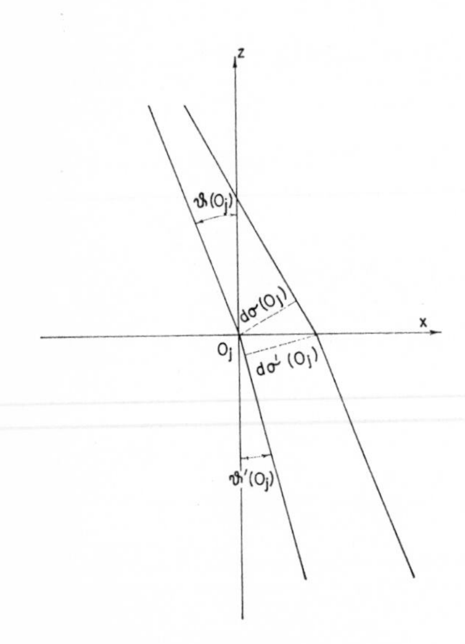

FIGURE 2.16 Trace of the cross-sectional area of a ray tube across the interface in the xz plane.

the ray at the depth z. Using Snell's law, $\sin \vartheta_j(z) = v_j(z) \sin \vartheta_0 / v_0$, we can write

$$r_j = \sin \vartheta_0 \, \tilde{r}_j, \tag{2.119}$$

where

$$\tilde{r}_j = \frac{1}{v_0} \left| \int_{z(O_{j-1})}^{z(O_j)} \frac{v_j(z) \, \mathrm{d}z}{[1 - \sin^2 \vartheta_0 \, v_j^2(z)/v_0^2]^{1/2}} \right|. \tag{2.120}$$

Note that similar formulae for r_j were also derived in an example given in section 2.2.1.

The above integral can be calculated simply for many velocity-depth functions $v_j(z)$ (see Kaufman, 1953). Some examples will be given in subsections 2.6.2 and 2.6.3.

If we know the analytic expression for $r_j(\vartheta_0)$, we can determine $\partial r_j / \partial \vartheta_0$ without difficulty. The final expression for the spreading function L is

$$L = \left[r \left(\sum_{j=1}^{k+1} \frac{\partial r_j}{\partial \vartheta_0} \right) \frac{\cos \vartheta(M)}{\sin \vartheta_0} \right]^{1/2} \prod_{j=1}^{k} \left(\frac{\cos \vartheta(O_j)}{\cos \vartheta'(O_j)} \right)^{1/2}. \tag{2.121}$$

This formula cannot be used when $r \to 0$, as then $\sin \vartheta_0 \to 0$. We can, however, simply rewrite it in the form

$$L = \left[\left(\sum_{j=1}^{k+1} \tilde{r}_j \right) \left(\sum_{j=1}^{k+1} \frac{\partial r_j}{\partial \vartheta_0} \right) \cos \vartheta(M) \right]^{1/2} \prod_{j=1}^{k} \left(\frac{\cos \vartheta(O_j)}{\cos \vartheta'(O_j)} \right)^{1/2}, \tag{2.122}$$

which is applicable for arbitrary values of the epicentral distance r.

For the travel time $\tau^{\mathrm{R}}(M)$, we can write

$$\tau^{\mathrm{R}}(M) = \sum_{j=1}^{k+1} \tau_j, \tag{2.123}$$

where τ_j is the time taken by the signal to travel from the point O_j to the point O_{j+1}, i.e.

$$\tau_j = \int_{O_{j-1}}^{O_j} \frac{\mathrm{d}s}{v_j}. \tag{2.124}$$

Using Snell's law, we can write (2.124) in the form

$$\tau_j = \left| \int_{z(O_{j-1})}^{z(O_j)} \frac{\mathrm{d}z}{v_j(z)[1 - \sin^2 \vartheta_0 \, v_j^2(z)/v_0^2]^{1/2}} \right|. \tag{2.125}$$

This formula was also derived in section 2.2.1. Using (2.106) to (2.109) we get the final expression for the displacement vector of the R-wave

$$\mathbf{W}^{R}(M) = U^{R}(M)\exp\{i\omega[t-\tau^{R}(M)]\}\mathbf{n}^{R}(M), \tag{2.126}$$

where

$$U^{R}(M) = \frac{(-1)^{\epsilon}}{L}\left(\frac{v_0\rho_0}{v(M)\rho(M)}\right)^{1/2}\prod_{j=1}^{k}\left(\frac{v'(O_j)\rho'(O_j)}{v(O_j)\rho(O_j)}\right)^{1/2} R_j, \tag{2.127}$$

$$\tau^{R}(M) = \sum_{j=1}^{k+1}\tau_j, \tag{2.128}$$

L being given by (2.121) or (2.122) and τ_j by (2.125).

For the r and z components of the displacement vector, we get from (2.126)

$$\begin{bmatrix} W_r^{R}(M) \\ W_z^{R}(M) \end{bmatrix} = U^{R}(M)\exp\{i\omega[t-\tau^{R}(M)]\}\begin{bmatrix} n_r^{R}(M) \\ n_z^{R}(M) \end{bmatrix}, \tag{2.129}$$

where $n_r^{R}(M)$ and $n_z^{R}(M)$ are the r and z components of the unit vector $\mathbf{n}^{R}(M)$ at the point M and are given by

$$n_r^{R}(M) = \sin\vartheta(M), \qquad n_z^{R}(M) = \pm\cos\vartheta(M), \tag{2.130}$$

if the R-wave has the character of a P-wave at the point M, or, if this wave is an SV-wave, by

$$n_r^{R}(M) = \cos\vartheta(M), \qquad n_z^{R}(M) = \mp\sin\vartheta(M). \tag{2.131}$$

In (2.130) and (2.131), the upper sign applies if the wave is propagating upwards, the lower one if it is propagating downwards through the point M.

The equations (2.126) to (2.128) may be used for SH-waves, too. For the φ-component of the displacement vector we have

$$W_\varphi^{R}(M) = U^{R}(M)\exp\{i\omega[t-\tau^{R}(M)]\}. \tag{2.132}$$

We may rewrite the expressions (2.129) in the form

$$\begin{aligned} W_r^{R}(M) &= A_r^{R}\exp[i\omega(t-\tau^{R})+i\chi_r^{R}], \\ W_z^{R}(M) &= A_z^{R}\exp[i\omega(t-\tau^{R})+i\chi_z^{R}], \end{aligned} \tag{2.133}$$

where A_r^{R} and A_z^{R} are the amplitudes of the horizontal and vertical components of displacement and χ_r^{R} and χ_z^{R} are their phase shifts. They are given by

$$\begin{bmatrix} A_r^{R} \\ A_z^{R} \end{bmatrix} = \frac{1}{|L|}\left(\frac{v_0\rho_0}{v(M)\rho(M)}\right)^{1/2}\prod_{j=1}^{k}\left(\frac{v'(O_j)\rho'(O_j)}{v(O_j)\rho(O_j)}\right)^{1/2}|R_j|\begin{bmatrix} |n_r^{R}| \\ |n_z^{R}| \end{bmatrix}, \tag{2.134}$$

$$\begin{bmatrix} \chi_r^{\mathrm{R}} \\ \chi_z^{\mathrm{R}} \end{bmatrix} = \sum_{j=1}^{k} \arg R_j - \arg L + \epsilon\pi + \begin{bmatrix} \arg n_r^{\mathrm{R}} \\ \arg n_z^{\mathrm{R}} \end{bmatrix}. \tag{2.135}$$

(See note (b) for the explanation of arg L.)

Notes

(a) If the jth element of the ray has a turning point within the layer, i.e. the points O_{j-1} and O_j lie on the same interface, the formulae for r_j and τ_j given above cannot be used. The expressions applicable in this case are (see figure 2.17):

$$r_j = \frac{2 \sin \vartheta_0}{v_0} \left| \int_{z(O_{j-1})}^{z^*} \frac{v_j(z) \mathrm{d}z}{[1 - \sin^2 \vartheta_0 v_j^2(z)/v_0^2]^{1/2}} \right|,$$
$$\tau_j = 2 \left| \int_{z(O_{j-1})}^{z^*} \frac{\mathrm{d}z}{v_j(z)[1 - \sin^2 \vartheta_0 v_j^2(z)/v_0^2]^{1/2}} \right|. \tag{2.136}$$

Here z^* is the z-coordinate of the turning point given by the relation, following from Snell's law,

$$v_j(z^*) = v_0/\sin \vartheta_0. \tag{2.137}$$

(b) As explained at the end of the section 2.5, we can always write for the spreading function L

$$L = |L|\exp(i \arg L), \qquad \arg L = -s\pi/2, \tag{2.138}$$

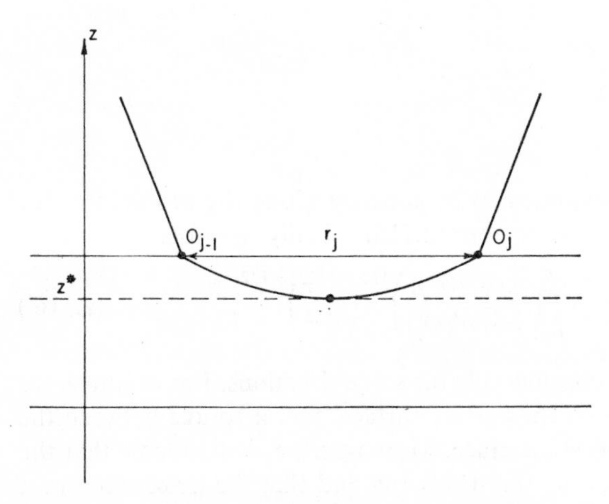

FIGURE 2.17 Turning point of the ray within a layer.

where s is the number of times the wave passed through a caustic. We do not need to know s for the determination of amplitudes (see (2.134)); this number is required only for determining the phase shifts. In the case of homogeneous layers with plane interfaces, s is always zero; i.e., $\arg L = 0$.

(c) If the receiver lies on the surface of the earth (free interface), the values of $n_r^{\mathrm{R}}(M)$ and $n_z^{\mathrm{R}}(M)$ in (2.129), (2.134), and (2.135) must be replaced by the coefficients of conversion.

(d) For P-waves (along the entire ray) and SH-waves, the value of ϵ is always zero. For a wave which propagates as SV along any element of the ray, $\epsilon = 0$ only if no SV element of the ray has a turning point within a layer. If any SV element of the ray has a turning point within the layer, $\epsilon \neq 0$ (see figure 2.12). In the case of homogeneous layers, ϵ always equals zero as no element of the ray can have a turning point within a layer.

(e) We can simply derive another expression for L, which is of considerable importance in the interpretation of seismic measurements. We shall assume that we know the time–distance curve $\tau = \tau(r)$ of the wave under consideration, along the profile in the direction of the r-axis (for $z = z_c =$ const.). We easily obtain the well-known formula

$$\frac{\mathrm{d}\tau(r)}{\mathrm{d}r} = p,$$

where p is the parameter of the ray, $p = \sin\vartheta_0/v_0 = \sin\vartheta(z_c)/v(z_c)$. Differentiating the above with respect to r, we find that

$$\frac{\mathrm{d}^2\tau}{\mathrm{d}r^2} = \frac{\cos\vartheta_0}{v_0}\frac{\mathrm{d}\vartheta_0}{\mathrm{d}r}$$

and, from this,

$$\frac{\mathrm{d}r}{\mathrm{d}\vartheta_0} = \frac{\cos\vartheta_0}{v_0}\left(\frac{\mathrm{d}^2\tau}{\mathrm{d}r^2}\right)^{-1}.$$

As the z-coordinate is assumed to be constant along the profile, we can insert this expression for $\partial r/\partial\vartheta_0$ into (2.116). Finally, we obtain

$$L = \left(\frac{r}{v_0}\cot\vartheta_0\cos\vartheta(z_c)\prod_{j=1}^{k}\frac{\cos\vartheta(O_j)}{\cos\vartheta'(O_j)}\right)^{1/2}\left(\frac{\mathrm{d}^2\tau}{\mathrm{d}r^2}\right)^{-1/2}, \qquad (2.116')$$

which can be simplified considerably for some situations. For example, we shall consider a profile on the earth's surface, and a source between the earth's surface and the first interface. Moreover, we shall assume that the wave is compressional along the whole ray and that the descending part of the ray is symmetrical with the ascending part. Then we have

$$\prod_{j=1}^{k} \frac{\cos \vartheta(O_j)}{\cos \vartheta'(O_j)} = 1,$$

and we obtain

$$L = [\cot \vartheta_0 \cos \vartheta(z_c)]^{1/2} \left(\frac{r}{v_0}\right)^{1/2} \bigg/ \left(\frac{d^2\tau}{dr^2}\right)^{1/2}. \tag{2.116''}$$

The same expression can be used when the wave is either SV or SH, along the whole ray.

Inserting the above equation into the expressions for the amplitudes of R-waves, (2.134), we see that the amplitudes are proportional to the factor $|d^2\tau/dr^2|^{1/2}$. Similar formulae for a spherically stratified earth's model were obtained by Bullen (1965, pp. 130–1), who gives a detailed seismological discussion of this fact.

2.6.2 *Homogeneous Layers*

We shall assume now that the velocity and density within individual layers are constant, i.e. $v_j(z) = v_j = \text{const.}$, $\rho_j(z) = \rho_j = \text{const.}$, $\vartheta_j(z) = \vartheta_j = \text{const.}$ Then $v'(O_{j-1}) = v(O_j)$, $\rho'(O_{j-1}) = \rho(O_j)$, $\vartheta'(O_{j-1}) = \vartheta(O_j)$. For r_j, $\tilde{r}_j$, and τ_j we have, from (2.119) and (2.125),

$$\left.\begin{aligned} r_j &= h_j \tan \vartheta_j, \\ \tilde{r}_j &= v_j h_j / v_1 \cos \vartheta_j, \end{aligned}\right\} \tag{2.139}$$

$$\tau_j = h_j / v_j \cos \vartheta_j. \tag{2.140}$$

The epicentral distance r and the travel time τ^R can be found by inserting (2.139) and (2.140) into (2.117) and (2.123).

From (2.139) it follows that

$$\frac{\partial r_j}{\partial \vartheta_0} = \frac{h_j \cos \vartheta_1 v_j}{\cos^3 \vartheta_j \, v_1}. \tag{2.141}$$

Inserting this into (2.121) yields

$$L = \left(r \cot \vartheta_1 \sum_{j=1}^{k+1} \frac{v_j h_j}{v_1} \frac{\cos \vartheta_1}{\cos^3 \vartheta_j}\right)^{1/2}, \tag{2.142}$$

or, using (2.122),

$$L = \frac{\cos \vartheta_1}{v_1} \left[\left(\sum_{j=1}^{k+1} \frac{v_j h_j}{\cos \vartheta_j}\right)\left(\sum_{j=1}^{k+1} \frac{v_j h_j}{\cos^3 \vartheta_j}\right)\right]^{1/2}. \tag{2.143}$$

The displacement vector of the R-wave $\mathbf{W}^R(M)$ is again given by (2.126), where $U^R(M)$ now becomes simply

$$U^{\mathrm{R}}(M) = \frac{1}{L} \prod_{j=1}^{k} R_j. \tag{2.144}$$

The expressions (2.129) for the r and z components of the displacement vector are applicable in the present situations as well.

The amplitudes of the horizontal and vertical components of the displacement vector and their phase shifts are now given by (see (2.134) and (2.135))

$$\begin{bmatrix} A_r^{\mathrm{R}} \\ A_z^{\mathrm{R}} \end{bmatrix} = \frac{1}{L} \prod_{j=1}^{k} |R_j| \begin{bmatrix} |n_r^{\mathrm{R}}| \\ |n_z^{\mathrm{R}}| \end{bmatrix}, \tag{2.145}$$

$$\begin{bmatrix} \chi_r^{\mathrm{R}} \\ \chi_z^{\mathrm{R}} \end{bmatrix} = \sum_{j=1}^{k} \arg R_j + \begin{bmatrix} \arg n_r^{\mathrm{R}} \\ \arg n_z^{\mathrm{R}} \end{bmatrix}. \tag{2.146}$$

For the *one-interface problem*, we obtain the simple expression

$$U^{\mathrm{R}}(M) = R_1/L, \tag{2.147}$$

where

$$L = \left[r \cot \vartheta_1 \left(\frac{h_1}{\cos^2 \vartheta_1} + \frac{v_2 h_2 \cos \vartheta_1}{v_1 \cos^3 \vartheta_2} \right) \right]^{1/2} \tag{2.148}$$

and R_1 is the R-coefficient of the appropriate type ($R_1 = R_{mn}$). Note that the expression for L is much simpler if $m = n$ (a case of reflected waves where incident and reflected waves are either both compressional or both shear). Then $\vartheta_2 = \vartheta_1$ and $v_2 = v_1$. Inserting this fact into (2.148) yields

$$L = \left[\frac{r(h_1+h_2)}{\sin \vartheta_1 \cos \vartheta_1} \right]^{1/2} = [r^2 + (h_1+h_2)^2]^{1/2}, \tag{2.149}$$

which indicates that the spreading function, in this case, is the distance from the image of the source.

The amplitudes and phase shifts of the horizontal and vertical components of displacement are given, for the one-interface problem, by

$$\begin{bmatrix} A_r^{\mathrm{R}} \\ A_z^{\mathrm{R}} \end{bmatrix} = \frac{|R_1|}{L} \begin{bmatrix} |n_r^{\mathrm{R}}| \\ |n_z^{\mathrm{R}}| \end{bmatrix}, \quad \begin{bmatrix} \chi_r^{\mathrm{R}} \\ \chi_z^{\mathrm{R}} \end{bmatrix} = \arg R_1 + \begin{bmatrix} \arg n_r^{\mathrm{R}} \\ \arg n_z^{\mathrm{R}} \end{bmatrix}. \tag{2.150}$$

Similar formulae can be easily written for SH-waves also.

2.6.3 *Constant Velocity Gradient inside Layers*

Let us assume that the variation of velocity in each layer is linear, i.e. the velocity $v_j(z)$ in the layer where the jth element of the ray lies is given by the relation

$$v_j(d) = v'(O_{j-1})\{1+b_j[d-d(O_{j-1})]\}, \tag{2.151}$$

where d is the depth (see equation (2.111)) and b_j is called the velocity gradient, which can be expressed as

$$b_j = \frac{v(O_j)-v'(O_{j-1})}{[d(O_j)-d(O_{j-1})]v'(O_{j-1})}. \tag{2.152}$$

Note that, strictly speaking, the velocity gradient is $b_j v'(O_{j-1})$; however, we shall call b_j the velocity gradient, for simplicity. If velocity increases with depth, $b_j > 0$. In the opposite case, $b_j < 0$. The dimension of b_j is (distance)$^{-1}$.

For r_j and τ_j we find, from (2.119) and (2.125), that

$$r_j = \frac{h_j\, v'(O_{j-1})}{[v(O_j)-v'(O_{j-1})]\sin\vartheta'(O_{j-1})}[\cos\vartheta'(O_{j-1})-\cos\vartheta(O_j)], \tag{2.153}$$

$$\tau_j = \frac{h_j}{v(O_j)-v'(O_{j-1})}\{\tanh^{-1}[\cos\vartheta'(O_{j-1})]-\tanh^{-1}[\cos\vartheta(O_j)]\} \tag{2.154}$$

The last equation can be rewritten as

$$\tau_j = \frac{h_j}{v(O_j)-v'(O_{j-1})}\ln\frac{\tan\frac{1}{2}\vartheta(O_j)}{\tan\frac{1}{2}\vartheta'(O_{j-1})}, \tag{2.155}$$

or

$$\tau_j = \frac{h_j}{v(O_j)-v'(O_{j-1})}\ln\frac{v(O_j)[1+\cos\vartheta'(O_{j-1})]}{v'(O_{j-1})[1+\cos\vartheta(O_j)]}. \tag{2.156}$$

If the jth element of the ray has a turning point within the layer (i.e. O_{j-1} and O_j lie on the same interface; see figure 2.17), r_j and τ_j are given by

$$r_j = 2\cot\vartheta'(O_{j-1})/|b_j|, \tag{2.157}$$

$$\tau_j = \frac{2}{|b_j|v'(O_{j-1})}\ln\frac{1+\cos\vartheta'(O_{j-1})}{\sin\vartheta'(O_{j-1})}. \tag{2.158}$$

These formulae follow from (2.153) and (2.156) if we insert $\cos\vartheta(O_j) = 0$, multiply by 2, and take into account (2.152).

For the determination of the spreading function L we must know $\partial r_j/\partial\vartheta_0$. From (2.153) we find that

$$\frac{\partial r_j}{\partial \vartheta_0} = \frac{h_j \cot \vartheta_0 \, v'(O_{j-1})}{[v(O_j) - v'(O_{j-1})] \sin \vartheta'(O_{j-1})} \left(\frac{1}{\cos \vartheta(O_j)} - \frac{1}{\cos \vartheta'(O_{j-1})} \right), \tag{2.159}$$

or, inserting (2.153),

$$\frac{\partial r_j}{\partial \vartheta_0} = \frac{r_j \cot \vartheta_0}{\cos \vartheta(O_j) \cos \vartheta'(O_{j-1})}. \tag{2.160}$$

If the ray has a turning point inside the jth element, we find, from (2.157), that

$$\frac{\partial r_j}{\partial \vartheta_0} = -\frac{2 \cos \vartheta_0 \, v'(O_{j-1})}{v_0 |b_j| \sin^2 \vartheta'(O_{j-1}) \cos \vartheta'(O_{j-1})}, \tag{2.161}$$

or

$$\frac{\partial r_j}{\partial \vartheta_0} = -\frac{r_j \cot \vartheta_0}{\cos^2 \vartheta'(O_{j-1})}. \tag{2.162}$$

Inserting the above expressions for r_j, τ_j, and $\partial r_j / \partial \vartheta_0$ into (2.117), (2.121), and (2.128), we find r, L, and τ^R. The displacement vector of the R-wave at the point M is then given by formulae (2.126) and (2.127), its displacement components by (2.129) or (2.132), and the amplitudes and phase shifts by (2.134) and (2.135), (see also (2.138)).

2.6.4 *Curved and Dipping Interfaces*

In this section, we shall again consider a medium consisting of *layers separated by smoothly curved interfaces of first order*. First, we shall find expressions for the continuation of the displacement vector of the R-wave generated by a point source. We recommend that the reader look carefully through the second part of section 2.3 (about the principal radii of curvature of the wave front of reflected and refracted waves) and through section 2.4.4 before starting this section. The notation of section 2.4.4 will be used here also.

We shall not investigate the general case, but only some simple situations which lead to formulae in compact form. First, we shall assume that the principal radii of curvature are continuous in the neighbourhood of the points O_j ($j = 1, 2, \ldots, k$; see figure 2.11), that they are large in comparison with the wavelength, that the thickness of individual layers is considerably larger than the wavelength, and that the velocity gradients are small in comparison with the frequency. Then the ray theory can be used, except in the vicinity of some singularities (caustics, critical points, boundary rays), and the formulae derived in sections 2.2 and 2.3 can be used successively, following the ray from one interface to another. However, the procedure remains generally rather cumbersome, particularly when the ray is a space curve and the layers are inhomogeneous. We shall investigate in detail one situation where the results are relatively simpler, and which is important in seismic prospecting and in seismology. This situation is characterized by the following assumptions:

(1) All the layers are homogeneous.

(2) The ray is a plane curve (it lies in a vertical plane $E_{\parallel}$).

(3) The angles ϕ_j between the vertical plane $E_{\parallel}$ and a principal normal section of the interface Σ_j at O_j ($j = 1, 2, \ldots, k$) are zero. (See more details about ϕ_j in section 2.3.)

The first two assumptions have a clear physical meaning. From the geological point of view, the third assumption means that the trend of geological structures is either parallel or perpendicular to the seismic profile. We appreciate the seriousness of these assumptions, but in practice situations which satisfy them are often found.

According to assumption 2, the planes of incidence at the points O_j ($j = 1, 2, \ldots, k$) coincide with the vertical plane $E_{\parallel}$. We shall denote the corresponding radii of curvature of Σ_j, of the wave front of the incident wave, and of the wave front of the reflected (refracted) wave at O_j by $R_{\parallel}^{(j)}$, $r_{\parallel}(O_j)$, and $r_{\parallel}'(O_j)$, respectively. The radii of curvature in the direction perpendicular to $E_{\parallel}$ will be denoted by $R_{\perp}^{(j)}$, $r_{\perp}(O_j)$, and $r_{\perp}'(O_j)$ (see section 2.3). According to assumption 3, the radii of curvature of the interface Σ_j at O_j ($j = 1, 2, \ldots, k$), $R_{\parallel}^{(j)}$ and $R_{\perp}^{(j)}$, are also the *principal* radii of curvature of Σ_j at O_j. The relation between the principal radii of wave fronts and the above-introduced radii of curvature of wave fronts is a little more complicated. When the normal principal cross-sections of the wave front incident on the first interface Σ_1 have an arbitrary orientation, the principal radii of the wave front at O_1 are generally different from $r_{\parallel}(O_1)$ and $r_{\perp}(O_1)$; and, consequently, the same holds for all points O_j ($j = 1, 2, \ldots, k$).

In the case of a *point source*, however, the wave front of the wave which is incident on the first interface is spherical. Any normal cross-section of a sphere is also a principal normal cross-section because the radius of curvature of the sphere is the same for any normal section. In this case, $r_{\parallel}(O_1)$ and $r_{\perp}(O_1)$ are also principal radii of curvature. We can easily see from (2.61) that, in this case, all $r_{\parallel}(O_j)$, $r_{\perp}(O_j)$, $r_{\parallel}'(O_j)$, and $r_{\perp}'(O_j)$ ($j = 1, 2, \ldots, k$) will also be principal radii of curvature. (In other words, the angles φ and φ' defined in section 2.3 are zero for all points O_j ($j = 1, 2, \ldots, k$).)

We shall not, however, start with the derivation of the formulae for the displacement vector of the R-wave generated by a point source. We shall first consider the continuation of the displacement vector of the R-wave along the ray, from the point M_0 to point M (see figure 2.11). We assume that $r_{\|}(M_0)$ and $r_{\perp}(M_0)$ are principal radii of the wave front of the incident wave at the point M_0 and, as in the case of a point source, we find that all $r_{\|}(O_j)$, $r_{\|}'(O_j)$, $r_{\perp}(O_j)$, and $r_{\perp}'(O_j)$ $(j = 1, 2, \ldots, k)$ are principal radii of curvature (i.e. $\varphi = \varphi' = 0$ for all O_j). Then we can use the simple procedure described in section 2.3 (see (2.61) and (2.61′)) for the successive determination of $r_{\|}(O_j)$, $r_{\|}'(O_j)$, $r_{\perp}(O_j)$, $r_{\perp}'(O_j)$ $(j = 1, 2, \ldots, k)$ if the values of $R_{\|}^{(j)}$, $R_{\perp}^{(j)}$ $(j = 1, 2, \ldots, k)$ and $r_{\|}(M_0)$, $r_{\perp}(M_0)$ are known. Since the jth element of the ray is a straight line,

$$\begin{aligned} r_{\|}(O_1) &= r_{\|}(M_0)+l_1, & r_{\perp}(O_1) &= r_{\perp}(M_0)+l_1, \\ r_{\|}(O_j) &= r_{\|}'(O_{j-1})+l_j, & r_{\perp}(O_j) &= r_{\perp}'(O_{j-1})+l_j \quad (j = 2, 3, \ldots), \end{aligned} \tag{2.163}$$

where l_j is the distance between the points O_j and O_{j-1}. The other relation connecting the principal radii $r_{\|}'(O_j)$ and $r_{\perp}'(O_j)$ of the R-waves with the principal radii $r_{\|}(O_j)$ and $r_{\perp}(O_j)$ of the incident wave were found in section 2.3 (see (2.61)). In our notation, we can write

$$r_{\|}'(O_j) = r_{\|}(O_j)/\Delta_{\|}^{(j)}, \qquad r_{\perp}'(O_j) = r_{\perp}(O_j)/\Delta_{\perp}^{(j)}, \tag{2.164}$$

where

$$\begin{aligned} \Delta_{\|}^{(j)} &= \frac{v_{j+1}\cos^2\vartheta(O_j)}{v_j\cos^2\vartheta'(O_j)} \\ &\quad + \frac{r_{\|}(O_j)}{R_{\|}^{(j)}\cos^2\vartheta'(O_j)}\left(\frac{v_{j+1}}{v_j}\cos\vartheta(O_j)\pm\cos\vartheta'(O_j)\right), \\ \Delta_{\perp}^{(j)} &= \frac{v_{j+1}}{v_j}+\frac{r_{\perp}(O_j)}{R_{\perp}^{(j)}}\left(\frac{v_{j+1}}{v_j}\cos\vartheta(O_j)\pm\cos\vartheta'(O_j)\right), \end{aligned} \tag{2.165}$$

$\vartheta(O_j)$ and $\vartheta'(O_j)$ being the acute angles between the ray and the local z_j-axis (normal to Σ_j) at the point O_j on the two sides and v_j the propagation velocity in the jth layer; the positive sign is for reflected waves at O_j and the negative sign for refracted waves at O_j.

If we now use (2.100), (2.28′), and (2.163) to (2.165), we find the final expression for $L(M, M_0)$, viz.,

$$L(M, M_0) = \left(\frac{[r_{\|}'(O_k)+l_{k+1}][r_{\perp}'(O_k)+l_{k+1}]}{r_{\|}(M_0)r_{\perp}(M_0)}\prod_{j=1}^{k}\Delta_{\|}^{(j)}\Delta_{\perp}^{(j)}\right)^{1/2}. \tag{2.166}$$

Inserting this into (2.99) gives us the final expression for the continuation of the displacement vector along the ray in the case of smoothly curved interfaces.

If we use (2.163) and (2.164), we can write

$$L(M, M_0) = \left\{\frac{1}{r_{\|}(M_0)r_{\perp}(M_0)}\left(\sum_{j=1}^{k+1} l_j \prod_{\nu=1}^{j-1} \Delta_{\|}^{(\nu)}\right)\left(\sum_{j=1}^{k+1} l_j \prod_{\nu=1}^{j-1} \Delta_{\perp}^{(\nu)}\right)\right\}^{1/2}, \tag{2.167}$$

where

$$\prod_{\nu=1}^{j-1} \Delta_{\|}^{(\nu)} = \prod_{\nu=1}^{j-1} \Delta_{\perp}^{(\nu)} = 1 \qquad \text{for } j = 1.$$

A similar procedure can be used to find the displacement vector of the R-wave generated by the point source lying at M_0. Assume that the wave generated by the source is given by (2.110). Then the displacement vector $\mathbf{W}^{\mathrm{R}}(M)$ of the R-wave at the point M is given by (see (2.106) to (2.108))

$$\mathbf{W}^{\mathrm{R}}(M) = U^{\mathrm{R}}(M)\exp\{i\omega[t-\tau^{\mathrm{R}}(M)]\}\mathbf{n}^{\mathrm{R}}(M), \tag{2.168}$$

where

$$U^{\mathrm{R}}(M) = \frac{(-1)^{\epsilon}}{L} \prod_{j=1}^{k} R_j, \tag{2.169}$$

$$L = \left([r_{\|}'(O_k)+l_{k+1}][r_{\perp}'(O_k)+l_{k+1}]\prod_{j=1}^{k} \Delta_{\|}^{(j)}\Delta_{\perp}^{(j)}\right)^{1/2}, \tag{2.170}$$

$$\tau^{\mathrm{R}}(M) = \sum_{j=1}^{k+1} (l_j/v_j). \tag{2.171}$$

Since the wave generated by the source, (2.110), propagates as a spherical wave in the first layer,

$$r_{\|}(O_1) = l_1, \qquad r_{\perp}(O_1) = l_1. \tag{2.172}$$

Using the expressions (2.163), (2.164), and (2.165), we can calculate $r_{\|}(O_j)$, $r_{\perp}(O_j)$, $r_{\|}'(O_j)$, and $r_{\perp}'(O_j)$ ($j = 1, 2, \ldots, k$), which we need for the determination of L by (2.170).

The other expression for L, similar to (2.167), is

$$L = \left[\left(\sum_{j=1}^{k+1} l_j \prod_{\nu=1}^{j-1} \Delta_{\|}^{(\nu)}\right)\left(\sum_{j=1}^{k+1} l_j \prod_{\nu=1}^{j-1} \Delta_{\perp}^{(\nu)}\right)\right]^{1/2}, \tag{2.173}$$

where

$$\prod_{\nu=1}^{j-1} \Delta_{\|}^{(\nu)} = \prod_{\nu=1}^{j-1} \Delta_{\perp}^{(\nu)} = 1 \qquad \text{for } j = 1.$$

Note that $R_{\|}^{(j)}$ is taken positive or negative depending on whether the interface Σ_j is convex or concave (at O_j) in the plane $E_{\|}$ for an observer situated on the side of the incident wave; and similarly for $R_\perp^{(j)}$ in the plane perpendicular to $E_\parallel$.

Now we shall briefly recapitulate the whole procedure. We assume that the following data are known: (1) position and type of source, (2) analytic description of the interfaces, (3) elastic parameters in all layers, (4) code of the wave under consideration. (When explicit expressions for the interfaces are not known, i.e. when the interfaces are given only in graphs or tables, we must simulate them by some smooth functions.) Then the procedure for finding the displacement vector of a multiply reflected and refracted wave $\mathbf{W}^\mathrm{R}$ (at the point M), described by the given code, is as follows:

(1) First, we must determine the parameters of the ray. This is only a kinematical problem and we shall not describe it here. As a result, we obtain the position of points O_j ($j = 1, 2, \ldots, k$), the values of l_j ($j = 1, 2, \ldots, k+1$), and the angles $\vartheta(O_j)$ and $\vartheta'(O_j)$ ($j = 1, 2, \ldots, k$). Using (2.171), we can determine $\tau^\mathrm{R}(M)$.

(2) We calculate $R_\parallel^{(j)}$ and $R_\perp^{(j)}$ at all points O_j from known formulae for the interfaces Σ_j.

(3) We calculate successively all $r_\parallel(O_j)$, $r_\perp(O_j)$, $r_\perp'(O_j)$, $r_\parallel'(O_j)$. First we put $r_\parallel(O_1) = l_1$, $r_\perp(O_1) = l_1$ and calculate $\Delta_\parallel^{(1)}$ and $\Delta_\perp^{(1)}$ from (2.165). From (2.164), we obtain $r_\parallel'(O_1)$ and $r_\perp'(O_1)$. Inserting into (2.163), we find the values of $r_\parallel(O_2)$ and $r_\perp(O_2)$. The procedure has to be repeated for all j ($j = 1, 2, \ldots, k$).

(4) We use the values determined in step 3 to determine L; see formula (2.170) or (2.173), and also (2.138).

(5) We calculate the R-coefficients at points O_j, R_j ($j = 1, 2, \ldots, k$), from (2.73) and determine ϵ from (2.96).

(6) Inserting into (2.169), we find $U^\mathrm{R}(M)$. The unit vector $\mathbf{n}^\mathrm{R}(M)$ can be determined by usual methods (see table 2.2).

A similar procedure has been described in detail by Alekseyev and Gel'chinskiy (1959). Marcinkovskaya and Krasavin (1968) have given some examples of numerical calculations.

Dipping Interfaces

If all the interfaces are plane, i.e. $R_\parallel^{(j)} = R_\perp^{(j)} = \infty$ for $j = 1, 2, \ldots, k$, the expressions (2.168) to (2.171) can be used. For L, however, we find, from (2.173) and (2.165), the simpler formula

$$L = \left[\left(\sum_{j=1}^{k+1} \frac{l_j v_j}{v_1}\right)\left(\sum_{j=1}^{k+1} \frac{l_j v_j}{v_1} \prod_{\nu=1}^{j-1} \frac{\cos^2\vartheta(O_\nu)}{\cos^2\vartheta'(O_\nu)}\right)\right]^{1/2}, \tag{2.174}$$

where the product term equals 1 for $j = 1$.

If the interfaces are plane and parallel to each other,

$$l_j = \frac{h_j}{\cos\vartheta_j}, \qquad \prod_{\nu=1}^{j-1} \frac{\cos^2\vartheta(O_\nu)}{\cos^2\vartheta'(O_\nu)} = \frac{\cos^2\vartheta_1}{\cos^2\vartheta_j}. \tag{2.175}$$

Then the expression (2.174) is equivalent to (2.143), as is to be expected from the geometry of the situation.

2.6.5 *Generalization for an Impulse Source*

We have assumed so far that the source is *harmonic*. Now we shall generalize the results obtained for a harmonic source to the case of an *impulse* source. The displacement vector of the impulse wave will be denoted by $\mathscr{W}$, instead of $\mathbf{W}$ which has been used for harmonic waves. We shall assume that the displacement vector of the impulse wave radiating from the source, at a point N in its neighbourhood, is given by

$$\mathscr{W}^0(N) = (1/R_0) f(t - R_0/v_0)\mathbf{n}(N), \tag{2.176}$$

where the symbols R_0, v_0, and $\mathbf{n}(N)$ have the same meaning as in (2.110). The spectrum of the function $f(t-R_0/v_0)$ must be of high frequency. In other respects, this function may be an arbitrary one as long as it satisfies the conditions permitting the use of the Fourier transform (viz., absolute integrability and the Dirichlet conditions). These conditions, are, however, always satisfied in the case of seismic impulses.

Using Fourier transforms, we can write

$$f(t) = \frac{1}{\pi}\,\mathrm{Re}\int_0^\infty S(\omega)\mathrm{e}^{i\omega t}\,\mathrm{d}\omega, \tag{2.177}$$

where $S(\omega)$ is the complex spectral function of the pulse $f(t)$, given by the expression

$$S(\omega) = \int_{-\infty}^{+\infty} f(t)\mathrm{e}^{-i\omega t}\,\mathrm{d}t. \tag{2.178}$$

We can write (2.176) in the form

$$\mathscr{W}^0(N) = \frac{1}{\pi}\,\mathrm{Re}\int_0^\infty S(\omega)\mathbf{W}^0(N)\,\mathrm{d}\omega, \tag{2.179}$$

where $\mathbf{W}^0(N)$ is the displacement vector of the incident harmonic wave, given by (2.110).

From the linear properties of the Fourier transform we know that we can write, for the displacement vector of an arbitrary type of R-wave, $\mathscr{W}^R(M)$, at any point M,

$$\mathscr{W}^R(M) = \frac{1}{\pi} \operatorname{Re} \int_0^\infty S(\omega) \mathbf{W}^R(M) \, d\omega, \tag{2.180}$$

where $\mathbf{W}^R$ is the displacement vector of the harmonic R-wave. The expressions for $\mathbf{W}^R$ are known from the previous sections. Generally, $\mathbf{W}^R(M)$ is given by the formula

$$\mathbf{W}^R(M) = U^R(M) \exp\{i\omega[t - \tau^R(M)]\} \mathbf{n}^R(M), \tag{2.181}$$

where $U^R(M)$, $\tau^R(M)$, and $\mathbf{n}^R(M)$ do not depend on frequency (see, e.g., (2.126) or (2.168)). Inserting (2.181) into (2.180) yields

$$\mathscr{W}^R(M) = \mathbf{n}^R(M) \frac{1}{\pi} \operatorname{Re} \left\{ U^R(M) \int_0^\infty S(\omega) \exp\{i\omega[t - \tau^R(M)]\} \, d\omega \right\}. \tag{2.182}$$

If $U^R(M)$ is real, we obtain very simply, from (2.182),

$$\mathscr{W}^R(M) = U^R(M) f(t - \tau^R(M)) \mathbf{n}^R(M). \tag{2.183}$$

Thus, if $U^R(M)$ is real, the form of the pulse of the R-wave is the same as the form of the impulse wave radiated by the source. It may, of course, differ in sign, as $U^R(M)$ can be negative, and also the components of $\mathbf{n}^R(M)$ can have different signs from those of $\mathbf{n}(M)$.

Now we shall consider that $U^R(M)$ is complex. From the physical point of view this happens, for example, when the wave is reflected from some interfaces beyond the critical angle (and the R-coefficients become complex), or when the R-wave passes through the caustics (and L becomes complex; see (2.138)). Then, we find from (2.182) that

$$\mathscr{W}^R(M) = |U^R(M)| \mathbf{n}^R(M) \left(\cos[\arg U^R(M)] \frac{1}{\pi} \operatorname{Re} \int_0^\infty S(\omega) \exp\{i\omega[t - \tau^R(M)]\} \, d\omega \right.$$
$$\left. - \sin[\arg U^R(M)] \frac{1}{\pi} \operatorname{Im} \int_0^\infty S(\omega) \exp\{i\omega[t - \tau^R(M)]\} \, d\omega \right).$$

If we define

$$g(t) = -\frac{1}{\pi} \operatorname{Im} \int_0^\infty S(\omega) e^{i\omega t} \, d\omega, \tag{2.184}$$

we obtain

$$\mathscr{W}^{\mathrm{R}}(M) = |U^{\mathrm{R}}(M)|\mathbf{n}^{\mathrm{R}}(M)\{f[t-\tau^{\mathrm{R}}(M)] \cos [\arg U^{\mathrm{R}}(M)] + g(t-\tau^{\mathrm{R}}(M)) \sin [\arg U^{\mathrm{R}}(M)]\}. \quad (2.185)$$

We now introduce the notation

$$f(t; \chi) = f(t) \cos \chi + g(t) \sin \chi, \quad (2.186)$$

where $f(t)$ and $g(t)$ are given by (2.177) and (2.184). Then the final result for the R-wave is

$$\mathscr{W}^{\mathrm{R}}(M) = |U^{\mathrm{R}}(M)| f(t-\tau^{\mathrm{R}}(M); \arg U^{\mathrm{R}}(M))\mathbf{n}^{\mathrm{R}}(M). \quad (2.187)$$

The values of $U^{\mathrm{R}}(M)$, $\tau^{\mathrm{R}}(M)$, and $\mathbf{n}^{\mathrm{R}}(M)$ have been found for many types of media in the preceding sections. Therefore, it is possible to calculate the pulse R-wave by using (2.178), (2.184), (2.186), and (2.187), when the incident pulse is given. From (2.187), of course, we obtain (2.183) if $\arg U^{\mathrm{R}}(M) = 0$. In general, however, the form of the signal of the R-wave may change with distance from the source, as $\arg U^{\mathrm{R}}(M)$ depends on the epicentral distance.

Only for a few functions $f(t)$ is it possible to write a convenient analytic expression for $g(t)$. In most cases we must use a computer to calculate it. The suitable procedures for numerical evaluation of $g(t)$ exploit the fact that it is the Hilbert transform of $f(t)$, i.e.

$$g(t) = \pi^{-1} \int_{-\infty}^{\infty} f(\xi)(\xi-t)^{-1}\mathrm{d}\xi$$

(where the integral is a Cauchy principal value). The functions $g(t)$ are also given graphically in some Russian books for a large variety of $f(t)$ (see, for example, Petrashen, 1957b).

The phase shift, $\arg U^{\mathrm{R}}(M)$, has a considerable influence on seismic signals. A pulse of the form $f(t)$ is deformed by the phase shift into a pulse of the form $f(t; \chi)$. The character of the function $f(t; \chi)$ has already been studied many times in the literature (Malinovskaya, 1957a, 1957b; Červený, 1966a; O'Brien, 1967) and we shall therefore deal only very briefly with the main conclusions. We are interested not only in the form of the signal but also in the change of its amplitude as a function of the phase shift.

The form of the pulse changes considerably under the influence of the phase shift, particularly for pulses with a small number of peaks. For pulses with many peaks, the changes are smaller. For sharp pulses, discontinuous at $t = t_0$ (such as step or exponential functions), the function

$f(t; \chi)$ reaches infinite values at $t = t_0$. If the amplitudes of individual peaks of the pulse are measured from the zero line, very marked changes are introduced. The amplitudes of the first peaks decrease with increasing phase shift, but the amplitudes of the last peaks increase and new peaks appear at the end of the pulse. The peaks of the function $f(t; \chi)$ shift in the time scale toward the onset of the pulse when χ increases. The sharpness of the pulse $f(t; \chi)$ also changes with χ.

The influence of the phase shift χ on the peak-to-trough amplitudes of $f(t; \chi)$ is, however, substantially smaller than its influence on amplitudes of individual peaks of the pulse measured from the zero line. This influence is especially small for long pulses (with many peaks). For example, for a sinusoidal pulse of the length of one period, the maximum change of the peak-to-trough amplitude is around 12 per cent, whereas for longer sinusoidal pulses it is always smaller than 5 per cent (Červený, 1966a). If we are content with an accuracy of up to 15 per cent, we need not take the influence of the phase shift on the peak-to-trough amplitudes into consideration. In the majority of cases in practical measurements by deep seismic sounding methods and in seismic prospecting, the number of peaks of seismic signals is usually larger than two. Then, when studying amplitudes (if, simultaneously, we are not interested in the form of the pulse), we need not deal with the character of the pulse at all. Of course, the peak-to-trough amplitudes must be measured in this case.

3
HEAD WAVES FROM A SINGLE INTERFACE

3.1
DESCRIPTION OF SOURCE AND MEDIA

In this chapter, we shall investigate the properties of head waves (H-waves) using the ray method. Basic equations and formulae pertaining to this method have been given in chapter 2. We shall study here the simplest problem involving head waves, viz., that of a plane interface between two homogeneous half-spaces. The point source may be of P, SV, or SH type and may be at an arbitrary point in the first or second medium. Results of this chapter are in accord with those obtained by the standard approximate evaluations (by the method of steepest descent, for example) of solutions found by wave methods (see chapter 7). The ray method, however, enables us to find the properties of H-waves in much more complicated media also, where the integral method becomes very difficult (or impossible) to use. In section 3.9 and in later chapters we shall encounter some of these situations.

The first investigators to obtain expressions for H-waves by means of the ray method were Friedrichs and Keller (1955); see also Friedlander (1958). They obtained formulae for an acoustic H-wave generated on a plane interface between two *fluids*. The case of H-waves generated on a plane interface between two *solid* media was first solved by Alekseyev and Gel'chinskiy (1958, 1961); see also Yanovskaya (n.d.). The computation matrix formulae for head wave coefficients (H-coefficients) based on their method were given by Yanovskaya (1966). No other papers seem to have been devoted to the ray method in relation to H-waves.

We assume that two perfectly elastic, isotropic and homogeneous half-spaces are in welded contact at a plane interface. The notation of section 2.3 is used for the elastic moduli, density, wave velocities, and their ratio, i.e. λ_i, μ_i, ρ_i, α_i, β_i, and γ_i ($i = 1, 2$). A point source of either P, SV, or SH type is situated at an arbitrary point in either medium at a distance h from the interface. The source is assumed to be symmetric, i.e. $g(\vartheta_0, \varphi_0) = 1$, and harmonic; but when we have the results for this source, they can be generalized to obtain the solution for an impulse source (section 3.7).

We introduce a Cartesian coordinate system (x, y, z) such that the interface is the plane $z = 0$ and the source, M_0, lies on the z-axis. The x-axis lies along the intersection of the plane of incidence with the interface. Its positive direction will be measured from the source, and the positive direction of the z-axis is in the first medium. We again introduce unit vectors at an arbitrary point of the ray, $\mathbf{n}_P$, $\mathbf{n}_{SV}$, and $\mathbf{n}_{SH}$, in the same directions as defined in section 2.3.

We shall assume that the displacement vector $\mathbf{W}^m$ of the wave emanating from the source is given by

$$\mathbf{W}^m = (1/R_0)\exp[i\omega(t - R_0/V_m)]\mathbf{n}_m, \tag{3.1}$$

where $\mathbf{n}_m$ may be $\mathbf{n}_P$, $\mathbf{n}_{SV}$, or $\mathbf{n}_{SH}$, depending on the type of source, and m specifies the nature of the source and the medium in which it is situated (see section 2.4.2). The components of the displacement vector $\mathbf{W}^m$ can be obtained using table 2.1 for P or SV waves, or immediately for SH-waves.

We shall also use a cylindrical coordinate system (r, z, φ) instead of the rectangular coordinate system (x, y, z). We assume that the r and z axes in the cylindrical coordinate system coincide with the x and z axes in the rectangular coordinate system. Then the unit coordinate vectors at an arbitrary point of the incident plane, $\mathbf{i}$, $\mathbf{j}$, $\mathbf{k}$, coincide with the unit coordinate vectors, $\mathbf{e}_r$, $\mathbf{e}_\varphi$, $\mathbf{e}_z$, at that point. The r, φ, and z components of the displacement vector in the cylindrical coordinate system will be, of course, the same as the x, y, and z components in the rectangular coordinate system. Note that r is the epicentral distance.

3.2 MATHEMATICAL PROCEDURE

In this section we shall derive expressions for head waves using the ray theory. We shall solve separately the cases of a P or SV source and an SH source.

As was shown in chapter 1, the origin of H-waves is connected with the wave that propagates along the interface. This wave, which propagates parallel to the interface, may be a critically reflected or refracted wave. In order to obtain expressions for head waves, therefore, we must know the properties of the R-waves propagating along the interface. We can use here the formulae derived in chapter 2.

3.2.1 *Source of P or SV Waves*

We assume that the displacement vector $\mathbf{W}^m$ of the wave which is incident on the interface is given by (3.1), where the index $m = 1, 2, 3$, or 4 characterizes the type and position of the source as in section 2.4.2. The properties and notation of the R-waves, generated when the wave (3.1) is incident on the interface, were given in chapter 2. We obtained the following expression for the displacement vector of the R-wave at the point $\bar{M}$ (see sections 2.4.2 and 2.6.2):

$$\mathbf{W}^{mn}(\bar{M}) = U^{mn}(\bar{M})\exp\{i\omega[t-\tau_{mn}(\bar{M})]\}\mathbf{n}_n(\bar{M}), \tag{3.2}$$

where

$$U^{mn}(\bar{M}) = R_{mn}/L, \tag{3.3}$$

$$L = \left\{r \cot \theta_m\left(\frac{h}{\cos^2 \theta_m}+\frac{V_n H \cos \theta_m}{V_m \cos^3 \theta_n}\right)\right\}^{1/2}, \tag{3.4}$$

$$\tau_{mn}(\bar{M}) = \frac{h}{V_m \cos \theta_m}+\frac{H}{V_n \cos \theta_n}. \tag{3.5}$$

The index n indicates the type of waves generated ($n = 1$ and 2 refer to P and SV waves in the first medium, $n = 3$ and 4 to P and SV waves in the second medium, respectively). θ_m and θ_n are angles of incidence and reflection (or refraction), V_m and V_n velocities of propagation of the incident and R-waves ($V_1 = \alpha_1$, $V_2 = \beta_1$, $V_3 = \alpha_2$, $V_4 = \beta_2$), h and H distances of the source and receiver from the interface, and R_{mn} the R-coefficient, given by (2.74). The components of the displacement vector $\mathbf{W}^{mn}$ can be obtained using table 2.2.

The formulae (3.3), (3.4), and (3.5) are parametric with parameters θ_m and θ_n. If we want to express U^{mn}, L, and τ_{mn} in terms of the epicentral distance r, we must add two more equations:

$$\sin \theta_m = V_m \sin \theta_n / V_n, \tag{3.6}$$

$$r = h \tan \theta_m + H \tan \theta_n. \tag{3.7}$$

We now use $\bar{l}$ to denote the length of the ray of the R-wave from the interface to the point $\bar{M}$,

$$\bar{l} = H/\cos \theta_n = [r(\bar{M}) - h \tan \theta_m]/\sin \theta_n \tag{3.8}$$

(see figure 3.1). Then we can write for $U^{mn}(\bar{M})$

$$U^{mn}(\bar{M}) = \cos \theta_n \tilde{U}^{mn}(\bar{M}), \tag{3.9}$$

where

$$\tilde{U}^{mn}(\bar{M}) = R_{mn}\left\{r \cot \theta_m \cos \theta_m \frac{V_n}{V_m}\left(\bar{l}+\frac{hV_m \cos^2 \theta_n}{V_n \cos^3 \theta_m}\right)\right\}^{-1/2}. \tag{3.10}$$

We shall now investigate the properties of the displacement vector $\mathbf{W}^{mn}(\bar{M})$ when the point $\bar{M}$ lies on the interface. In order to distinguish this from other situations, we shall label the point $\bar{M}$ as A if it lies on the interface. At the point A, then, we obtain (see figure 3.2 and (2.74″) for Θ)

$$\theta_n = \tfrac{1}{2}\pi, \qquad \cos\theta_n = 0, \qquad \sin\theta_n = 1, \qquad \Theta = 1/V_n. \tag{3.11}$$

The angle of incidence θ_m for which (3.11) holds is the critical angle, which will be denoted by $\theta_m{}^*$. From (3.6) we obtain

$$\sin\theta_m{}^* = V_m/V_n. \tag{3.12}$$

Similarly, we introduce (see 2.74)

$$R_{mn}^* = R_{mn}(\theta_m{}^*) = (R_{mn})_{\Theta=1/V_n}. \tag{3.13}$$

It is clear from (3.12) that a necessary condition for the ray to be parallel to the interface is

$$V_m/V_n < 1. \tag{3.14}$$

This, of course, is also a necessary condition for generation of H-waves from the wave $\mathbf{W}^{mn}$.

The ray of the incident wave corresponding to the critical angle $\theta_m{}^*$ will be called the critical ray, and the horizontal distance between the source and the point of intersection of the critical ray with the interface (see figure 3.2) is the interface critical distance r_{mn}^*, given by

$$r_{mn}^* = h\tan\theta_m{}^* = hV_m(V_n{}^2 - V_m{}^2)^{-1/2}. \tag{3.15}$$

The ray of the R-wave $\mathbf{W}^{mn}$ can be parallel to the interface only at distances $r > r_{mn}^*$. H-waves can, therefore, be generated by the wave $\mathbf{W}^{mn}$ only at distances $r > r_{mn}^*$.

The formula (3.2) gives us the zeroth approximation of the R-wave $\mathbf{W}^{mn}$. We can easily see that U^{mn} is zero at the point A (see (3.9) for

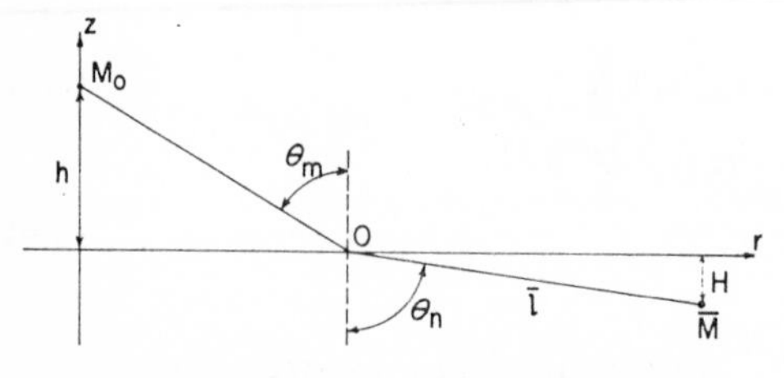

FIGURE 3.1 A ray of the R-wave starting from the source (M_0) and arriving at $\bar{M}$.

$\theta_n = \frac{1}{2}\pi$). The first non-vanishing term in the ray series of the R-wave at the interface will be the term for $k = 1$. We can therefore write

$$\mathbf{W}^{mn}(A) = (1/i\omega)\mathbf{W}_1{}^{mn}(A)\exp\{i\omega[t-\tau_{mn}(A)]\}. \tag{3.16}$$

We take into account the fact that the ray of the wave $\mathbf{W}^{mn}(A)$ is parallel to the interface, i.e. $\mathbf{n}_P = \mathbf{i}$, $\mathbf{n}_{SV} = (-1)^{\varepsilon_n}\mathbf{k}$, where ε_n is given by (2.52). Then we can write for $\mathbf{W}_1{}^{mn}(A)$ (see (2.31) and (2.45))

$$\mathbf{W}_1{}^{mn}(A) = U_1{}^{mn}(A)\mathbf{i} - V_n \frac{\partial U^{mn}(A)}{\partial z}\mathbf{k} \tag{3.17}$$

for $n = 1$ and 3, and

$$\mathbf{W}_1{}^{mn}(A) = (-1)^{\varepsilon_n}\left(V_n \frac{\partial U^{mn}(A)}{\partial z}\mathbf{i} + U_1{}^{mn}(A)\mathbf{k}\right) \tag{3.17'}$$

for $n = 2$ and 4. In the above expressions, $U_1{}^{mn}\mathbf{i}$ and $U_1{}^{mn}\mathbf{k}$ are the principal components of $\mathbf{W}_1{}^{mn}$ and the other two terms are the additional components. The y-component has been omitted in (3.17). We cannot determine $U_1{}^{mn}(A)$, as we do not know the initial data for it. However, we shall not need to know it in discussing the properties of head waves. (Head waves generally do not depend on it, as will be shown later.)

When the wave $\mathbf{W}^{mn}$ propagates along a boundary, the interface conditions (2.57) must be satisfied. We shall investigate these conditions at a general point A where $\mathbf{W}^{mn}$ now plays the role of an incident wave. It was shown in section 2.3 that four new waves must arise. We shall indicate the type of waves generated by the index p (corresponding to ν in section 2.3). As in the case of m and n, p may be 1, 2, 3, or 4 ($p = 1$ for a compressional wave in the first medium, $p = 2$ for a shear wave in the first medium, and $p = 3$ and 4 for compressional and shear waves in the second medium).

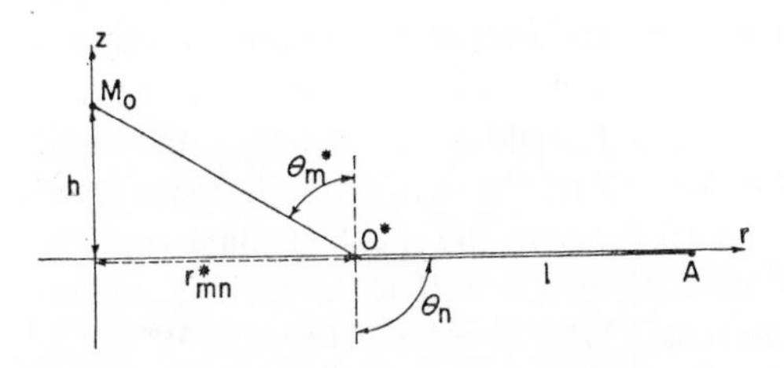

FIGURE 3.2 A critically refracted ray; $\theta_m{}^*$ is the critical angle and r^*_{mn} is the interface critical distance.

Then we can write for the displacement vectors of the generated waves:

$$\mathbf{W}^{mnp}(A) = \exp\{i\omega[t-\tau_{mnp}(A)]\}\sum_{k=0}^{\infty}(i\omega)^{-k}\mathbf{W}_k^{mnp}(A), \tag{3.18}$$

$p = 1, 2, 3, 4$ (see (2.47)). Here m and n are the indices characterizing the incident wave $\mathbf{W}^{mn}$. If we denote the velocity of propagation of the wave $\mathbf{W}^{mnp}$ by V_p ($V_1 = \alpha_1$, $V_2 = \beta_1$, $V_3 = \alpha_2$, $V_4 = \beta_2$), we can write the eikonal equations for the phase function τ_{mnp}, see (2.49), as

$$(\nabla\tau_{mnp})^2 = (1/V_p)^2, \tag{3.19}$$

with the initial condition

$$\tau_{mnp}(A) = \tau_{mn}(A). \tag{3.19'}$$

The acute angle between the ray of the pth wave and the z-axis is $\theta_p{}^*$. (We can write the asterisk here for the same reasons as in the case of $\theta_m{}^*$ given by (3.12).) From (2.51) follows Snell's law

$$\sin\theta_p{}^* = V_p\sin\theta_n/V_n. \tag{3.20}$$

As $\sin\theta_n = 1$ in our case, we obtain

$$\sin\theta_p{}^* = V_p/V_n. \tag{3.20'}$$

If $V_p > V_n$, we find from (3.20′) and (2.53) that $\sin\theta_p{}^* > 1$ and $\cos\theta_p{}^* = -i(\sin^2\theta_p{}^*-1)^{1/2}$. Then the phase function τ_{mnp} is complex for $z \neq 0$, and the amplitude of the wave $\mathbf{W}^{mnp}$ decreases exponentially with distance from the interface. Waves with these properties are known as *inhomogeneous head waves* or *head waves of the surface type* (Červený, 1957a; Zaitsev, 1959). We shall not pay attention to these head waves in this book.

It follows from (3.20′) that the regular (homogeneous) head waves $\mathbf{W}^{mnp}$ exist only if

$$V_p < V_n. \tag{3.21}$$

Expressions (3.14) and (3.21) give us the necessary velocity conditions which must be fulfilled for the homogeneous head wave $\mathbf{W}^{mnp}$ to exist.

If we change the position of the point A along the interface, the angle $\theta_p{}^*$ does not change (see (3.20′)). Therefore, the rays of the homogeneous head wave $\mathbf{W}^{mnp}$ are parallel to each other in the incident plane and the wave fronts τ_{mnp} = const. are conical in space. The exceptions are inhomogeneous head waves and the wave $\mathbf{W}^{mnp}$ ($n = p$). The wave $\mathbf{W}^{mnn}$ has the character of a reflected wave (with the angle of reflection $\theta_p{}^* = \frac{1}{2}\pi$; see (3.20′)). This wave will, of course, travel along the interface with the same velocity as the wave $\mathbf{W}^{mn}$, viz., V_n. Since we are interested only in properties of H-waves, we shall not discuss this reflected wave in any detail and shall concentrate our attention on $\mathbf{W}^{mnp}$ when $n \neq p$.

Figure 3.3 illustrates the various waves that arise when we assume that $\alpha_2 > \beta_2 > \alpha_1 > \beta_1$ and that the displacement vector of the wave incident on the interface at the point A (with an angle of incidence $\theta_n = \frac{1}{2}\pi$) is $\mathbf{W}^{m3}$, i.e. this wave is a compressional wave propagating in the second medium (since $n = 3$). The generated waves are $\mathbf{W}^{m31}$ (P-wave in the first medium), $\mathbf{W}^{m32}$ (SV-wave in the first medium), $\mathbf{W}^{m33}$ (P-wave in the second medium), and $\mathbf{W}^{m34}$ (SV-wave in the second medium). The waves $\mathbf{W}^{m31}$, $\mathbf{W}^{m32}$, and $\mathbf{W}^{m34}$ are homogeneous H-waves, and $\mathbf{W}^{m33}$ is a reflected wave.

We shall now apply equations I, II, III, and IV from the general set of interface conditions (2.57) for the determination of the coefficients in the ray series (3.18). We shall follow the procedure described in section 2.3.1. First we must determine the zeroth coefficients $\mathbf{W}_0{}^{mnp}$ from (2.57) for $k = 0$. It was shown above that the displacement vector of the incident wave $\mathbf{W}^{mn}$ is zero at the point A. It follows that $\mathbf{W}_0{}^{mnp}(A)$ also vanishes (see (2.75)). From the formulae for continuation of the displacement

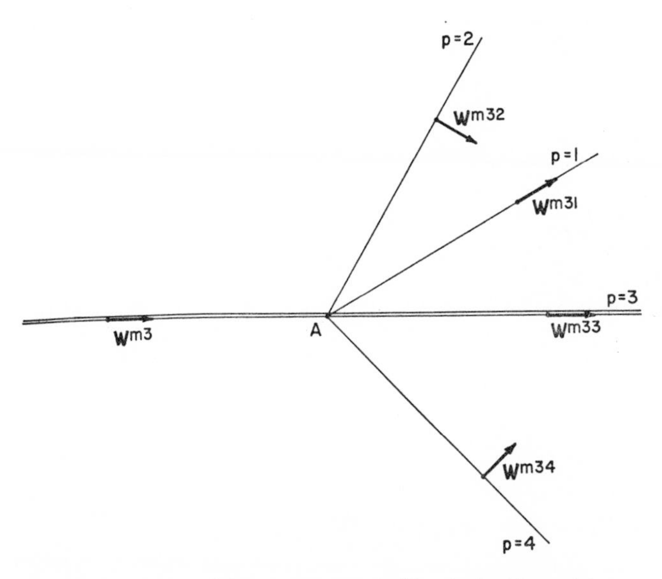

FIGURE 3.3 Rays of the waves generated by the incidence (at point A) of a P-wave propagating parallel to the interface when $\alpha_2 > \beta_2 > \alpha_1 > \beta_1$. The arrows indicate positive directions of displacement for the various waves.

vector along a ray, (2.64′), it follows that $\mathbf{W}_0^{mnp}$ equals zero not only on the interface but everywhere.

Now we shall solve (2.57) for $k = 1$ to find $\mathbf{W}_1^{mnp}(A)$. Since $\mathbf{W}_0^{mnp}$ equals zero, values of F_{i4}, given by (2.58″), equal zero, also. The additional components of $\mathbf{W}_1^{mnp}$ can generally be found by differentiating $\mathbf{W}_0^{mnp}$. Hence, the additional components of $\mathbf{W}_1^{mnp}$ vanish and $F_{i3} = 0$ (see (2.58″)). $\mathbf{W}_1^{mnp}$ are represented only by the principal components and we can write the following expressions for $\mathbf{W}^{mnp}$ (if we neglect higher terms):

$$\mathbf{W}^{mnp}(A) = (1/i\omega)U^{mnp}(A)\exp\{i\omega[t-\tau_{mnp}(A)]\}\mathbf{n}_p^*(A), \tag{3.22}$$

where the unit vector $\mathbf{n}_p^*$ may be $\mathbf{n}_\mathrm{P}^*$ or $\mathbf{n}_\mathrm{SV}^*$, depending on the type of wave. (For $p = 1$ and 3, $\mathbf{n}_p^* = \mathbf{n}_\mathrm{P}^*$, and for $p = 2$ and 4 $\mathbf{n}_p^* = \mathbf{n}_\mathrm{SV}^*$). In the notation of section 2.3, $U^{mn1} = W_{1,\mathrm{P}}^1$, $U^{mn2} = W_{1,\mathrm{SV}}^2$, $U^{mn3} = W_{1,\mathrm{P}}^3$, $U^{mn4} = W_{1,\mathrm{SV}}^4$.

Now we must find F_{i1} and F_{i2} from the expressions for the incident wave $\mathbf{W}^{mn}$, i.e. from (3.2), (3.16), (3.17), and (3.17′). For F_{i2} we find easily from (3.2) and (2.58′) that

$$F_{12} = F_{22} = F_{32} = 0, \qquad F_{42} = (-1)^{\varepsilon_n+1}\mu_{\varepsilon_n}\frac{\partial U^{mn}}{\partial z}, \tag{3.23}$$

for $n = 1$ and 3, and

$$F_{12} = F_{22} = F_{42} = 0, \qquad F_{32} = -(\lambda_{\varepsilon_n}+2\mu_{\varepsilon_n})\frac{\partial U^{mn}}{\partial z}, \tag{3.23′}$$

for $n = 2$ and 4.

For the determination of F_{i1}, we can use (3.17), (3.17′), (2.58), and the simple relations

$$\frac{\partial\tau_{mn}}{\partial x} = \frac{\sin\theta_n}{V_n} = \frac{1}{V_n}, \qquad \frac{\partial\tau_{mn}}{\partial z} = 0. \tag{3.24}$$

We obtain

$$F_{11} = (-1)^{\varepsilon_n}U_1^{mn}, \qquad F_{31} = (-1)^{\varepsilon_n}(\lambda_{\varepsilon_n}/V_n)U_1^{mn},$$
$$F_{21} = (-1)^{\varepsilon_n+1}V_n\frac{\partial U^{mn}}{\partial z}, \qquad F_{41} = (-1)^{\varepsilon_n+1}\mu_{\varepsilon_n}\frac{\partial U^{mn}}{\partial z}, \tag{3.25}$$

for $n = 1$ and 3, and

$$F_{11} = V_n \frac{\partial U^{mn}}{\partial z}, \qquad F_{31} = \lambda_{\varepsilon_n} \frac{\partial U^{mn}}{\partial z},$$
$$F_{21} = U_1^{mn}, \qquad F_{41} = (\mu_{\varepsilon_n}/V_n) U_1^{mn} \tag{3.25'}$$

for $n = 2$ and 4.

Inserting (3.23) through (3.25′) into (2.57′), we obtain from (2.57) a set of four linear equations for U^{mnp} $(p = 1, 2, 3, 4)$

$$\sum_{p=1}^{4} a_{ip}^* U^{mnp} = F_i^{mn} \qquad (i = 1, 2, 3, 4), \tag{3.26}$$

where the coefficients a_{ip}^* are given by the matrix

$$A^* = \begin{bmatrix} \sin\theta_1^* & \cos\theta_2^* & -\sin\theta_3^* & -\cos\theta_4^* \\ \cos\theta_1^* & -\sin\theta_2^* & \cos\theta_3^* & -\sin\theta_4^* \\ \rho_1\alpha_1 \cos 2\theta_2^* & -\rho_1\alpha_1\gamma_1 \sin 2\theta_2^* & -\rho_2\alpha_2 \cos 2\theta_4^* & \rho_2\alpha_2\gamma_2 \sin 2\theta_4^* \\ \rho_1\alpha_1\gamma_1^2 \sin 2\theta_1^* & \rho_1\alpha_1\gamma_1 \cos 2\theta_2^* & \rho_2\alpha_2\gamma_2^2 \sin 2\theta_3^* & \rho_2\alpha_2\gamma_2 \cos 2\theta_4^* \end{bmatrix} \tag{3.27}$$

and F_i^{mn} are given by

$$\begin{aligned} F_1^{mn} &= (-1)^{\varepsilon_n} U_1^{mn}, \\ F_2^{mn} &= F^{mn}, \\ F_3^{mn} &= (-1)^{\varepsilon_n} (\lambda_{\varepsilon_n}/V_n) U_1^{mn}, \\ F_4^{mn} &= 2\rho_{\varepsilon_n} \gamma_{\varepsilon_n}^2 V_n F^{mn}, \end{aligned} \tag{3.28}$$

for $n = 1$ and 3, and

$$\begin{aligned} F_1^{mn} &= (-1)^{\varepsilon_n+1} F^{mn}, \\ F_2^{mn} &= U_1^{mn}, \\ F_3^{mn} &= (-1)^{\varepsilon_n} 2\rho_{\varepsilon_n} \gamma_{\varepsilon_n} \alpha_{\varepsilon_n} F^{mn}, \\ F_4^{mn} &= (\mu_{\varepsilon_n}/V_n) U_1^{mn}, \end{aligned} \tag{3.28'}$$

for $n = 2$ and 4. The function F^{mn} in the above formulae is given by the relation

$$F^{mn} = (-1)^{\varepsilon_n+1} V_n \frac{\partial U^{mn}}{\partial z}. \tag{3.29}$$

If we introduce, in the derivatives $\partial U^{mn}/\partial z$, the distance of the receiver from the interface, H, instead of z we can write

$$\frac{\partial U^{mn}}{\partial z} = (-1)^{\varepsilon_n+1} \frac{\partial U^{mn}}{\partial H}. \tag{3.29'}$$

Inserting this into (3.29) yields

$$F^{mn} = V_n \frac{\partial U^{mn}}{\partial H}. \tag{3.30}$$

We can see from (3.27) and (2.70) that the matrix A is identical with the matrix A^* when θ_j^* are substituted for θ_j. The expressions for F_i^{mn}, (3.28) and (3.28′), depend on F^{mn} and U_1^{mn}. We can divide F_i^{mn} into two parts, one depending only on F^{mn} and the other only on U_1^{mn}. We can write

$$F_i^{mn} = \hat{F}_i^{mn} + \tilde{F}_i^{mn}, \tag{3.31}$$

where

$$\begin{aligned} \hat{F}_1^{mn} &= 0, & \tilde{F}_1^{mn} &= (-1)^{\varepsilon_n} U_1^{mn}, \\ \hat{F}_2^{mn} &= F^{mn}, & \tilde{F}_2^{mn} &= 0, \\ \hat{F}_3^{mn} &= 0, & \tilde{F}_3^{mn} &= (-1)^{\varepsilon_n} (\lambda_{\varepsilon_n}/V_n) U_1^{mn}, \\ \hat{F}_4^{mn} &= 2\rho_{\varepsilon_n} \gamma_{\varepsilon_n}^2 V_n F^{mn}, & \tilde{F}_4^{mn} &= 0. \end{aligned} \tag{3.32}$$

for $n = 1$ and 3, and

$$\begin{aligned} \hat{F}_1^{mn} &= (-1)^{\varepsilon_n+1} F^{mn}, & \tilde{F}_1^{mn} &= 0, \\ \hat{F}_2^{mn} &= 0, & \tilde{F}_2^{mn} &= U_1^{mn}, \\ \hat{F}_3^{mn} &= (-1)^{\varepsilon_n} 2\rho_{\varepsilon_n} \gamma_{\varepsilon_n} \alpha_{\varepsilon_n} F^{mn}, & \tilde{F}_3^{mn} &= 0, \\ \hat{F}_4^{mn} &= 0, & \tilde{F}_4^{mn} &= (\mu_{\varepsilon_n}/V_n) U_1^{mn}, \end{aligned} \tag{3.33}$$

for $n = 2$ and 4.

Then we can divide the solution U^{mnp} of the set of equations (3.26) into two parts, viz.,

$$U^{mnp} = \hat{U}^{mnp} + \tilde{U}^{mnp}, \tag{3.34}$$

where $\hat{U}^{mnp}$ and $\tilde{U}^{mnp}$ correspond to the functions $\hat{F}_i^{mn}$ and $\tilde{F}_i^{mn}$ in the right-hand side of (3.26), respectively.

Comparing the expressions for $\tilde{F}_i^{mn}$ with the coefficients a_{ip}^* of the matrix A^* (3.27) we can see that, for $\theta_n^* = \frac{1}{2}\pi$,

$$\tilde{F}_i^{mn} = b_{ni}^* U_1^{mn} (-1)^{n+1}, \tag{3.35}$$

where

$$b_{ni}^* = (-1)^i a_{in}^*. \tag{3.36}$$

The system of four linear equations for $\tilde{U}^{mnp}$ is therefore identical with (2.69), where $\theta_n = \theta_n{}^*$ is substituted. The solution of this system is

$$\tilde{U}^{mnp} = \{R_{np}\}_{\theta_n=\pi/2} U_1{}^{mn}(-1)^{n+1}, \tag{3.37}$$

where R_{np} are the R-coefficients, given by (2.74). We can easily see from (2.74) that for $n \neq p$ all R_{np} vanish for $\theta_n = \frac{1}{2}\pi$ (i.e., for $P_n = 0$ in the notation of section 2.4.2), and, therefore, $\tilde{U}^{mnp} = 0$ for $n \neq p$. The only non-vanishing R_{np} are those for $n = p$. As was mentioned earlier, we shall not be interested in properties of the wave $\mathbf{W}^{mnn}$ $(n = p)$, since this wave is not an H-wave but a reflected wave propagating along the interface. Thus, we can write finally for $n \neq p$ the system of equations (3.26), where $\hat{F}_i^{mn}$ is substituted for F_i^{mn}. The solution of this system of equations is

$$U^{mnp}(A) = M_{np}F^{mn}(A), \tag{3.38}$$

where F^{mn} is given by (3.30) and M_{np} is the solution of a set of four linear equations

$$\sum_{p=1}^{4} a_{ip}^* M_{np} = c_{ni} \qquad (i = 1, 2, 3, 4), \tag{3.39}$$

where the coefficients a_{ip}^* are given by the matrix A^* in (3.27) and the coefficients c_{ni} are given by the column matrices

$$c_1 = \begin{bmatrix} 0 \\ 1 \\ 0 \\ 2\rho_1\alpha_1\gamma_1{}^2 \end{bmatrix}, \quad c_2 = \begin{bmatrix} 1 \\ 0 \\ -2\rho_1\alpha_1\gamma_1 \\ 0 \end{bmatrix}, \quad c_3 = \begin{bmatrix} 0 \\ 1 \\ 0 \\ 2\rho_2\alpha_2\gamma_2{}^2 \end{bmatrix}, \quad c_4 = \begin{bmatrix} -1 \\ 0 \\ 2\rho_2\gamma_2\alpha_2 \\ 0 \end{bmatrix}. \tag{3.40}$$

If we want to obtain numerical values of M_{np}, it is most convenient to calculate them directly from the set of equations (3.39). (We must not forget that a_{ip}^* may be complex.) In some cases, however, we need to know the analytic expressions for M_{np}. If we use

$$\Theta = \frac{\sin\theta_1{}^*}{V_1} = \frac{\sin\theta_2{}^*}{V_2} = \frac{\sin\theta_3{}^*}{V_3} = \frac{\sin\theta_4{}^*}{V_4} \tag{3.41}$$

and

$$P_i = (1 - V_i{}^2\Theta^2)^{1/2} = \cos\theta_i{}^* \qquad (i = 1, 2, 3, 4), \tag{3.42}$$

we obtain

$$\begin{aligned}
M_{12} &= -\alpha_1\{\Theta D^{-1}[P_3P_4q(\rho_1+q\Theta^2)\\
&\qquad +\beta_2\alpha_2(\rho_2-q\Theta^2)(\rho_2-\rho_1-q\Theta^2)]\}_{\Theta=1/\alpha_1},\\
M_{13} &= \alpha_1\rho_1\{D^{-1}[\beta_2P_2(\rho_2-q\Theta^2)+\beta_1P_4(\rho_1+q\Theta^2)]\}_{\Theta=1/\alpha_1},\\
M_{14} &= -\alpha_1\rho_1\{\Theta D^{-1}[qP_2P_3-\alpha_2\beta_1(\rho_2-\rho_1-q\Theta^2)]\}_{\Theta=1/\alpha_1},\\
M_{23} &= -\rho_1\beta_1\{\Theta D^{-1}[qP_1P_4-\alpha_1\beta_2(\rho_2-\rho_1-q\Theta^2)]\}_{\Theta=1/\beta_1},\\
M_{24} &= -\rho_1\beta_1\{D^{-1}[(\rho_1+q\Theta^2)\alpha_1P_3+(\rho_2-q\Theta^2)\alpha_2P_1]\}_{\Theta=1/\beta_1},\\
M_{31} &= \alpha_2\rho_2\{D^{-1}[\beta_1P_4(\rho_1+q\Theta^2)+\beta_2P_2(\rho_2-q\Theta^2)]\}_{\Theta=1/\alpha_2},\\
M_{32} &= \alpha_2\rho_2\{\Theta D^{-1}[P_1P_4q-\alpha_1\beta_2(\rho_2-\rho_1-q\Theta^2)]\}_{\Theta=1/\alpha_2},\\
M_{34} &= \alpha_2\{\Theta D^{-1}[P_2P_1q(\rho_2-q\Theta^2)\\
&\qquad +\beta_1\alpha_1(\rho_1+q\Theta^2)(\rho_2-\rho_1-q\Theta^2)]\}_{\Theta=1/\alpha_2},\\
M_{41} &= \beta_2\rho_2\{\Theta D^{-1}[qP_2P_3-(\rho_2-\rho_1-q\Theta^2)\alpha_2\beta_1]\}_{\Theta=1/\beta_2},\\
M_{42} &= -\beta_2\rho_2\{D^{-1}[\alpha_2P_1(\rho_2-q\Theta^2)+\alpha_1P_3(\rho_1+q\Theta^2)]\}_{\Theta=1/\beta_2},
\end{aligned} \tag{3.43}$$

where

$$\begin{aligned}
D &= \alpha_1\alpha_2\beta_1\beta_2\Theta^2(\rho_2-\rho_1-q\Theta^2)^2+P_1P_2\alpha_2\beta_2(\rho_2-q\Theta^2)^2\\
&\quad +\rho_1\rho_2(P_1P_4\beta_1\alpha_2+P_2P_3\beta_2\alpha_1)+P_3P_4\alpha_1\beta_1(\rho_1+q\Theta^2)^2\\
&\quad +P_1P_2P_3P_4\Theta^2q^2
\end{aligned} \tag{3.44}$$

and

$$q = 2(\rho_2\beta_2{}^2-\rho_1\beta_1{}^2).$$

We have not given the formulae for M_{np} when $n = p$ since the wave $\mathbf{W}^{mnp}$ is not an H-wave in this case. Since the compressional wave velocity is always higher than the shear wave velocity, the condition (3.21) is not satisfied for $n = 2$, $p = 1$ and for $n = 4$, $p = 3$. From this it follows that the waves $\mathbf{W}^{m21}$ and $\mathbf{W}^{m43}$ are inhomogeneous head waves. Therefore, the formulae for M_{21} and M_{43} are not given in (3.43).

Comparing the formulae for M_{np} and the R-coefficients, R_{np} (see (2.74)), we can see that

$$M_{np} = \tfrac{1}{2}(-1)^{n+1}(R_{np}/P_n)_{\Theta=1/V_n}. \tag{3.45}$$

This is a very important expression which enables us to write M_{np} without solving the set of equations (3.39) if we know the formulae for the R-coefficients. Of course, this formula could have been obtained directly by comparing the set of equations (3.39) with the corresponding set of equations for the R-coefficients (2.73).

If we insert (3.38) into (3.22), we obtain for the displacement vector of the head wave at the point A

$$\mathbf{W}^{mnp}(A) = \frac{1}{i\omega}V_nM_{np}\frac{\partial U^{mn}(A)}{\partial H}\exp\{i\omega[t-\tau_{mnp}(A)]\}\mathbf{n}_p{}^*(A), \tag{3.46}$$

where $\partial U^{mn}(A)/\partial H$ denotes $\partial U^{mn}/\partial H$ at the point A. This equation is a general formula for head waves, independent of the nature of the wave $\mathbf{W}^{mn}$. Inserting different expressions for U^{mn}, we obtain formulae for H-waves propagating in different media.

In the case of a spherical wave incident on a plane interface between two homogeneous media, U^{mn} is given by (3.9). We use the simple relation

$$\frac{\partial U^{mn}(A)}{\partial H} = -\frac{1}{l}\left(\frac{\partial U^{mn}}{\partial \theta_n}\right)_{\theta_n=\pi/2}, \tag{3.47}$$

where l is the distance between the point A and the interface critical point (see figure 3.2),

$$l = r(A) - r^*_{mn} = r(A) - h \tan \theta_m{}^* = (\bar{l})_{\theta_n=\pi/2}. \tag{3.47'}$$

Then we obtain from (3.9)

$$\frac{\partial U^{mn}(A)}{\partial H} = -\frac{1}{l}\left(\frac{\partial(\cos \theta_n \tilde{U}^{mn})}{\partial \theta_n}\right)_{\theta_n=\pi/2} = \frac{1}{l}(\tilde{U}^{mn})_{\theta_n=\pi/2}.$$

Using (3.10), we have finally

$$\frac{\partial U^{mn}(A)}{\partial H} = \frac{R^*_{mn} \tan \theta_m{}^*}{r^{1/2} l^{3/2}}, \tag{3.48}$$

where R^*_{mn} is defined by (3.13) and r is the epicentral distance of the point A.

Inserting (3.48) into (3.46) yields

$$\mathbf{W}^{mnp}(A) = \frac{V_n M_{np} R^*_{mn} \tan \theta_m{}^*}{i\omega r^{1/2} l^{3/2}} \exp\{i\omega[t - \tau_{mnp}(A)]\}\mathbf{n}_p{}^*(A). \tag{3.49}$$

We now introduce the head wave coefficient (H-coefficient) Γ_{mnp} by

$$\Gamma_{mnp} = R^*_{mn} M_{np}. \tag{3.50}$$

H-coefficients are very important in the investigation of H-waves. We shall discuss them more in the following sections.

Inserting (3.50) into (3.49) yields

$$\mathbf{W}^{mnp}(A) = \frac{V_n \Gamma_{mnp} \tan \theta_m{}^*}{i\omega r^{1/2} l^{3/2}} \exp\{i\omega[t - \tau_{mnp}(A)]\}\mathbf{n}_p{}^*(A). \tag{3.51}$$

This is the final expression for the displacement vector of an arbitrary type of H-wave at the point A *lying on the interface.* If we want to know the displacement vector of an H-wave at a general point M lying on the ray of the H-wave, at a distance H from the interface, we use the equation for the continuation of a displacement vector along a ray given in chapter 2. Thus we obtain

$$\mathbf{W}^{mnp}(M) = \left(\frac{\mathrm{d}\sigma(A)}{\mathrm{d}\sigma(M)}\right)^{1/2} \frac{V_n\Gamma_{mnp}\tan\theta_m^*}{i\omega r^{1/2}l^{3/2}} \exp\{i\omega[t-\tau_{mnp}(M)]\}\mathbf{n}_p^*(M), \tag{3.52}$$

where $\mathrm{d}\sigma(A)$ and $\mathrm{d}\sigma(M)$ are the cross-sectional areas of an elementary ray tube of the wave under consideration at the points A and M, respectively. As the head wave is conical, its rays are parallel in the rz plane and spreading only occurs in the φ direction. We obtain easily

$$\frac{\mathrm{d}\sigma(A)}{\mathrm{d}\sigma(M)} = \frac{r(A)}{r(M)}. \tag{3.53}$$

Inserting (3.53) into (3.52), we obtain

$$\mathbf{W}^{mnp}(M) = \frac{V_n\Gamma_{mnp}\tan\theta_m^*}{i\omega[r(M)]^{1/2}l^{3/2}}\exp\{i\omega[t-\tau_{mnp}(M)]\}\mathbf{n}_p^*(M), \tag{3.54}$$

where $r(M)$ is the horizontal distance of the receiver M from the source. We can write for it

$$r(M) = r(A)+H\tan\theta_p^*. \tag{3.55}$$

The value of l is given in (3.47′). We can also write for it, inserting (3.55) into (3.47′),

$$l = r(M)-r^*_{mnp}, \tag{3.56}$$

where

$$r^*_{mnp} = h\tan\theta_m^*+H\tan\theta_p^* = \frac{hV_m}{[V_n^2-V_m^2]^{1/2}}+\frac{HV_p}{[V_n^2-V_p^2]^{1/2}}. \tag{3.57}$$

r^*_{mnp} is usually called the *critical distance*, and the point which lies at the critical distance from the source (for a given H) is called the *critical point*. Since the head wave can arise only if the ray of the wave $\mathbf{W}^{mn}$ is parallel to the interface, l must be greater than zero. Therefore, head waves exist only if $r > r^*_{mnp}$ for a given location of the source and a given H. Inserting (3.56) into (3.54), we can write finally

$$\mathbf{W}^{mnp}(M) = \frac{V_n\Gamma_{mnp}\tan\theta_m^*}{i\omega r^{1/2}(r-r^*_{mnp})^{3/2}} \exp\{i\omega[t-\tau_{mnp}(M)]\}\mathbf{n}_p^*(M), \tag{3.58}$$

where r is the horizontal distance of the point M from the source, and r^*_{mnp} is given by (3.57). For the travel time τ_{mnp} we can write

$$\tau_{mnp} = \frac{r-r^*_{mnp}}{V_n} + \frac{h}{V_m[1-(V_m/V_n)^2]^{1/2}} + \frac{H}{V_p[1-(V_p/V_n)^2]^{1/2}}, \tag{3.59}$$

or, using (3.57),

$$\tau_{mnp} = \frac{r}{V_n} + \frac{h}{V_m}\left[1-\left(\frac{V_m}{V_n}\right)^2\right]^{1/2} + \frac{H}{V_p}\left[1-\left(\frac{V_p}{V_n}\right)^2\right]^{1/2}. \tag{3.60}$$

The expression (3.58) characterizes the displacement vector of homogeneous H-waves only if all of the following conditions are satisfied:

(a) $m \neq n$. If $m = n$, the H-wave cannot arise since none of the rays of the R-wave can be parallel to the interface, except when the source is right on the interface – a case which we shall not consider in this book.

(b) $n \neq p$. If $n = p$, the wave $\mathbf{W}^{mnp}$ is not an H-wave, but a wave propagating along the interface.

(c) $V_m/V_n < 1$ and $V_p/V_n < 1$. If $V_m/V_n > 1$, no head waves arise. On the other hand, if $V_m/V_n < 1$ but $V_p/V_n > 1$, we obtain the so-called inhomogeneous H-waves.

(d) $r > r^*_{mnp}$ (see formula (3.57) for r^*_{mnp}). The waves exist only beyond the critical point.

For the horizontal and vertical components of the displacement vector $\mathbf{W}^{mnp}$, we get from (3.58)

$$\begin{bmatrix} W_r^{mnp} \\ W_z^{mnp} \end{bmatrix} = \frac{V_n\Gamma_{mnp}\tan\theta_m^*}{i\omega r^{1/2}(r-r^*_{mnp})^{3/2}}\exp[i\omega(t-\tau_{mnp})]\begin{bmatrix} n^*_{pr} \\ n^*_{pz} \end{bmatrix}, \tag{3.61}$$

where n^*_{pr} and n^*_{pz} are projections of the unit vector $\mathbf{n}_p{}^*$ on r (or x) and z axes. They do not depend on the position of the point M and are the same along the whole ray. They are given in table 3.1, where $\sin\theta_p{}^* = V_p/V_n$.

We can write (3.61) in another form:

$$\begin{aligned} W_r^{mnp} &= A_r^{mnp}\exp[i\omega(t-\tau_{mnp})+i\chi_r^{mnp}], \\ W_z^{mnp} &= A_z^{mnp}\exp[i\omega(t-\tau_{mnp})+i\chi_z^{mnp}], \end{aligned} \tag{3.62}$$

TABLE 3.1

Projections of $\mathbf{n}_p{}^*$ on r and z axes

	$p = 1$	$p = 2$	$p = 3$	$p = 4$
n^*_{pr}	$\sin\theta_1{}^*$	$\cos\theta_2{}^*$	$\sin\theta_3{}^*$	$\cos\theta_4{}^*$
n^*_{pz}	$\cos\theta_1{}^*$	$-\sin\theta_2{}^*$	$-\cos\theta_3{}^*$	$\sin\theta_4{}^*$

where A_r^{mnp} and A_z^{mnp} are amplitudes of the horizontal and the vertical components of the displacement vector of the H-wave, and χ_r^{mnp} and χ_z^{mnp} are their phase shifts. They are given by

$$\begin{bmatrix} A_r^{mnp} \\ A_z^{mnp} \end{bmatrix} = \frac{V_n \tan \theta_m^*}{\omega r^{1/2}(r-r_{mnp}^*)^{3/2}} |\Gamma_{mnp}| \begin{bmatrix} |n_{pr}^*| \\ |n_{pz}^*| \end{bmatrix}, \tag{3.63}$$

$$\begin{bmatrix} \chi_r^{mnp} \\ \chi_z^{mnp} \end{bmatrix} = \arg \Gamma_{mnp} - \tfrac{1}{2}\pi + \begin{bmatrix} \arg n_{pr}^* \\ \arg n_{pz}^* \end{bmatrix}, \tag{3.64}$$

We see from (3.63) that for $r \to r_{mnp}^*$ (i.e. when the receiver approaches the critical point) the amplitudes tend to infinity. The ray formulae for H-waves, however, are not exact in the close vicinity of the critical point. This special case will be discussed later in chapter 7.

Special Cases

(1) *Interface between two liquids* If both the media are liquids (i.e. $\gamma_1 = \gamma_2 = 0$), no shear waves can exist. Therefore, neither m, n, nor p can assume the value 2 or 4. We obtain only one type of H-wave, viz., $\mathbf{W}^{131}$ when the source is situated in the first medium and $\mathbf{W}^{313}$ when the source lies in the second medium. For a given interface, only one of these waves can exist, because the necessary condition for the existence of the wave $\mathbf{W}^{131}$ is $\alpha_2 > \alpha_1$, whereas the necessary condition for the existence of the wave $\mathbf{W}^{313}$ is $\alpha_2 < \alpha_1$.

The expressions (3.58), (3.61), and (3.62) apply to both these waves. For calculating Γ_{131} and Γ_{313}, we can again use the general formula (3.50). R_{mn}^* and M_{np}, however, cannot be calculated from the sets of equations (2.73) and (3.39) which become unstable when both γ_1 and γ_2 vanish. The interface conditions for two liquids in contact are given by two equations, corresponding to the second and the third row of the above-mentioned sets of equations.

From these two equations, we find that

$$R_{13}^* = 2\rho_1\alpha_1/\rho_2\alpha_2, \qquad R_{31}^* = 2\rho_2\alpha_2/\rho_1\alpha_1. \tag{3.65}$$

Similarly, we obtain two equations for M_{np} corresponding to the second and third row of (3.39), viz.,

$$\begin{aligned} \cos\theta_1^* M_{n1} + \cos\theta_3^* M_{n3} &= 1, \\ \rho_1\alpha_1 M_{n1} - \rho_2\alpha_2 M_{n3} &= 0. \end{aligned} \tag{3.66}$$

The first equation is quite sufficient for the determination of M_{13} and M_{31}. For $n = 1$, $\cos\theta_1^* = 0$, and we obtain

$$M_{13} = \frac{1}{\cos \theta_3{}^*} = \frac{1}{[1-(\alpha_2/\alpha_1)^2]^{1/2}}. \tag{3.67}$$

For $n = 3$, $\cos \theta_3{}^* = 0$, and we get

$$M_{31} = \frac{1}{\cos \theta_1{}^*} = \frac{1}{[1-(\alpha_1/\alpha_2)^2]^{1/2}}. \tag{3.67'}$$

From (3.65), (3.67), and (3.67′) we obtain the H-coefficients, Γ_{131} and Γ_{313}, for the case of two liquids in contact:

$$\Gamma_{131} = \frac{2\rho_1\alpha_1}{\rho_2\alpha_2[1-(\alpha_1/\alpha_2)^2]^{1/2}}, \qquad \Gamma_{313} = \frac{2\rho_2\alpha_2}{\rho_1\alpha_1[1-(\alpha_2/\alpha_1)^2]^{1/2}}. \tag{3.68}$$

Both H-coefficients are, of course, symmetric.

The formulae (3.65) through (3.68) can also be obtained directly by taking the limit $\beta_1 \to 0$ and $\beta_2 \to 0$ in the corresponding expressions for the solid–solid interface.

(2) *Interface between liquid and solid* All formulae given for a solid–solid interface are applicable to the situation when one of the media is liquid. Naturally, we must substitute $\gamma_1 = 0$ (or $\gamma_2 = 0$), and we must not attempt to calculate the waves propagating as shear waves in the liquid medium.

(3) *Free interface* Suppose that one of the half-spaces, say the first one, is a vacuum. Only one type of H-wave can be generated on the free interface, viz. the H-wave $\mathbf{W}^{434}$. All formulae given for a solid–solid interface are applicable to this situation.

3.2.2 *Source of SH-Waves*

The analysis of H-waves caused by an SH-source is very similar to that for a P or SV source. The displacement vector $\mathbf{W}^m$ of an SH-wave incident on the interface is given by (3.1) (where $m = 2$ or 4, $\mathbf{n}_m = \mathbf{n}_{SH}$). The expression for R-waves was derived in section 2.4.2B. The zeroth approximation for the R-wave, $\mathbf{W}^{mn}$ ($n = 2$ or 4) at the point $\bar{M}$, is the same as equations (3.2) to (3.10) except that for $\mathbf{n}_n(\bar{M})$ we must substitute $\mathbf{n}_{SH}(\bar{M})$ and instead of R_{mn} we must use R_{mn}^{SH}, the R-coefficients for SH-waves given in (2.80). As before, we investigate properties of the R-wave $\mathbf{W}^{mn}(\bar{M})$ when the point $\bar{M}$ lies on the interface, i.e. $\theta_n = \frac{1}{2}\pi$. We shall again designate the point $\bar{M}$ lying on the interface by A. The point $\bar{M}$ can lie on the interface (i.e. we have a wave propagating parallel to the boundary) only if $m \neq n$ and $V_n > V_m$.

The displacement vector $\mathbf{W}^{mn}(A)$ vanishes in the zero-order approximation; see (3.9) for $\theta_n = \frac{1}{2}\pi$. If we ignore the x and z components, which we do not need further, we can write the leading coefficient in the ray series as

$$\mathbf{W}_1^{mn}(A) = U_1^{mn}(A)\mathbf{j}. \tag{3.69}$$

Two new waves with displacement vectors $\mathbf{W}^{mnp}$ can arise at A. The index p indicates the type of waves generated ($p = 2$ for an SH-wave in the first medium and $p = 4$ for an SH-wave in the second medium). All the indices, m, n, and p, can take only the values 2 and 4 with the same restrictions as before.

The ray series expression for the displacement vector $\mathbf{W}^{mnp}$ is given by (3.18), where the phase function τ_{mnp} can be determined from (3.19) and (3.19′). The coefficients $\mathbf{W}_k^{mnp}$ can be determined from equations v and vi of the general set of interface conditions (2.57). The zeroth term, $\mathbf{W}_0^{mnp}$, vanishes since the zeroth term of the incident wave is zero at the point A. We can therefore write

$$\mathbf{W}^{mnp}(A) = (1/i\omega)U^{mnp}(A)\exp\{i\omega[t-\tau_{mnp}(A)]\}\mathbf{n}_{\mathrm{SH}}. \tag{3.70}$$

Now we shall solve equations v and vi from (2.57) for $k = 1$. Using the same procedure as in the case of P and SV waves, we find that

$$\begin{gathered} F_{52} = F_{53} = F_{54} = F_{61} = F_{63} = F_{64} = 0, \\ F_{51} = (-1)^{\varepsilon_n} U_1^{mn}, \qquad F_{62} = (-1)^{\varepsilon_n+1}\mu_{\varepsilon_n}\frac{\partial U^{mn}}{\partial z}. \end{gathered} \tag{3.71}$$

Inserting (3.71) into (2.57) and (2.57′) yields

$$\begin{aligned} U^{mn2} \qquad\qquad - U^{mn4} \qquad &= (-1)^{\varepsilon_n} U_1^{mn}, \\ U^{mn2}\rho_1\beta_1\cos\theta_2^* + U^{mn4}\rho_2\beta_2\cos\theta_4^* &= \rho_{\varepsilon_n}\beta_{\varepsilon_n}F^{mn}, \end{aligned} \tag{3.72}$$

where F^{mn} is given by (3.30) and the angles θ_p^* are given by

$$\sin\theta_p^* = V_p\sin\theta_n/V_n = V_p/V_n, \tag{3.73}$$

(as $\theta_n = \frac{1}{2}\pi$).

We see from (3.72) that, for $n \neq p$, the second equation is sufficient to give the solution, which can be written in the form

$$U^{mnp}(A) = M_{np}^{\mathrm{SH}}F^{mn}(A), \tag{3.74}$$

where

$$M_{24}^{\mathrm{SH}} = \frac{\rho_1\beta_1}{\rho_2\beta_2\cos\theta_4^*} = \frac{\rho_1\beta_1}{\rho_2\beta_2[1-(\beta_2/\beta_1)^2]^{1/2}}, \tag{3.75}$$

$$M_{42}^{\mathrm{SH}} = \frac{\rho_2\beta_2}{\rho_1\beta_1 \cos\theta_2{}^*} = \frac{\rho_2\beta_2}{\rho_1\beta_1[1-(\beta_1/\beta_2)^2]^{1/2}}. \tag{3.76}$$

For the determination of U^{m22} or U^{m44}, we need to know $U_1{}^{mn}$. However, we are not interested in these waves since they are not H-waves.

Inserting (3.74) into (3.70) yields

$$\mathbf{W}^{mnp}(A) = \frac{1}{i\omega} M_{np}^{\mathrm{SH}} V_n \frac{\partial U^{mn}(A)}{\partial H} \exp\{i\omega[t-\tau_{mnp}(A)]\}\mathbf{n}_{\mathrm{SH}}, \tag{3.77}$$

which is a general formula for SH H-waves. Inserting different expressions for U^{mn}, we obtain formulae for H-waves propagating in different media.

Using the same procedure for calculating $\partial U^{mn}/\partial H$ in the present case as was used for P and SV waves, we obtain

$$\mathbf{W}^{mnp}(A) = \frac{V_n \Gamma_{mnp}^{\mathrm{SH}} \tan\theta_m{}^*}{i\omega r^{1/2} l^{3/2}} \exp\{i\omega[t-\tau_{mnp}(A)]\}\mathbf{n}_{\mathrm{SH}}, \tag{3.78}$$

where we have introduced the H-coefficients

$$\Gamma_{mnp}^{\mathrm{SH}} = M_{np}^{\mathrm{SH}} R_{mn}^{\mathrm{SH}*}. \tag{3.79}$$

As was mentioned earlier, a necessary condition for H-waves is $m \neq n$ and $n \neq p$. Therefore, for SH-waves, we must have $m = p$, since m, n, and p can have one of two possible values, viz., 2 or 4. Therefore, we can write

$$\mathbf{W}^{mnm}(A) = \frac{V_n \Gamma_{mnm}^{\mathrm{SH}} \tan\theta_m{}^*}{i\omega r^{1/2} l^{3/2}} \exp\{i\omega[t-\tau_{mnm}(A)]\}\mathbf{n}_{\mathrm{SH}}. \tag{3.80}$$

Only two different $\Gamma_{mnm}^{\mathrm{SH}}$ can exist, viz., $\Gamma_{242}^{\mathrm{SH}}$ and $\Gamma_{424}^{\mathrm{SH}}$. We obtain them from (3.79). We can see from (2.80) that

$$R_{24}^{\mathrm{SH}*} = R_{42}^{\mathrm{SH}*} = 2. \tag{3.81}$$

From (3.75), (3.76), (3.79), and (3.81) we get

$$\begin{aligned} \Gamma_{242}^{\mathrm{SH}} &= \frac{2\rho_2\beta_2}{\rho_1\beta_1[1-(\beta_1/\beta_2)^2]^{1/2}}, \\ \Gamma_{424}^{\mathrm{SH}} &= \frac{2\rho_1\beta_1}{\rho_2\beta_2[1-(\beta_2/\beta_1)^2]^{1/2}}. \end{aligned} \tag{3.82}$$

Using the same procedure as in the case of P and SV waves, we can write for the displacement vector of an H-wave at an arbitrary point M lying on the ray of the wave under consideration

$$\mathbf{W}^{mnm}(M) = \frac{V_n \Gamma^{SH}_{mnm} \tan \theta_m^*}{i\omega r^{1/2}(r-r^*_{mnm})^{3/2}} \exp\{i\omega[t-\tau_{mnm}(M)]\}\mathbf{n}_{SH}. \tag{3.83}$$

This formula is the same as the expression (3.58) for determining P or SV H-waves, but the unit vector $\mathbf{n}_{SH}$ and the H-coefficient Γ^{SH}_{mnm} differ. Values of r^*_{mnm} and τ_{mnm} in (3.83) may be calculated using (3.57) and (3.60). These values are, of course, the same as for SV-waves.

We can write the φ-component of the SH H-wave in the form

$$W_\varphi^{mnm} = \frac{V_n \Gamma^{SH}_{mnm} \tan \theta_m^*}{i\omega r^{1/2}(r-r^*_{mnm})^{3/2}} \exp\{i\omega[t-\tau_{mnm}(M)]\}, \tag{3.84}$$

or

$$W_\varphi^{mnm} = A_\varphi^{mnm} \exp[i\omega(t-\tau_{mnm})+i\chi_\varphi^{mnm}], \tag{3.85}$$

where the amplitude A_φ^{mnm} is given by

$$A_\varphi^{mnm} = \frac{V_n \tan \theta_m^* \Gamma^{SH}_{mnm}}{\omega r^{1/2}(r-r^*_{mnm})^{3/2}}, \tag{3.86}$$

and the phase shift χ_φ^{mnm} is constant and is given by

$$\chi_\varphi^{mnm} = -\tfrac{1}{2}\pi. \tag{3.87}$$

Note that Γ^{SH}_{mnm} is always real and positive.

For a given interface, only one SH H-wave can exist. If, for example, $\beta_2 > \beta_1$ and the source lies in the first medium, the H-wave $\mathbf{W}^{242}$ exists. If the source lies in the second medium, and if $\beta_2 < \beta_1$, we obtain only $\mathbf{W}^{424}$. In other situations no head wave exists.

3.3
SUMMARY OF RESULTS

In this section we shall summarize the formulae for all possible types of H-waves that can be generated by a single interface. In all the formulae $V_1 = \alpha_1$, $V_2 = \beta_1$, $V_3 = \alpha_2$, $V_4 = \beta_2$.

3.3.1 *P and SV Head Waves*

For a solid–solid interface, twenty-six types of H-waves can exist, as given schematically in figure 3.4. Of course, all these H-waves cannot exist simultaneously; their number depends on the type and position of the source and on the elastic parameters of the interface. For example, if a source of compressional waves lies in the first medium, the maximum number of H-waves generated by the interface is *five*, and if a source of shear waves lies in the first medium, the maximum number generated is *six* (see figure 3.7).

We can write the general formula for the displacement vector of an arbitrary H-wave, (3.58), in the form

$$\mathbf{W}^{mnp} = \frac{D_{mnp}}{i\omega r^{1/2}(r - r^*_{mnp})^{3/2}} \exp[i\omega(t - \tau_{mnp})]\mathbf{n}_p^*, \tag{3.88}$$

from which the r and z components can be obtained:

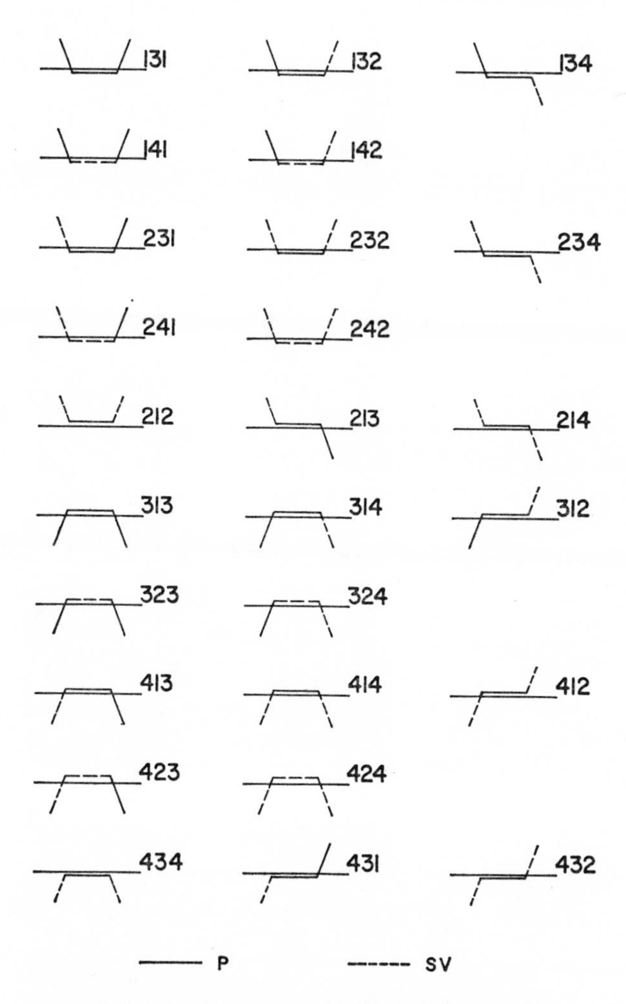

FIGURE 3.4 Schematic representation of all the possible types of P and SV head waves for a single solid–solid interface. All these waves cannot exist simultaneously. Each type of head wave is characterized by three numbers, *mnp*.

TABLE 3.2
Summary of the various formulae for head waves for a single interface

Type mnp	Exists if	Travel time τ_{mnp}	Critical distance r^*_{mnp}	D_{mnp}	D_{mnp} real if	n^*_{pr}	n^*_{pz}
131	$\alpha_2 > \alpha_1$	$\frac{r}{\alpha_2}+\frac{h+H}{\alpha_1}\left[1-\left(\frac{\alpha_1}{\alpha_2}\right)^2\right]^{1/2}$	$\frac{(h+H)\alpha_1}{[\alpha_2^2-\alpha_1^2]^{1/2}}$	$\frac{\alpha_1\alpha_2}{[\alpha_2^2-\alpha_1^2]^{1/2}}\Gamma_{131}$	always	$\frac{\alpha_1}{\alpha_2}$	$\left[1-\left(\frac{\alpha_1}{\alpha_2}\right)^2\right]^{1/2}$
132	$\alpha_2 > \alpha_1$	$\frac{r}{\alpha_2}+\frac{h}{\alpha_1}\left[1-\left(\frac{\alpha_1}{\alpha_2}\right)^2\right]^{1/2}+\frac{H}{\beta_1}\left[1-\left(\frac{\beta_1}{\alpha_2}\right)^2\right]^{1/2}$	$\frac{h\alpha_1}{[\alpha_2^2-\alpha_1^2]^{1/2}}+\frac{H\beta_1}{[\alpha_2^2-\beta_1^2]^{1/2}}$	$\frac{\alpha_1\alpha_2}{[\alpha_2^2-\alpha_1^2]^{1/2}}\Gamma_{132}$	always	$\left[1-\left(\frac{\beta_1}{\alpha_2}\right)^2\right]^{1/2}$	$-\frac{\beta_1}{\alpha_2}$
134	$\alpha_2 > \alpha_1$	$\frac{r}{\alpha_2}+\frac{h}{\alpha_1}\left[1-\left(\frac{\alpha_1}{\alpha_2}\right)^2\right]^{1/2}+\frac{H}{\beta_2}\left[1-\left(\frac{\beta_2}{\alpha_2}\right)^2\right]^{1/2}$	$\frac{h\alpha_1}{[\alpha_2^2-\alpha_1^2]^{1/2}}+\frac{H\beta_2}{[\alpha_2^2-\beta_2^2]^{1/2}}$	$\frac{\alpha_1\alpha_2}{[\alpha_2^2-\alpha_1^2]^{1/2}}\Gamma_{134}$	always	$\left[1-\left(\frac{\beta_2}{\alpha_2}\right)^2\right]^{1/2}$	$\frac{\beta_2}{\alpha_2}$
141	$\beta_2 > \alpha_1$	$\frac{r}{\beta_2}+\frac{h+H}{\alpha_1}\left[1-\left(\frac{\alpha_1}{\beta_2}\right)^2\right]^{1/2}$	$\frac{(h+H)\alpha_1}{[\beta_2^2-\alpha_1^2]^{1/2}}$	$\frac{\alpha_1\beta_2}{[\beta_2^2-\alpha_1^2]^{1/2}}\Gamma_{141}$	never	$\frac{\alpha_1}{\beta_2}$	$\left[1-\left(\frac{\alpha_1}{\beta_2}\right)^2\right]^{1/2}$
142	$\beta_2 > \alpha_1$	$\frac{r}{\beta_2}+\frac{h}{\alpha_1}\left[1-\left(\frac{\alpha_1}{\beta_2}\right)^2\right]^{1/2}+\frac{H}{\beta_1}\left[1-\left(\frac{\beta_1}{\beta_2}\right)^2\right]^{1/2}$	$\frac{h\alpha_1}{[\beta_2^2-\alpha_1^2]^{1/2}}+\frac{H\beta_1}{[\beta_2^2-\beta_1^2]^{1/2}}$	$\frac{\alpha_1\beta_2}{[\beta_2^2-\alpha_1^2]^{1/2}}\Gamma_{142}$	never	$\left[1-\left(\frac{\beta_1}{\beta_2}\right)^2\right]^{1/2}$	$-\frac{\beta_1}{\beta_2}$
231	$\alpha_2 > \alpha_1$	$\frac{r}{\alpha_2}+\frac{h}{\beta_1}\left[1-\left(\frac{\beta_1}{\alpha_2}\right)^2\right]^{1/2}+\frac{H}{\alpha_1}\left[1-\left(\frac{\alpha_1}{\alpha_2}\right)^2\right]^{1/2}$	$\frac{h\beta_1}{[\alpha_2^2-\beta_1^2]^{1/2}}+\frac{H\alpha_1}{[\alpha_2^2-\alpha_1^2]^{1/2}}$	$\frac{\beta_1\alpha_2}{[\alpha_2^2-\beta_1^2]^{1/2}}\Gamma_{231}$	always	$\frac{\alpha_1}{\alpha_2}$	$\left[1-\left(\frac{\alpha_1}{\alpha_2}\right)^2\right]^{1/2}$
232	$\alpha_2 > \beta_1$	$\frac{r}{\alpha_2}+\frac{h+H}{\beta_1}\left[1-\left(\frac{\beta_1}{\alpha_2}\right)^2\right]^{1/2}$	$\frac{(h+H)\beta_1}{[\alpha_2^2-\beta_1^2]^{1/2}}$	$\frac{\beta_1\alpha_2}{[\alpha_2^2-\beta_1^2]^{1/2}}\Gamma_{232}$	$\alpha_2 > \alpha_1$	$\left[1-\left(\frac{\beta_1}{\alpha_2}\right)^2\right]^{1/2}$	$-\frac{\beta_1}{\alpha_2}$
234	$\alpha_2 > \beta_1$	$\frac{r}{\alpha_2}+\frac{h}{\beta_1}\left[1-\left(\frac{\beta_1}{\alpha_2}\right)^2\right]^{1/2}+\frac{H}{\beta_2}\left[1-\left(\frac{\beta_2}{\alpha_2}\right)^2\right]^{1/2}$	$\frac{h\beta_1}{[\alpha_2^2-\beta_1^2]^{1/2}}+\frac{H\beta_2}{[\alpha_2^2-\beta_2^2]^{1/2}}$	$\frac{\beta_1\alpha_2}{[\alpha_2^2-\beta_1^2]^{1/2}}\Gamma_{234}$	$\alpha_2 > \alpha_1$	$\left[1-\left(\frac{\beta_2}{\alpha_2}\right)^2\right]^{1/2}$	$\frac{\beta_2}{\alpha_2}$

241	$\beta_2 > \alpha_1$	$\frac{r}{\beta_2}+\frac{h}{\beta_1}\left[1-\left(\frac{\beta_1}{\beta_2}\right)^2\right]^{1/2}+\frac{H}{\alpha_1}\left[1-\left(\frac{\alpha_1}{\beta_2}\right)^2\right]^{1/2}$	$\frac{h\beta_1}{[\beta_2{}^2-\beta_1{}^2]^{1/2}}+\frac{H\alpha_1}{[\beta_2{}^2-\alpha_1{}^2]^{1/2}}$	$\frac{\beta_1\beta_2}{[\beta_2{}^2-\beta_1{}^2]^{1/2}}\Gamma_{241}$	never	$\frac{\alpha_1}{\beta_2}$	$\left[1-\left(\frac{\alpha_1}{\beta_2}\right)^2\right]^{1/2}$
242	$\beta_2 > \beta_1$	$\frac{r}{\beta_2}+\frac{h+H}{\beta_1}\left[1-\left(\frac{\beta_1}{\beta_2}\right)^2\right]^{1/2}$	$\frac{(h+H)\beta_1}{[\beta_2{}^2-\beta_1{}^2]^{1/2}}$	$\frac{\beta_1\beta_2}{[\beta_2{}^2-\beta_1{}^2]^{1/2}}\Gamma_{242}$	never	$\left[1-\left(\frac{\beta_1}{\beta_2}\right)^2\right]^{1/2}$	$-\frac{\beta_1}{\beta_2}$
212	always	$\frac{r}{\alpha_1}+\frac{h+H}{\beta_1}\left[1-\left(\frac{\beta_1}{\alpha_1}\right)^2\right]^{1/2}$	$\frac{(h+H)\beta_1}{[\alpha_1{}^2-\beta_1{}^2]^{1/2}}$	$\frac{\beta_1\alpha_1}{[\alpha_1{}^2-\beta_1{}^2]^{1/2}}\Gamma_{212}$	$\alpha_1 > \alpha_2$	$\left[1-\left(\frac{\beta_1}{\alpha_1}\right)^2\right]^{1/2}$	$-\frac{\beta_1}{\alpha_1}$
213	$\alpha_1 > \alpha_2$	$\frac{r}{\alpha_1}+\frac{h}{\beta_1}\left[1-\left(\frac{\beta_1}{\alpha_1}\right)^2\right]^{1/2}+\frac{H}{\alpha_2}\left[1-\left(\frac{\alpha_2}{\alpha_1}\right)^2\right]^{1/2}$	$\frac{h\beta_1}{[\alpha_1{}^2-\beta_1{}^2]^{1/2}}+\frac{H\alpha_2}{[\alpha_1{}^2-\alpha_2{}^2]^{1/2}}$	$\frac{\beta_1\alpha_1}{[\alpha_1{}^2-\beta_1{}^2]^{1/2}}\Gamma_{213}$	always	$\frac{\alpha_2}{\alpha_1}$	$-\left[1-\left(\frac{\alpha_2}{\alpha_1}\right)^2\right]^{1/2}$
214	$\alpha_1 > \beta_2$	$\frac{r}{\alpha_1}+\frac{h}{\beta_1}\left[1-\left(\frac{\beta_1}{\alpha_1}\right)^2\right]^{1/2}+\frac{H}{\beta_2}\left[1-\left(\frac{\beta_2}{\alpha_1}\right)^2\right]^{1/2}$	$\frac{h\beta_1}{[\alpha_1{}^2-\beta_1{}^2]^{1/2}}+\frac{H\beta_2}{[\alpha_1{}^2-\beta_2{}^2]^{1/2}}$	$\frac{\beta_1\alpha_1}{[\alpha_1{}^2-\beta_1{}^2]^{1/2}}\Gamma_{214}$	$\alpha_1 > \alpha_2$	$\left[1-\left(\frac{\beta_2}{\alpha_1}\right)^2\right]^{1/2}$	$\frac{\beta_2}{\alpha_1}$
313	$\alpha_1 > \alpha_2$	$\frac{r}{\alpha_1}+\frac{h+H}{\alpha_2}\left[1-\left(\frac{\alpha_2}{\alpha_1}\right)^2\right]^{1/2}$	$\frac{(h+H)\alpha_2}{[\alpha_1{}^2-\alpha_2{}^2]^{1/2}}$	$\frac{\alpha_1\alpha_2}{[\alpha_1{}^2-\alpha_2{}^2]^{1/2}}\Gamma_{313}$	always	$\frac{\alpha_2}{\alpha_1}$	$-\left[1-\left(\frac{\alpha_2}{\alpha_1}\right)^2\right]^{1/2}$
314	$\alpha_1 > \alpha_2$	$\frac{r}{\alpha_1}+\frac{h}{\alpha_2}\left[1-\left(\frac{\alpha_2}{\alpha_1}\right)^2\right]^{1/2}+\frac{H}{\beta_2}\left[1-\left(\frac{\beta_2}{\alpha_1}\right)^2\right]^{1/2}$	$\frac{h\alpha_2}{[\alpha_1{}^2-\alpha_2{}^2]^{1/2}}+\frac{H\beta_2}{[\alpha_1{}^2-\beta_2{}^2]^{1/2}}$	$\frac{\alpha_1\alpha_2}{[\alpha_1{}^2-\alpha_2{}^2]^{1/2}}\Gamma_{314}$	always	$\left[1-\left(\frac{\beta_2}{\alpha_1}\right)^2\right]^{1/2}$	$\frac{\beta_2}{\alpha_1}$
312	$\alpha_1 > \alpha_2$	$\frac{r}{\alpha_1}+\frac{h}{\alpha_2}\left[1-\left(\frac{\alpha_2}{\alpha_1}\right)^2\right]^{1/2}+\frac{H}{\beta_1}\left[1-\left(\frac{\beta_1}{\alpha_1}\right)^2\right]^{1/2}$	$\frac{h\alpha_2}{[\alpha_1{}^2-\alpha_2{}^2]^{1/2}}+\frac{H\beta_1}{[\alpha_1{}^2-\beta_1{}^2]^{1/2}}$	$\frac{\alpha_1\alpha_2}{[\alpha_1{}^2-\alpha_2{}^2]^{1/2}}\Gamma_{312}$	always	$\left[1-\left(\frac{\beta_1}{\alpha_1}\right)^2\right]^{1/2}$	$-\frac{\beta_1}{\alpha_1}$
323	$\beta_1 > \alpha_2$	$\frac{r}{\beta_1}+\frac{h+H}{\alpha_2}\left[1-\left(\frac{\alpha_2}{\beta_1}\right)^2\right]^{1/2}$	$\frac{(h+H)\alpha_2}{[\beta_1{}^2-\alpha_2{}^2]^{1/2}}$	$\frac{\beta_1\alpha_2}{[\beta_1{}^2-\alpha_2{}^2]^{1/2}}\Gamma_{323}$	never	$\frac{\alpha_2}{\beta_1}$	$-\left[1-\left(\frac{\alpha_2}{\beta_1}\right)^2\right]^{1/2}$
324	$\beta_1 > \alpha_2$	$\frac{r}{\beta_1}+\frac{h}{\alpha_2}\left[1-\left(\frac{\alpha_2}{\beta_1}\right)^2\right]^{1/2}+\frac{H}{\beta_2}\left[1-\left(\frac{\beta_2}{\beta_1}\right)^2\right]^{1/2}$	$\frac{h\alpha_2}{[\beta_1{}^2-\alpha_2{}^2]^{1/2}}+\frac{H\beta_2}{[\beta_1{}^2-\beta_2{}^2]^{1/2}}$	$\frac{\beta_1\alpha_2}{[\beta_1{}^2-\alpha_2{}^2]^{1/2}}\Gamma_{324}$	never	$\left[1-\left(\frac{\beta_2}{\beta_1}\right)^2\right]^{1/2}$	$\frac{\beta_2}{\beta_1}$

TABLE 3.2 (concluded)

Type mnp	Exists if	Travel time τ_{mnp}	Critical distance r^*_{mnp}	D_{mnp}	D_{mnp} real if	n^*_{pr}	n^*_{pz}
413	$\alpha_1 > \alpha_2$	$\frac{r}{\alpha_1}+\frac{h}{\beta_2}\left[1-\left(\frac{\beta_2}{\alpha_1}\right)^2\right]^{1/2}+\frac{H}{\alpha_2}\left[1-\left(\frac{\alpha_2}{\alpha_1}\right)^2\right]^{1/2}$	$\frac{h\beta_2}{[\alpha_1{}^2-\beta_2{}^2]^{1/2}}+\frac{H\alpha_2}{[\alpha_1{}^2-\alpha_2{}^2]^{1/2}}$	$\frac{\alpha_1\beta_2}{[\alpha_1{}^2-\beta_2{}^2]^{1/2}}\Gamma_{413}$	always	$\frac{\alpha_2}{\alpha_1}$	$-\left[1-\left(\frac{\alpha_2}{\alpha_1}\right)^2\right]^{1/2}$
414	$\alpha_1 > \beta_2$	$\frac{r}{\alpha_1}+\frac{h+H}{\beta_2}\left[1-\left(\frac{\beta_2}{\alpha_1}\right)^2\right]^{1/2}$	$\frac{(h+H)\beta_2}{[\alpha_1{}^2-\beta_2{}^2]^{1/2}}$	$\frac{\alpha_1\beta_2}{[\alpha_1{}^2-\beta_2{}^2]^{1/2}}\Gamma_{414}$	$\alpha_1 > \alpha_2$	$\left[1-\left(\frac{\beta_2}{\alpha_1}\right)^2\right]^{1/2}$	$\frac{\beta_2}{\alpha_1}$
412	$\alpha_1 > \beta_2$	$\frac{r}{\alpha_1}+\frac{h}{\beta_2}\left[1-\left(\frac{\beta_2}{\alpha_1}\right)^2\right]^{1/2}+\frac{H}{\beta_1}\left[1-\left(\frac{\beta_1}{\alpha_1}\right)^2\right]^{1/2}$	$\frac{h\beta_2}{[\alpha_1{}^2-\beta_2{}^2]^{1/2}}+\frac{H\beta_1}{[\alpha_1{}^2-\beta_1{}^2]^{1/2}}$	$\frac{\alpha_1\beta_2}{[\alpha_1{}^2-\beta_2{}^2]^{1/2}}\Gamma_{412}$	$\alpha_1 > \alpha_2$	$\left[1-\left(\frac{\beta_1}{\alpha_1}\right)^2\right]^{1/2}$	$-\frac{\beta_1}{\alpha_1}$
423	$\beta_1 > \alpha_2$	$\frac{r}{\beta_1}+\frac{h}{\beta_2}\left[1-\left(\frac{\beta_2}{\beta_1}\right)^2\right]^{1/2}+\frac{H}{\alpha_2}\left[1-\left(\frac{\alpha_2}{\beta_1}\right)^2\right]^{1/2}$	$\frac{h\beta_2}{[\beta_1{}^2-\beta_2{}^2]^{1/2}}+\frac{H\alpha_2}{[\beta_1{}^2-\alpha_2{}^2]^{1/2}}$	$\frac{\beta_1\beta_2}{[\beta_1{}^2-\beta_2{}^2]^{1/2}}\Gamma_{423}$	never	$\frac{\alpha_2}{\beta_1}$	$-\left[1-\left(\frac{\alpha_2}{\beta_1}\right)^2\right]^{1/2}$
424	$\beta_1 > \beta_2$	$\frac{r}{\beta_1}+\frac{h+H}{\beta_2}\left[1-\left(\frac{\beta_2}{\beta_1}\right)^2\right]^{1/2}$	$\frac{(h+H)\beta_2}{[\beta_1{}^2-\beta_2{}^2]^{1/2}}$	$\frac{\beta_1\beta_2}{[\beta_1{}^2-\beta_2{}^2]^{1/2}}\Gamma_{424}$	never	$\left[1-\left(\frac{\beta_2}{\beta_1}\right)^2\right]^{1/2}$	$\frac{\beta_2}{\beta_1}$
434	always	$\frac{r}{\alpha_2}+\frac{h+H}{\beta_2}\left[1-\left(\frac{\beta_2}{\alpha_2}\right)^2\right]^{1/2}$	$\frac{(h+H)\beta_2}{[\alpha_2{}^2-\beta_2{}^2]^{1/2}}$	$\frac{\beta_2\alpha_2}{[\alpha_2{}^2-\beta_2{}^2]^{1/2}}\Gamma_{434}$	$\alpha_2 > \alpha_1$	$\left[1-\left(\frac{\beta_2}{\alpha_2}\right)^2\right]^{1/2}$	$\frac{\beta_2}{\alpha_2}$
431	$\alpha_2 > \alpha_1$	$\frac{r}{\alpha_2}+\frac{h}{\beta_2}\left[1-\left(\frac{\beta_2}{\alpha_2}\right)^2\right]^{1/2}+\frac{H}{\alpha_1}\left[1-\left(\frac{\alpha_1}{\alpha_2}\right)^2\right]^{1/2}$	$\frac{h\beta_2}{[\alpha_2{}^2-\beta_2{}^2]^{1/2}}+\frac{H\alpha_1}{[\alpha_2{}^2-\alpha_1{}^2]^{1/2}}$	$\frac{\beta_2\alpha_2}{[\alpha_2{}^2-\beta_2{}^2]^{1/2}}\Gamma_{431}$	always	$\frac{\alpha_1}{\alpha_2}$	$\left[1-\left(\frac{\alpha_1}{\alpha_2}\right)^2\right]^{1/2}$
432	$\alpha_2 > \beta_1$	$\frac{r}{\alpha_2}+\frac{h}{\beta_2}\left[1-\left(\frac{\beta_2}{\alpha_2}\right)^2\right]^{1/2}+\frac{H}{\beta_1}\left[1-\left(\frac{\beta_1}{\alpha_2}\right)^2\right]^{1/2}$	$\frac{h\beta_2}{[\alpha_2{}^2-\beta_2{}^2]^{1/2}}+\frac{H\beta_1}{[\alpha_2{}^2-\beta_1{}^2]^{1/2}}$	$\frac{\beta_2\alpha_2}{[\alpha_2{}^2-\beta_2{}^2]^{1/2}}\Gamma_{432}$	$\alpha_2 > \alpha_1$	$\left[1-\left(\frac{\beta_1}{\alpha_2}\right)^2\right]^{1/2}$	$-\frac{\beta_1}{\alpha_2}$

$$\begin{bmatrix} W_r^{mnp} \\ W_z^{mnp} \end{bmatrix} = \frac{D_{mnp}}{i\omega r^{1/2}(r-r_{mnp}^*)^{3/2}} \exp[i\omega(t-\tau_{mnp})] \begin{bmatrix} n_{pr}^* \\ n_{pz}^* \end{bmatrix}. \tag{3.89}$$

Here D_{mnp} is a function which does not depend on the position of the source or receiver (for a given type of wave) or on the frequency of the wave. It is given by

$$D_{mnp} = V_n \Gamma_{mnp} \tan \theta_m^* = \frac{V_n V_m}{(V_n^2 - V_m^2)^{1/2}} \Gamma_{mnp}, \tag{3.90}$$

where Γ_{mnp} is given by (3.50). (Detailed expressions for Γ_{mnp} will be given in section 3.5.) r_{mnp}^* is the critical distance and is given by

$$r_{mnp}^* = \frac{hV_m}{(V_n^2 - V_m^2)^{1/2}} + \frac{HV_p}{(V_n^2 - V_p^2)^{1/2}}, \tag{3.91}$$

and τ_{mnp} is the travel time of the H-wave, given by

$$\tau_{mnp} = \frac{r}{V_n} + \frac{h}{V_m}[1-(V_m/V_n)^2]^{1/2} + \frac{H}{V_p}[1-(V_p/V_n)^2]^{1/2}, \tag{3.92}$$

where r is the horizontal distance between the receiver and the source and h and H are the distances of the source and the receiver from the interface, respectively. $\mathbf{n}_p^*$ is a unit vector associated with the wave ($\mathbf{n}_p^* = \mathbf{n}_P$ for $p = 1$ and 3, $\mathbf{n}_p^* = \mathbf{n}_{SV}$ for $p = 2$ and 4). n_{pr}^* and n_{pz}^* are projections of this vector on the r and z axes, respectively. They are given in table 3.1.

The wave (3.88) is an H-wave only if $m \neq n$, $n \neq p$, $V_m < V_n$, and $V_p < V_n$, and exists only for distances $r > r_{mnp}^*$.

In table 3.2 we give formulae for all twenty-six possible types of H-waves. Note that the sixth column shows whether the value of D_{mnp} is real or complex. For ten H-waves D_{mnp} is always real, for eight others it is always complex, and for the remaining eight types it may be real or complex depending on the velocity relations on the interface. More detailed discussion will be given in section 3.5. We can write the following expressions for the amplitudes of the horizontal and vertical components of the displacement vector of the H-wave, A_r^{mnp} and A_z^{mnp}, and the phase shifts, χ_r^{mnp} and χ_z^{mnp}:

$$\begin{bmatrix} A_r^{mnp} \\ A_z^{mnp} \end{bmatrix} = \frac{|D_{mnp}|}{\omega r^{1/2}(r-r_{mnp}^*)^{3/2}} \begin{bmatrix} |n_{pr}^*| \\ |n_{pz}^*| \end{bmatrix}, \tag{3.93}$$

$$\begin{bmatrix} \chi_r^{mnp} \\ \chi_z^{mnp} \end{bmatrix} = -\tfrac{1}{2}\pi + \arg D_{mnp} + \begin{bmatrix} \arg n_{pr}^* \\ \arg n_{pz}^* \end{bmatrix}, \tag{3.94}$$

where $\arg n_{pr}^*$ and $\arg n_{pz}^*$ are either 0 or π, and $\arg D_{mnp} = \arg \Gamma_{mnp}$. χ_r^{mnp} and χ_z^{mnp} are constant, for a given H-wave, and do not depend on the position of the source or the receiver or on the frequency of the wave.

The expressions given in table 3.2 can be used for an arbitrary type of interface (solid–solid, solid–liquid, liquid–liquid, free boundary) if the relevant head wave $\mathbf{W}^{mnp}$ exists, but the expressions for the H-coefficients, Γ_{mnp}, may be different in different cases.

3.3.2 *SH Waves*

Only one of two possible SH H-waves, $\mathbf{W}^{242}$ and $\mathbf{W}^{424}$, can exist for a given interface ($\mathbf{W}^{242}$ for $\beta_1 < \beta_2$, $\mathbf{W}^{424}$ for $\beta_2 < \beta_1$). For the displacement vector of these waves, we can write (see (3.83))

$$\mathbf{W}^{mnm} = \frac{D_{mnm}}{i\omega r^{1/2}(r-r^*_{mnm})^{3/2}} \exp[i\omega(t-\tau_{mnm})]\mathbf{n}_{SH}, \tag{3.95}$$

and for the φ-component,

$$W_\varphi{}^{mnm} = \frac{D_{mnm}}{i\omega r^{1/2}(r-r^*_{mnm})^{3/2}} \exp[i\omega(t-\tau_{mnm})], \tag{3.96}$$

where D_{mnm}, r^*_{mnm}, and τ_{mnm} are again given by (3.90) to (3.92). For the special case of $mnm = 242$ or 424, we can use table 3.2 to find values of D_{mnm}, r^*_{mnm}, and τ_{mnm}. Now, of course, Γ_{242} and Γ_{424} have to be replaced by Γ^{SH}_{242} and Γ^{SH}_{424} (which are always real and positive). For the amplitude ($A_\varphi{}^{mnm}$) of the φ-component of the displacement vector, and for the phase shift ($\chi_\varphi{}^{mnm}$), we obtain

$$A_\varphi{}^{mnm} = \frac{D_{mnm}}{\omega r^{1/2}(r-r^*_{mnm})^{3/2}}, \qquad \chi_\varphi{}^{mnm} = -\tfrac{1}{2}\pi. \tag{3.97}$$

The expression for Γ^{SH}_{mnm} is given in section 3.5.1F.

3.3.3 *Accuracy of the Formulae for Head Waves*

The ray method, which we used for the derivation of H-waves, is only an approximate one; it does not give us exact results. The accuracy of the results we have obtained is greater for higher frequencies. The ray method lacks precision in the neighbourhood of the singular points of the phase function. For H-waves, the singular point is the critical point. Immediately at the critical point, we obtain H-waves of infinite amplitude (see (3.63), for $r = r^*_{mnp}$). In the neighbourhood of the critical point, the formulae for H-waves are obviously inaccurate (the *neighbourhood* being larger for smaller frequencies).

Using the ray method, it is not possible to appreciate the degree of inaccuracy and the dimensions of the region around the critical point in which the results are in error. In order to do this we must compare them with calculations based on more exact methods, as will be done in chapter 7. We can only conclude that in the neighbourhood of the critical point we must use our results with caution. Note that the formulae for R-waves are also inaccurate in the neighbourhood of the critical points. The H-wave $\mathbf{W}^{mnp}$ does not exist in the neighbourhood of the critical point separately from other waves; it always interferes here with the R-wave $\mathbf{W}^{mp}$ (see the time-distance curves in figure 3.5, and a more detailed analysis in section 3.8). The resulting interference wave $\mathbf{W}^{mp}+\mathbf{W}^{mnp}$ has properties which are quite different from those of the simple H-wave, and will therefore be studied separately in section 3.8. The pure H-waves $\mathbf{W}^{mnp}$ exist separately only beyond the interference zone of both waves. It is found from comparisons with exact results that the H-wave formulae based on the ray method are usually sufficiently accurate beyond the interference zone. For example, when $\gamma_1 \sim \gamma_2$ and $\rho_1 \sim \rho_2$, the inaccuracy of the ray formula for a longitudinal H-wave $\mathbf{W}^{131}$ is usually not larger than 5 per cent beyond this zone.

3.4
CLASSIFICATION OF HEAD WAVES

Every H-wave $\mathbf{W}^{mnp}$ is closely connected with two R-waves, $\mathbf{W}^{mn}$ and $\mathbf{W}^{mp}$. In previous sections, we have mentioned the wave $\mathbf{W}^{mn}$ very often (see (3.2) to (3.5) and figure 3.5). This wave propagates along the interface and causes the H-wave. Its wave front has a common point (A in figure 3.5) on the interface with the wave front of the H-wave. The R-wave $\mathbf{W}^{mp}$ has one common ray (O^*A^*) with the H-wave, which is called the *critical ray*. For this ray, $\sin\theta_p = \sin\theta_p{}^* = V_p/V_n$. This R-wave propagates in the same medium as the H-wave and is of the same type, i.e. it is a compressional wave if $\mathbf{W}^{mnp}$ is compressional, and it is a shear wave if $\mathbf{W}^{mnp}$ is shear. The wave fronts of the H-wave $\mathbf{W}^{mnp}$ and the R-wave $\mathbf{W}^{mp}$ are tangent at the point A^* lying on this critical ray (see figure 3.5). The horizontal distance of the point A^* from the source is the critical distance r^*_{mnp} given by (3.91). The H-wave exists only at distances r greater than r^*_{mnp}. The trace of the wave front of the H-wave is the straight line A^*A. (In three dimensions, the wave front will constitute the frustum of a cone.) At the critical distance, the travel times of both waves $\mathbf{W}^{mnp}$ and $\mathbf{W}^{mp}$ are the same. The time-distance curve of the R-wave $\mathbf{W}^{mp}$ has a hyperbolic shape whereas that of the H-wave $\mathbf{W}^{mnp}$ is a straight line. The latter curve is tangential to the former at the critical distance as shown in figure 3.5.

A given R-wave $\mathbf{W}^{mp}$ can have more than one critical ray associated with different kinds of $\mathbf{W}^{mnp}$. The maximum number is two if $\mathbf{W}^{mp}$ is a compressional wave and three if it is a shear wave.

H-waves can be classified into four groups according to the type of $\mathbf{W}^{mn}$ and $\mathbf{W}^{mp}$ (see figure 3.6):

(1) *Head waves of first kind*: $\mathbf{W}^{mn}$ is a refracted wave and $\mathbf{W}^{mp}$ is a reflected wave.

(2) *Head waves of second kind*: Both $\mathbf{W}^{mn}$ and $\mathbf{W}^{mp}$ are refracted waves.

(3) *Head waves of third kind*: $\mathbf{W}^{mn}$ is a reflected wave and $\mathbf{W}^{mp}$ is a refracted wave.

(4) *Head waves of fourth kind*: Both $\mathbf{W}^{mn}$ and $\mathbf{W}^{mp}$ are reflected waves.

This basis for classification was first used by Petrashen (1959) although his terminology is different from that adopted here. Pod''yapol'skiy's (1966a) classification is based on the characteristics of $\mathbf{W}^{mn}$ and $\mathbf{W}^{np}$ (rather than those of $\mathbf{W}^{mn}$ and $\mathbf{W}^{mp}$). Thus he calls the above four groups

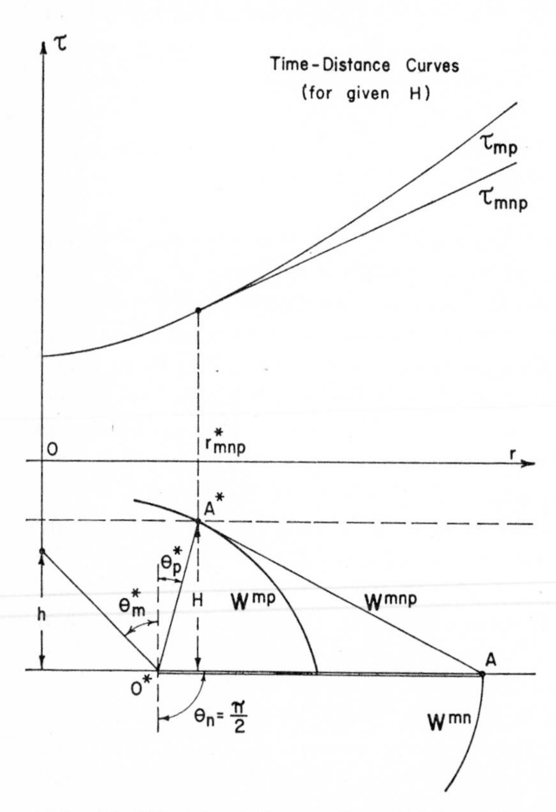

FIGURE 3.5 Wave front diagram (lower half) and time-distance curves (upper half) for the reflected wave $\mathbf{W}^{mp}$ and head wave $\mathbf{W}^{mnp}$.

twice-refracted, refracted-reflected, reflected-refracted, and twice-reflected, respectively.

The twenty-six possible H-waves fall in the following groups (see figure 3.4):

Head waves of first kind: 131, 132, 141, 142, 231, 232, 241, 242, 313, 314, 323, 324, 413, 414, 423, 424 (16 waves).

Head waves of second kind: 134, 234, 312, 412 (4 waves).

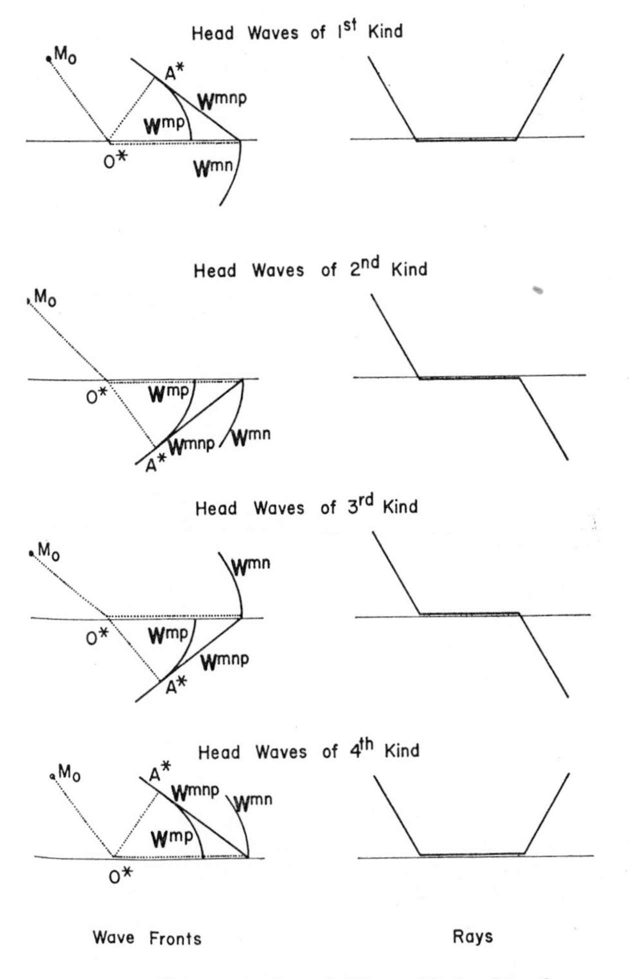

FIGURE 3.6 Characterization of different kinds of head waves ($\mathbf{W}^{mnp}$) based on the types of two R-waves, $\mathbf{W}^{mp}$ and $\mathbf{W}^{mn}$.

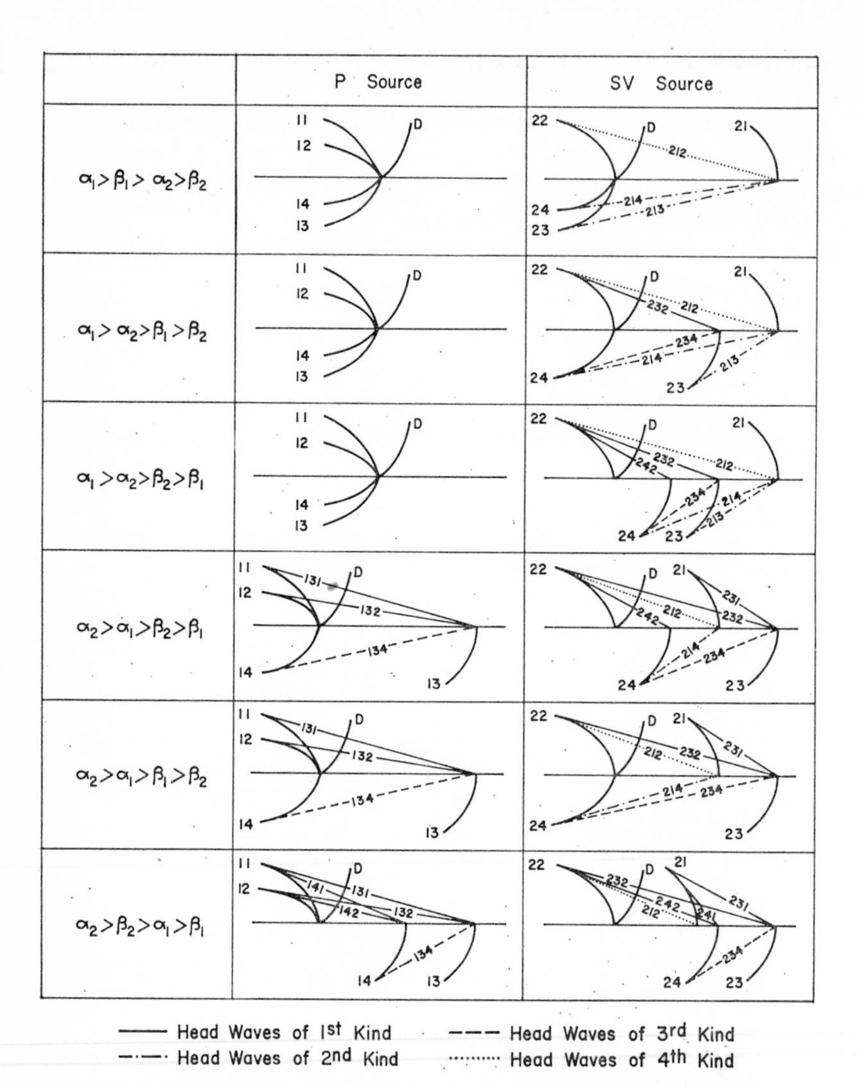

FIGURE 3.7 Various kinds of H-waves and R-waves for all the possible velocity distributions for a single interface when the source lies in the first medium. Each type of head wave is characterized by three numbers, *mnp*, and the R-wave by two numbers, *mn*. Also the direct wave (D) is presented. The wave fronts of R and D waves are represented by thick solid lines.

Head waves of third kind: 213, 214, 431, 432 (4 waves).
Head waves of fourth kind: 212, 434 (2 waves).

The most important H-waves in seismic prospecting and in seismic investigations of the earth's crust are those of the first kind, which have, therefore, been called *basic head waves*. The number of H-waves generated by an interface depends on the type of source and on the distribution of the wave velocities. Six types of velocity distributions are possible:

(1) $\alpha_1 > \beta_1 > \alpha_2 > \beta_2$, (4) $\alpha_2 > \alpha_1 > \beta_2 > \beta_1$,

(2) $\alpha_1 > \alpha_2 > \beta_1 > \beta_2$, (5) $\alpha_2 > \alpha_1 > \beta_1 > \beta_2$,

(3) $\alpha_1 > \alpha_2 > \beta_2 > \beta_1$, (6) $\alpha_2 > \beta_2 > \alpha_1 > \beta_1$.

Figure 3.7 shows the wave fronts of all the possible H-waves for the above six situations, separately for a P-source and an S-source lying in the first medium ($m = 1, 2$). For a P or an SV source lying in the second medium ($m = 3$ or 4), we can easily obtain a similar picture. The wave fronts of the R-waves as well as of the direct wave (D) are also given in the figure, which is an adaptation of a figure given by Petrashen (1959).

3.5 HEAD WAVE COEFFICIENTS

3.5.1 *Formulae for Head Wave Coefficients*

The H-wave coefficients, Γ_{mnp}, are generally given by (3.50), i.e.

$$\Gamma_{mnp} = R^*_{mn} M_{np}, \tag{3.98}$$

where R^*_{mn} is the solution of the set of equations (2.73) (for $\sin\theta_i = \sin\theta_i^* = V_i/V_n$) and M_{np} is the solution of the set of equations (3.39).

It is very convenient to use these two sets of equations to determine Γ_{mnp}. However, we often need to know explicit formulae for Γ_{mnp}. These formulae, for different types of interface, are given below.

A Solid–Solid Interface

For the twenty-six possible Γ_{mnp}, we obtain

$$\begin{aligned}
\Gamma_{131} &= 2\alpha_1\alpha_2\rho_1\rho_2\{P_1D^{-2}(\beta_2P_2X+\beta_1P_4Y)^2\}_{\Theta=1/\alpha_2},\\
\Gamma_{132} &= 2\alpha_1\rho_1\rho_2\{P_1D^{-2}(\beta_2P_2X+\beta_1P_4Y)(qP_1P_4-\alpha_1\beta_2Z)\}_{\Theta=1/\alpha_2},\\
\Gamma_{134} &= 2\alpha_1\rho_1\{P_1D^{-2}(\beta_2P_2X+\beta_1P_4Y)(qP_1P_2X+\beta_1\alpha_1YZ)\}_{\Theta=1/\alpha_2},\\
\Gamma_{141} &= -2\alpha_1\rho_1\rho_2\{P_1\Theta D^{-2}(qP_2P_3-\alpha_2\beta_1Z)^2\}_{\Theta=1/\beta_2},
\end{aligned} \tag{3.99}$$

$$\Gamma_{142} = 2\alpha_1\rho_1\rho_2\{P_1D^{-2}(qP_2P_3-\alpha_2\beta_1Z)(\alpha_2P_1X+\alpha_1P_3Y)\}_{\Theta=1/\beta_2},$$
$$\Gamma_{231} = 2\beta_1\rho_1\rho_2\{P_2D^{-2}(qP_1P_4-\alpha_1\beta_2Z)(\beta_1P_4Y+\beta_2P_2X)\}_{\Theta=1/\alpha_2},$$
$$\Gamma_{232} = 2\beta_1\rho_1\rho_2\{P_2\Theta D^{-2}(qP_1P_4-\alpha_1\beta_2Z)^2\}_{\Theta=1/\alpha_2},$$
$$\Gamma_{234} = 2\beta_1\rho_1\{\Theta P_2D^{-2}(qP_1P_4-\alpha_1\beta_2Z)(qP_1P_2X+\alpha_1\beta_1YZ)\}_{\Theta=1/\alpha_2},$$
$$\Gamma_{241} = 2\beta_1\rho_1\rho_2\{P_2D^{-2}(\alpha_1P_3Y+\alpha_2P_1X)(qP_2P_3-\beta_1\alpha_2Z)\}_{\Theta=1/\beta_2},$$
$$\Gamma_{242} = -2\beta_1\beta_2\rho_1\rho_2\{P_2D^{-2}(\alpha_1P_3Y+\alpha_2P_1X)^2\}_{\Theta=1/\beta_2},$$
$$\Gamma_{212} = 2\beta_1\{\Theta P_2D^{-2}(qP_3P_4Y+\alpha_2\beta_2XZ)^2\}_{\Theta=1/\alpha_1},$$
$$\Gamma_{213} = -2\beta_1\rho_1\{P_2D^{-2}(qP_3P_4Y+\alpha_2\beta_2XZ)(\beta_2P_2X+\beta_1P_4Y)\}_{\Theta=1/\alpha_1},$$
$$\Gamma_{214} = 2\beta_1\rho_1\{\Theta P_2D^{-2}(qP_3P_4Y+\alpha_2\beta_2XZ)(qP_2P_3-\alpha_2\beta_1Z)\}_{\Theta=1/\alpha_1},$$
$$\Gamma_{313} = 2\alpha_1\alpha_2\rho_1\rho_2\{P_3D^{-2}(\beta_1P_4Y+\beta_2P_2X)^2\}_{\Theta=1/\alpha_1},$$
$$\Gamma_{314} = -2\alpha_2\rho_1\rho_2\{P_3D^{-2}(\beta_1P_4Y+\beta_2P_2X)(qP_2P_3-\alpha_2\beta_1Z)\}_{\Theta=1/\alpha_1},$$
$$\Gamma_{312} = -2\alpha_2\rho_2\{P_3D^{-2}(\beta_1P_4Y+\beta_2P_2X)(qP_3P_4Y+\alpha_2\beta_2XZ)\}_{\Theta=1/\alpha_1},$$
$$\Gamma_{323} = -2\alpha_2\rho_1\rho_2\{\Theta P_3D^{-2}(qP_1P_4-\alpha_1\beta_2Z)^2\}_{\Theta=1/\beta_1},$$
$$\Gamma_{324} = -2\alpha_2\rho_1\rho_2\{P_3D^{-2}(qP_4P_1-\alpha_1\beta_2Z)(\alpha_1P_3Y+\alpha_2P_1X)\}_{\Theta=1/\beta_1},$$
$$\Gamma_{413} = -2\beta_2\rho_1\rho_2\{P_4D^{-2}(qP_2P_3-\alpha_2\beta_1Z)(\beta_2P_2X+\beta_1P_4Y)\}_{\Theta=1/\alpha_1},$$
$$\Gamma_{414} = 2\beta_2\rho_1\rho_2\{\Theta P_4D^{-2}(qP_2P_3-\alpha_2\beta_1Z)^2\}_{\Theta=1/\alpha_1},$$
$$\Gamma_{412} = 2\beta_2\rho_2\{\Theta P_4D^{-2}(qP_2P_3-\alpha_2\beta_1Z)(qP_3P_4Y+\beta_2\alpha_2XZ)\}_{\Theta=1/\alpha_1},$$
$$\Gamma_{423} = -2\beta_2\rho_1\rho_2\{P_4D^{-2}(\alpha_2P_1X+\alpha_1P_3Y)(qP_1P_4-\alpha_1\beta_2Z)\}_{\Theta=1/\beta_1},$$
$$\Gamma_{424} = -2\beta_1\beta_2\rho_1\rho_2\{P_4D^{-2}(\alpha_2P_1X+\alpha_1P_3Y)^2\}_{\Theta=1/\beta_1},$$
$$\Gamma_{434} = 2\beta_2\{\Theta P_4D^{-2}(qP_1P_2X+\alpha_1\beta_1YZ)^2\}_{\Theta=1/\alpha_2},$$
$$\Gamma_{431} = 2\beta_2\rho_2\{P_4D^{-2}(qP_1P_2X+\alpha_1\beta_1YZ)(\beta_1P_4Y+\beta_2P_2X)\}_{\Theta=1/\alpha_2},$$
$$\Gamma_{432} = 2\beta_2\rho_2\{\Theta P_4D^{-2}(qP_1P_2X+\alpha_1\beta_1YZ)(qP_1P_4-\alpha_1\beta_2Z)\}_{\Theta=1/\alpha_2},$$

(3.99) cont.

where

$$q = 2(\rho_2\beta_2{}^2-\rho_1\beta_1{}^2),$$
$$P_1 = (1-\alpha_1{}^2\Theta^2)^{1/2}, \qquad P_2 = (1-\beta_1{}^2\Theta^2)^{1/2},$$
$$P_3 = (1-\alpha_2{}^2\Theta^2)^{1/2}, \qquad P_4 = (1-\beta_2{}^2\Theta^2)^{1/2},$$
$$X = \rho_2-q\Theta^2, \qquad Y = \rho_1+q\Theta^2, \qquad Z = \rho_2-\rho_1-q\Theta^2, \tag{3.100}$$
$$D = \alpha_1\alpha_2\beta_1\beta_2\Theta^2Z^2+\alpha_2\beta_2P_1P_2X^2+\rho_1\rho_2(\alpha_2\beta_1P_1P_4+\alpha_1\beta_2P_2P_3) +\alpha_1\beta_1P_3P_4Y^2+q^2\Theta^2P_1P_2P_3P_4.$$

If P_j is imaginary, we must take the negative value of the radical, i.e.,

$$P_j = -i[(V_j\Theta)^2-1]^{1/2} \qquad \text{for } V_j\Theta > 1. \tag{3.101}$$

Note that the H-coefficients Γ_{131}, Γ_{132}, Γ_{134}, Γ_{231}, Γ_{213}, Γ_{313}, Γ_{314}, Γ_{312}, Γ_{413}, Γ_{431} are always real, and the coefficients Γ_{141}, Γ_{142}, Γ_{241}, and Γ_{242} are always complex because P_3 is imaginary. Similarly, Γ_{323}, Γ_{324}, Γ_{423}, and Γ_{424} are always complex because P_1 is imaginary. The remaining Γ_{mnp} may be real or complex, depending on the distribution of the wave velocities. The coefficients Γ_{232}, Γ_{234}, Γ_{434}, and Γ_{432} are real if $\alpha_2 > \alpha_1$ (if $\alpha_2 < \alpha_1$, they are complex, since P_1 is then imaginary), and the coefficients Γ_{214}, Γ_{414}, Γ_{212}, and Γ_{412} are real if $\alpha_1 > \alpha_2$ (if $\alpha_1 < \alpha_2$, they are complex, since P_3 is then imaginary).

B Liquid–Solid Interface

If the first medium is liquid and the other solid, nine H-waves can, in general, be generated on the interface: 131, 134, 141, 313, 314, 413, 414, 434, and 431. The corresponding H-coefficients can be calculated again from the general sets of equations if we substitute $\gamma_1 = 0$. The explicit formulae for a liquid–solid interface are as follows:

$$\begin{aligned}
\Gamma_{131} &= 2\alpha_1\rho_1/\{\alpha_2\rho_2[1-(\alpha_1/\alpha_2)^2]^{1/2}(1-2\beta_2{}^2/\alpha_2{}^2)^2\},\\
\Gamma_{134} &= 4\alpha_1\beta_2\rho_1/[\alpha_2{}^2\rho_2(1-2\beta_2{}^2/\alpha_2{}^2)^2],\\
\Gamma_{141} &= 8\alpha_1\beta_2\rho_1\rho_2[1-(\alpha_1/\beta_2)^2]^{1/2}(\alpha_2{}^2/\beta_2{}^2-1)/\{\alpha_2\rho_2[1-(\alpha_1/\beta_2)^2]^{1/2}\\
&\qquad -i\rho_1\alpha_1(\alpha_2{}^2/\beta_2{}^2-1)^{1/2}\}^2,\\
\Gamma_{313} &= 2\alpha_2\rho_2(1-2\beta_2{}^2/\alpha_1{}^2)^2/\{\rho_1\alpha_1[1-(\alpha_2/\alpha_1)^2]^{1/2}\}\\
\Gamma_{314} &= -4\alpha_2\beta_2\rho_2(1-2\beta_2{}^2/\alpha_1{}^2)/\rho_1\alpha_1{}^2,\\
\Gamma_{413} &= -4\beta_2{}^2\rho_2[1-(\beta_2/\alpha_1)^2]^{1/2}(1-2\beta_2{}^2/\alpha_1{}^2)/\{\rho_1\alpha_1{}^2[1-(\alpha_2/\alpha_1)^2]^{1/2}\},\\
\Gamma_{414} &= 8\beta_2{}^3\rho_2[1-(\beta_2/\alpha_1)^2]^{1/2}/\alpha_1{}^3\rho_1,\\
\Gamma_{434} &= 8\beta_2{}^3[1-(\beta_2/\alpha_2)^2]^{1/2}/[\alpha_2{}^3(1-2\beta_2{}^2/\alpha_2{}^2)^2],\\
\Gamma_{431} &= 4\beta_2{}^2[1-(\beta_2/\alpha_2)^2]^{1/2}/\{\alpha_2{}^2[1-(\alpha_1/\alpha_2)^2]^{1/2}(1-2\beta_2{}^2/\alpha_2{}^2)^2\}.
\end{aligned} \tag{3.102}$$

All Γ_{mnp} are real except for Γ_{141}, which is always complex.

C Solid–Liquid Interface

If the first medium is solid and the other liquid, nine H-waves can arise: 131, 132, 231, 232, 212, 213, 313, 312, 323. The corresponding H-wave coefficients can be calculated from the general sets of equations (substituting $\gamma_2 = 0$). The explicit formulae for this case are the following:

$$
\begin{aligned}
\Gamma_{131} &= 2\alpha_1\rho_1(1-2\beta_1{}^2/\alpha_2{}^2)^2/\{\rho_2\alpha_2[1-(\alpha_1/\alpha_2)^2]^{1/2}\},\\
\Gamma_{132} &= -4\alpha_1\beta_1\rho_1(1-2\beta_1{}^2/\alpha_2{}^2)/\rho_2\alpha_2{}^2,\\
\Gamma_{231} &= -4\rho_1\beta_1{}^2[1-(\beta_1/\alpha_2)^2]^{1/2}(1-2\beta_1{}^2/\alpha_2{}^2)/\{\rho_2\alpha_2{}^2[1-(\alpha_1/\alpha_2)^2]^{1/2}\},\\
\Gamma_{232} &= 8\beta_1{}^3\rho_1[1-(\beta_1/\alpha_2)^2]^{1/2}/\rho_2\alpha_2{}^3,\\
\Gamma_{212} &= 8\beta_1{}^3[1-(\beta_1/\alpha_1)^2]^{1/2}/[\alpha_1{}^3(1-2\beta_1{}^2/\alpha_1{}^2)^2],\\
\Gamma_{213} &= 4\beta_1{}^2[1-(\beta_1/\alpha_1)^2]^{1/2}/\{\alpha_1{}^2[1-(\alpha_2/\alpha_1)^2]^{1/2}(1-2\beta_1{}^2/\alpha_1{}^2)^2\},\\
\Gamma_{313} &= 2\alpha_2\rho_2/\{\alpha_1\rho_1[1-(\alpha_2/\alpha_1)^2]^{1/2}(1-2\beta_1{}^2/\alpha_1{}^2)^2\},\\
\Gamma_{312} &= 4\beta_1\alpha_2\rho_2/[\alpha_1{}^2\rho_1(1-2\beta_1{}^2/\alpha_1{}^2)^2],\\
\Gamma_{323} &= -8\rho_1\rho_2\beta_1\alpha_2[1-(\alpha_2/\beta_1)^2]^{1/2}[1-(\alpha_1/\beta_1)^2]/\{\alpha_1\rho_1[1-(\alpha_2/\beta_1)^2]^{1/2}\\
&\qquad -i\rho_2\alpha_2[(\alpha_1/\beta_1)^2-1]^{1/2}\}.
\end{aligned}
\tag{3.103}
$$

All Γ_{mnp} are real except for Γ_{323}, which is always complex.

D Liquid–Liquid Interface

In this case only the H-wave $\mathbf{W}^{131}$ can exist when $\alpha_1 < \alpha_2$, and the H-wave $\mathbf{W}^{313}$ when $\alpha_2 < \alpha_1$. The corresponding H-coefficients cannot be calculated from the general sets of equations, which become unstable for $\gamma_1 = \gamma_2 = 0$, but must be determined from the liquid–liquid interface conditions directly. The formulae are

$$
\Gamma_{131} = \frac{2\rho_1\alpha_1}{\rho_2\alpha_2[1-(\alpha_1/\alpha_2)^2]^{1/2}}, \qquad \Gamma_{313} = \frac{2\rho_2\alpha_2}{\rho_1\alpha_1[1-(\alpha_2/\alpha_1)^2]^{1/2}}. \tag{3.104}
$$

Both coefficients are real and positive.

E Free Interface

Assume that the first medium is a vacuum. In this case only one H-wave can be generated on the interface, viz., $\mathbf{W}^{434}$. The formula for the corresponding coefficient Γ_{434} may be obtained from the general sets of equations if we substitute $\rho_1 = 0$, $\alpha_1 = 0$ (and any arbitrary value for γ_1). The explicit formula is

$$
\Gamma_{434} = 8\beta_2{}^3[1-(\beta_2/\alpha_2)^2]^{1/2}/\alpha_2{}^3(1-2\beta_2{}^2/\alpha_2{}^2)^2. \tag{3.105}
$$

Similarly, if the second medium is a vacuum, the H-wave $\mathbf{W}^{212}$ can exist. The H-coefficient Γ_{212} is given by

$$
\Gamma_{212} = 8\beta_1{}^3[1-(\beta_1/\alpha_1)^2]^{1/2}/\alpha_1{}^3(1-2\beta_1{}^2/\alpha_1{}^2)^2. \tag{3.105$'$}
$$

Both of the coefficients are real and positive.

F SH Head Wave Coefficients

For SH-waves, only one of the two possible H-waves $\mathbf{W}^{242}$ and $\mathbf{W}^{424}$ can exist ($\mathbf{W}^{242}$ if $\beta_1 < \beta_2$ and $\mathbf{W}^{424}$ if $\beta_1 > \beta_2$). The formulae for the corresponding H-coefficients are

$$\Gamma_{242}^{SH} = \frac{2\rho_2\beta_2}{\rho_1\beta_1[1-(\beta_1/\beta_2)^2]^{1/2}}, \qquad \Gamma_{424}^{SH} = \frac{2\rho_1\beta_1}{\rho_2\beta_2[1-(\beta_2/\beta_1)^2]^{1/2}}. \quad (3.106)$$

Both coefficients are real and positive.

3.5.2 *Relationship between Head Wave Coefficients and Coefficients of Reflection and Refraction*

We shall give here three formulae connecting H-coefficients and R-coefficients.

(1) Inserting (3.45) into (3.98) we obtain the first formula:

$$\Gamma_{mnp} = \tfrac{1}{2}(-1)^{n+1}(R_{mn}\,R_{np}/P_n)_{\Theta=1/V_n}. \quad (3.107)$$

A similar expression for the H-coefficient was given earlier by Pod''yapol'-skiy (1966a). We can easily see that this formula is applicable to all the cases we have considered. Note that $P_n = 0$ for $\Theta = 1/V_n$, but that this creates no problem since the expression for R_{np} contains a factor P_n which cancels the P_n in the denominator.

(2) The second formula connecting Γ_{mnp} with the R-coefficients, R_{mp}, is

$$\Gamma_{mnp} = -(\mathrm{d}R_{mp}/\mathrm{d}P_n)_{\Theta=1/V_n}. \quad (3.108)$$

This expression does not follow directly from our presentation, but for individual cases we can see that it is correct. It may be obtained directly if we use the Sommerfeld integral method for the derivation of H-waves. It also follows directly from the general equations for R-coefficients given by Pod''yapol'skiy (1959b).

(3) A third formula for Γ_{mnp} is also used often. If we write R_{mp} in the form

$$R_{mp} = \frac{R_{mp}^{(2)} + \bar{R}_{mp}^{(2)}P_n}{R_{mp}^{(1)} + \bar{R}_{mp}^{(1)}P_n}, \quad (3.109)$$

where the functions $R_{mp}^{(1)}$, $\bar{R}_{mp}^{(1)}$, $R_{mp}^{(2)}$, and $\bar{R}_{mp}^{(2)}$ do not contain the radical P_n, we obtain from (3.108)

$$\Gamma_{mnp} = [(\bar{R}_{mp}^{(1)}R_{mp}^{(2)} - R_{mp}^{(1)}\bar{R}_{mp}^{(2)})/(R_{mp}^{(1)})^2]_{\Theta=1/V_n}. \quad (3.110)$$

This expression is most commonly used in papers dealing with H-waves by the integral methods (Muskat, 1933; Heelan, 1953; Červený, 1957a; and others).

3.5.3 *Formulae Connecting Different Head Wave Coefficients*

We can see from (3.99) that the H-coefficients Γ_{mnp} and Γ_{pnm} are connected by the formula

$$\rho_{\varepsilon_p} V_p[1-(V_p/V_n)^2]^{1/2}\Gamma_{mnp} = \rho_{\varepsilon_m} V_m[1-(V_m/V_n)^2]^{1/2}\Gamma_{pnm}. \tag{3.111}$$

A similar expression was given by Yanovskaya (1966). In this equation $\varepsilon_i = 1$ for $i = 1, 2$ and $\varepsilon_i = 2$ for $i = 3, 4$. The formula follows directly from (3.107) and from the properties of the R-coefficients given in section 2.4.2, also.

The other formula connecting H-coefficients is quite clear from the physical properties of head waves. We see from figure 3.4 that the twenty-six possible head waves can be divided into two groups: the first group (of thirteen waves) corresponds to the source in the first medium, and the second group corresponds to the source in the second medium. If we interchange the parameters of the media (i.e. take the first medium as the second and the second as the first), the expressions for the H-coefficients of the two groups must interchange too. We shall define $\bar{m}$ as follows: $\bar{m} = 3, 4, 1$ or 2 if $m = 1, 2, 3$ or 4, respectively. Similarly we define $\bar{n}$ and $\bar{p}$. Then we can write

$$\Gamma_{mnp}(\alpha_1, \beta_1, \rho_1, \alpha_2, \beta_2, \rho_2) = \Gamma_{\bar{m}\bar{n}\bar{p}}(\alpha_2, \beta_2, \rho_2, \alpha_1, \beta_1, \rho_1). \tag{3.111$'$}$$

This expression also follows directly from (3.107) and the properties of R-coefficients given in section 2.4.2.

Some authors have used *potential* H-coefficients instead of the *displacement* H-coefficients used in our work. The potential H-coefficients $\tilde{\Gamma}_{mnp}$ are usually defined as follows:

$$\tilde{\Gamma}_{mnp} = (V_p/V_m)\Gamma_{mnp}. \tag{3.112}$$

They are used very often in Russian papers, and an extensive tabulation of them can be found in Petrashen (1957b).

Note that comparisons of the formulae for head waves with those given by other authors must be made with caution since the head wave coefficients very often differ by a multiplication factor. For example, the coefficients A_{10}^* introduced by Zvolinskiy (1958) and X by Heelan (1953) are connected with the coefficient Γ_{131} defined here by the relation

$$\Gamma_{131} = (\alpha_1/\alpha_2)A_{10}^* = [1-(\alpha_1/\alpha_2)^2]^{1/2}X.$$

Note that the definition of the H-coefficients given in this book corresponds completely with that given by Pod"yapol'skiy (1966a).

3.5.4 *Numerical Values of Head Wave Coefficients*

The head wave coefficients depend on the elastic parameters, i.e. α_1, β_1, ρ_1, α_2, β_2, ρ_2. However, it can be easily seen from (3.99) that the dependence is only on four independent ratios of these parameters. These ratios are ρ_1/ρ_2, α_1/α_2, γ_1 $(= \beta_1/\alpha_1)$, γ_2 $(= \beta_2/\alpha_2)$ or ρ_1/ρ_2, β_1/β_2, γ_1, γ_2, or ρ_1/ρ_2, α_1/α_2, α_1/β_2, γ_1. It would be very simple to write the formulae for H-coefficients in terms of these four parameters, but the formulae (3.99) are more convenient when calculating head waves propagating in layered media (where each layer is characterized by three elastic parameters, viz., α_i, β_i, ρ_i). We can also use, as a standard subroutine for the calculation of H-coefficients, the formula (3.50) (and the sets of equations (2.73) and (3.39)), which is considerably shorter than the program based on formulae (3.99).

Numerical values of the H-coefficients for basic head waves of the type 131, 132, 141, 142, 231, 232, 241, and 242 were calculated by Smirnova for large variations in the parameters involved (see Petrashen, 1957b, vol. 2). A few curves, based on her tables, are given in Malinovskaya (1957a), Keilis-Borok (1960), and O'Brien (1967). Smirnova, in her tables, gives values of potential H-coefficients $\tilde{\Gamma}$, and in order to use her data in our formulae for H-waves (based on displacement H-coefficients), we must use (3.112). (Note that what we have called H-coefficients have also been called the *coefficients of the formation of head waves* by Smirnova, Pod"yapol'skiy, and many other Russian workers; the *coefficients of head waves* by Zvolinskiy (1957); and the *amplitude coefficients of head waves* by O'Brien (1967).) Except for the references mentioned above, numerical values of H-coefficients appear only rarely in books and papers devoted to head waves.

Numerical values of various head wave coefficients are given in figures 3.8 to 3.22. Those for a solid–solid interface are given in figures 3.8 to 3.17, for a liquid–solid interface in figures 3.18 and 3.19, for a liquid–liquid interface in figure 3.20, for a free interface in figure 3.21, and for SH-waves in figure 3.22.

A Solid–Solid Interface (figures 3.8 to 3.17)

As was shown earlier, the H-coefficients depend on four parameters, viz., α_1/α_2, γ_1 $(= \beta_1/\alpha_1)$, γ_2 $(= \beta_2/\alpha_2)$, and ρ_1/ρ_2. The most significant variation is with the parameter α_1/α_2, which is the one we have considered in the diagrams, where its value has been taken to change from 0 to 1. Note that the dependence of the coefficients is often rather complicated when α_1/α_2 is very close to one $(0.97 < \alpha_1/\alpha_2 < 1)$. In some of the following figures, therefore, values of the H-coefficients are not presented for α_1/α_2 approach-

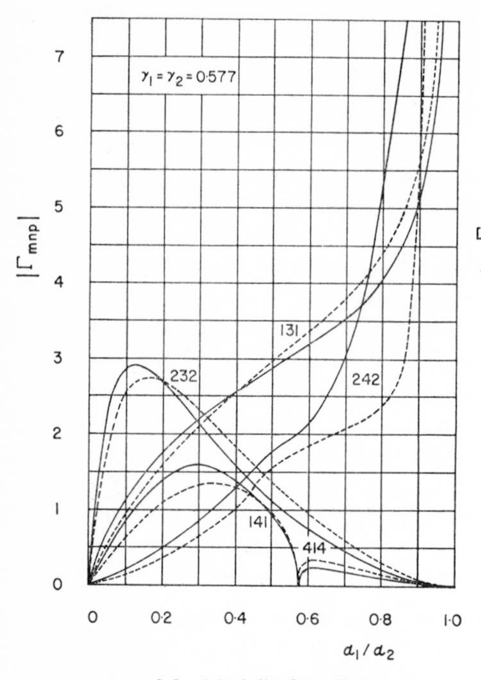

FIGURE 3.8 Moduli of coefficients of head waves of the first kind versus α_1/α_2 when $m = p$. The solid lines are for $\rho_1/\rho_2 = 1.0$ and the dashed for $\rho_1/\rho_2 = 0.7$. $\gamma_1 = \gamma_2 = 0.577$.

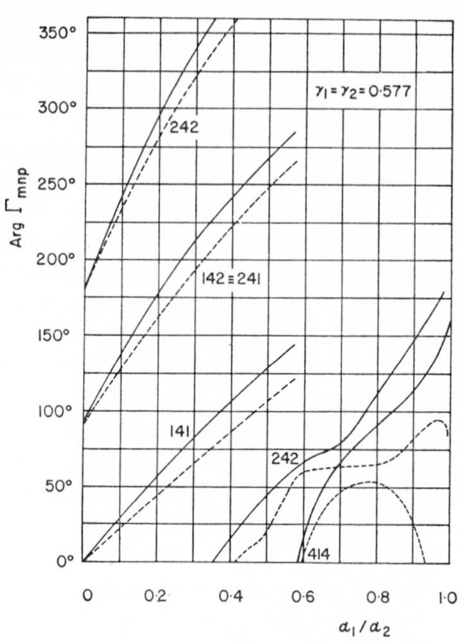

FIGURE 3.10 Arguments of coefficients of head waves of the first kind vs. α_1/α_2. The solid lines are for $\rho_1/\rho_2 = 1.0$ and the dashed for $\rho_1/\rho_2 = 0.7$. $\gamma_1 = \gamma_2 = 0.577$.

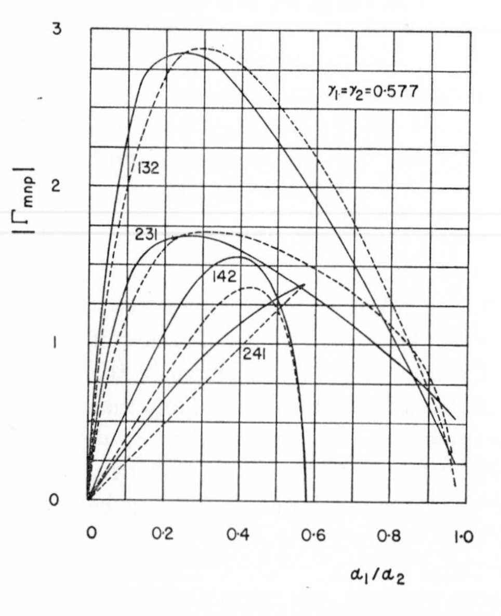

FIGURE 3.9 Moduli of coefficients of head waves of the first kind versus α_1/α_2 when $m \neq p$. The solid lines are for $\rho_1/\rho_2 = 1.0$ and the dashed for $\rho_1/\rho_2 = 0.7$. $\gamma_1 = \gamma_2 = 0.577$.

ing unity, and the general conclusions of this section do not take into account the case of α_1/α_2 very close to one. The dependence of H-coefficients on ρ_1/ρ_2 is usually not large; therefore, only two values of ρ_1/ρ_2 are considered, 0.7 and 1.0. The dashed curves are for $\rho_1/\rho_2 = 0.7$ and the solid for $\rho_1/\rho_2 = 1.0$. The values of γ_1 and γ_2 vary from figure to figure.

(1) *Head waves of the first kind* (figures 3.8 to 3.14) Head waves of the first kind (basic head waves) are the most important head waves in seismic prospecting and in seismic investigations of the earth's crust. Therefore, numerical values of the corresponding H-coefficients are given here in much more detail than for the other types.

In figures 3.8 and 3.9, where $\gamma_1 = \gamma_2 = 1/\sqrt{3}$, the moduli of the H-coefficients Γ_{131}, Γ_{132}, Γ_{231}, Γ_{232}, Γ_{141}, Γ_{142}, Γ_{241}, Γ_{242}, and Γ_{414} are given. The head waves 141, 142, and 241 exist only when $\alpha_1/\beta_2 < 1$, i.e. when $\alpha_1/\alpha_2 < \gamma_2$ $(= 0.577)$. The head wave 414 exists when $\alpha_1/\beta_2 > 1$, i.e. for $\alpha_1/\alpha_2 > \gamma_2$ $(= 0.577)$. No other basic head waves exist for $\alpha_1/\alpha_2 < 1$. When $\alpha_1/\alpha_2 > 1$, the H-coefficients can be determined from the above figures using formula (3.111′).

We can see from these figures that $|\Gamma_{131}|$, $|\Gamma_{242}|$, and $|\Gamma_{241}|$ increase with increasing α_1/α_2, whereas $|\Gamma_{132}|$, $|\Gamma_{141}|$, $|\Gamma_{232}|$, $|\Gamma_{231}|$, and $|\Gamma_{142}|$ increase for small α_1/α_2, attain a maximum value, and then decrease. The maximum lies at very small α_1/α_2 ($\sim$0.1–0.2) for $|\Gamma_{232}|$ and at α_1/α_2 $\sim$0.2–0.5 for $|\Gamma_{231}|$, $|\Gamma_{142}|$, $|\Gamma_{141}|$, and $|\Gamma_{132}|$. The modulus of the head wave coefficient Γ_{414} (which exists only for $\alpha_1/\alpha_2 > \gamma_2$) increases very rapidly initially, has a maximum at about $\alpha_1/\alpha_2 \sim 0.6$–$0.65$, and decreases for larger α_1/α_2. The highest head wave coefficients for small velocity contrast $(0.6 < \alpha_1/\alpha_2 < 1.0)$ correspond to the waves 131 and 242, but for large velocity contrast $(0 < \alpha_1/\alpha_2 < 0.5)$ other H-coefficients are comparable with these or even larger (e.g., 232 and 132). The effect of variations in the value of ρ_1/ρ_2 on the moduli of these coefficients is usually small.

The H-coefficients Γ_{131}, Γ_{132}, Γ_{231}, and Γ_{232} are real (for $\gamma_1 = \gamma_2$ they are all positive), and the coefficients Γ_{141}, Γ_{142}, Γ_{241}, Γ_{242}, and Γ_{414} are complex. The arguments are given in figure 3.10.

It is not possible to appreciate the influence of γ_1 and γ_2 on the head wave coefficients from the above figures; therefore, values of some important head wave coefficients are shown in figures 3.11 to 3.14 for $\gamma_1 = 0.3, 0.4, 0.5$, and 0.6 and for $\gamma_2 = 0.3, 0.4, 0.5$, and 0.6. The numbers near the curves always denote the value of γ_2, and the solid line is again for $\rho_1/\rho_2 = 1$ and the dashed line for $\rho_1/\rho_2 = 0.7$.

In figure 3.11 values of Γ_{131} (which are always real and positive) are given. The influence of ρ_1/ρ_2 and γ_1 is seen to be not large. The influence

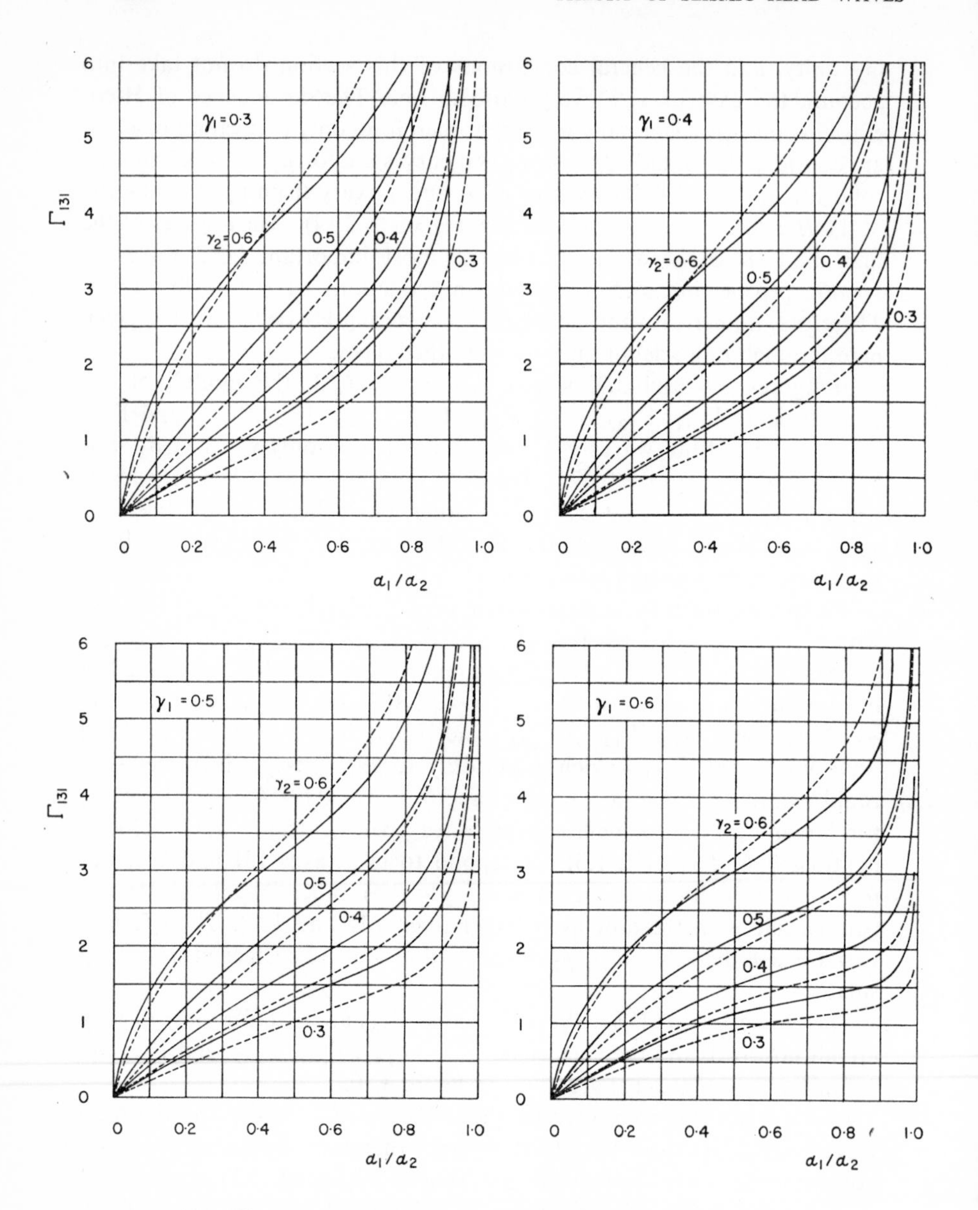

FIGURE 3.11 Γ_{131} as a function of α_1/α_2 for different values of γ_1 and γ_2. The solid lines are for $\rho_1/\rho_2 = 1.0$ and the dashed for $\rho_1/\rho_2 = 0.7$.

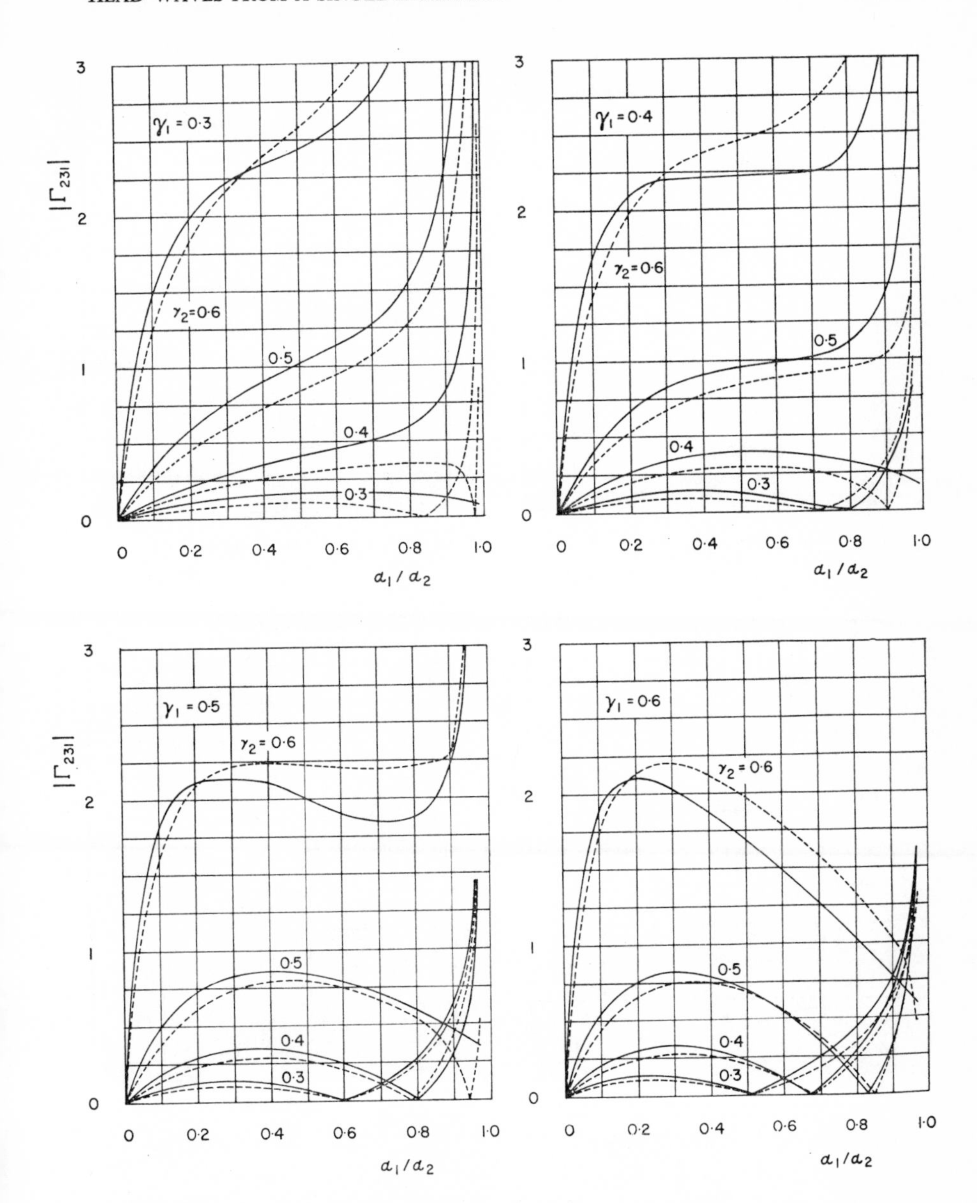

FIGURE 3.12 $|\Gamma_{231}|$ as a function of α_1/α_2 for different values of γ_1 and γ_2. The solid lines are for $\rho_1/\rho_2 = 1.0$ and the dashed for $\rho_1/\rho_2 = 0.7$.

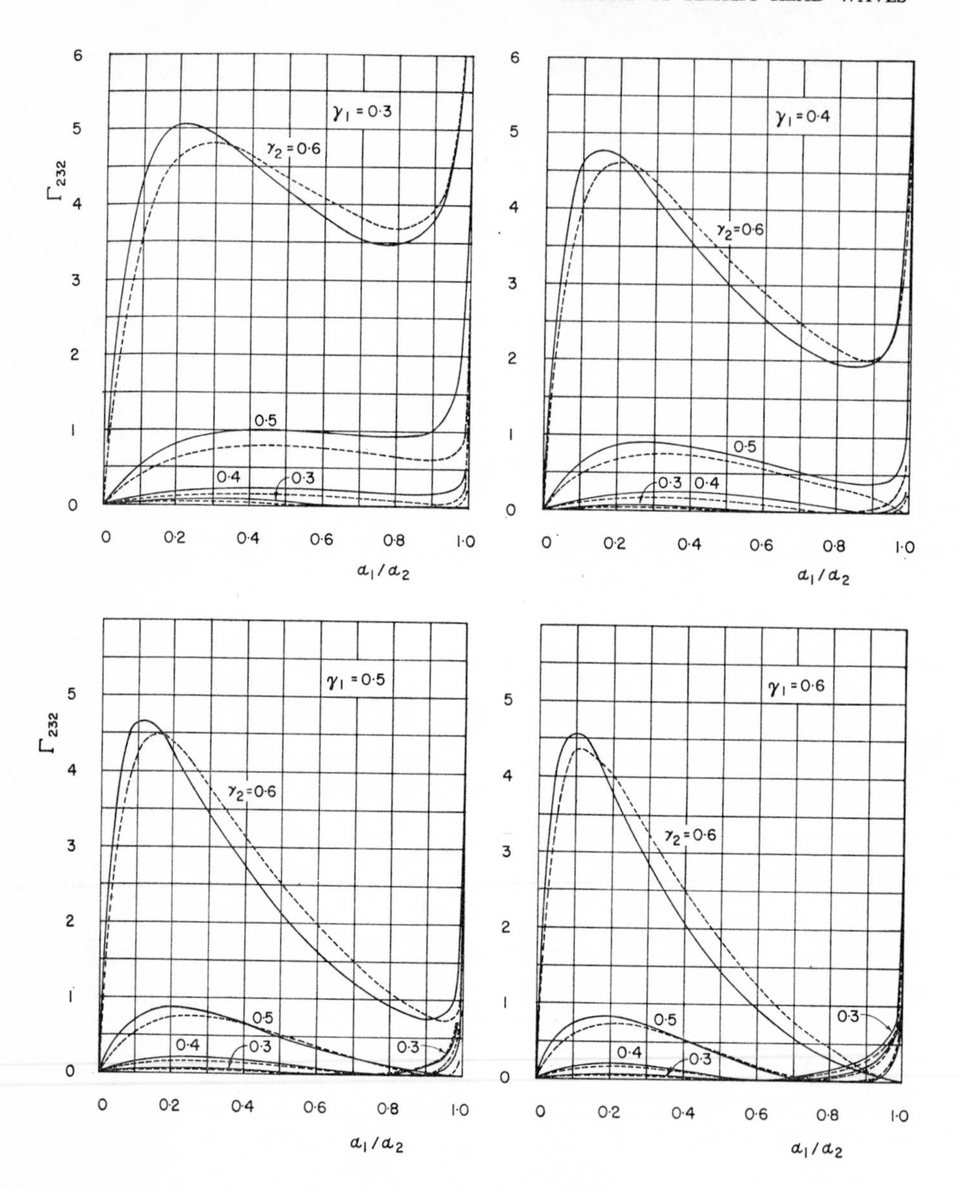

FIGURE 3.13 Γ_{232} as a function of α_1/α_2 for different values of γ_1 and γ_2. The solid lines are for $\rho_1/\rho_2 = 1.0$ and the dashed for $\rho_1/\rho_2 = 0.7$.

of γ_2, however, is more considerable; the higher the value of γ_2, the larger are the values of Γ_{131}. For any combination of γ_1 and γ_2, Γ_{131} is an increasing function of α_1/α_2. For $\alpha_1/\alpha_2 \to 1$, it increases as $[1-(\alpha_1/\alpha_2)^2]^{-1/2}$. Some exceptions may appear for α_1/α_2 very close to one when $\gamma_1 \neq \gamma_2$ and/or $\rho_1/\rho_2 \neq 1$. Then Γ_{131} may have a maximum for some α_1/α_2 very close to one. (It is not possible to see these maxima in figure 3.11 as they are very close to one, but their existence is clear from formula (3.99) for

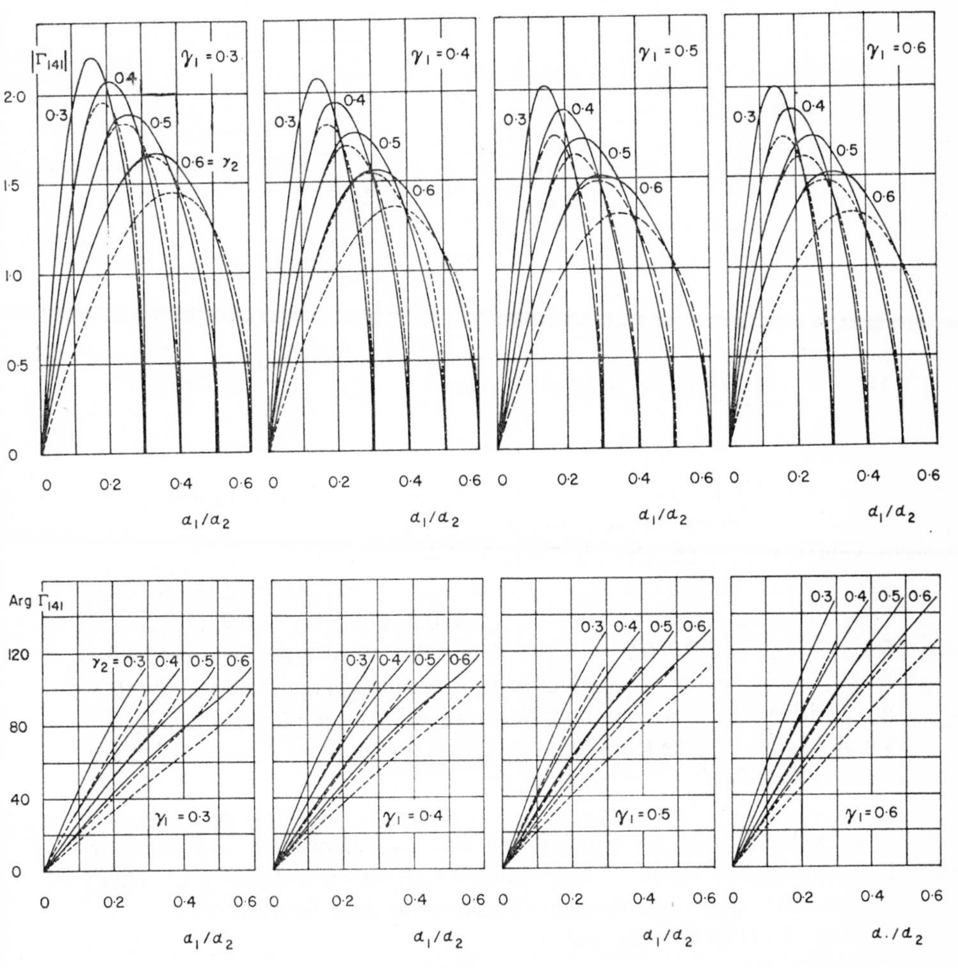

FIGURE 3.14 $|\Gamma_{141}|$ (top) and arg Γ_{141} (bottom) as a function of α_1/α_2 for different values of γ_1 and γ_2. The solid lines are for $\rho_1/\rho_2 = 1.0$ and the dashed for $\rho_1/\rho_2 = 0.7$.

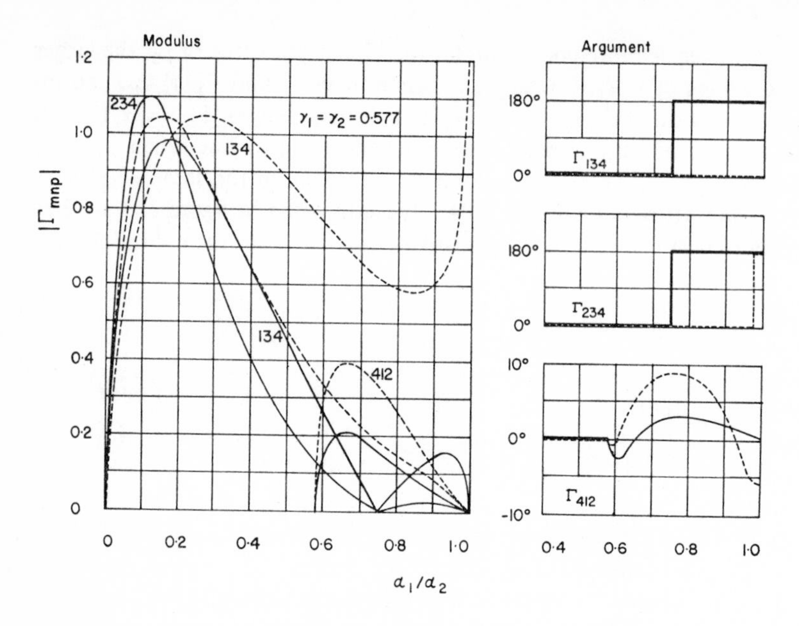

FIGURE 3.15 Coefficients of head waves of the second kind versus α_1/α_2. The solid lines are for $\rho_1/\rho_2 = 1.0$ and the dashed for $\rho_1/\rho_2 = 0.7$. $\gamma_1 = \gamma_2 = 0.577$.

Γ_{131}.) As this maximum lies very near to one, it has no practical importance (see more details in Červený, 1957b).

Values of Γ_{231} are given in figure 3.12 for the same values of γ_1 and γ_2 as in figure 3.11. We can again see that Γ_{231} is very sensitive to the value of γ_2. For example, for $\gamma_1 = 0.3$, $\rho_1/\rho_2 = 0.7$, and $\alpha_1/\alpha_2 = 0.4$, $\Gamma_{231} = 2.4$ for $\gamma_2 = 0.6$ and 0.1 for $\gamma_2 = 0.3$. The influence of other parameters (γ_1 and ρ_1/ρ_2) is smaller. The values of Γ_{231} are always real, and usually they are positive, although they may be negative for large α_1/α_2 (especially for small γ_2). We can see from figure 3.12 that for small γ_2, Γ_{231} vanishes for some value of α_1/α_2. For example, for $\gamma_1 = 0.5$ and $\rho_1/\rho_2 = 1$, $\Gamma_{231} = 0$ for $\alpha_1/\alpha_2 = 0.6$ when $\gamma_2 = 0.3$, and for $\alpha_1/\alpha_2 = 0.8$ when $\gamma_2 = 0.4$. Up to the value of α_1/α_2 for which it vanishes, Γ_{231} is positive, and then, for larger α_1/α_2, it is negative. This applies to all curves in figure 3.12. Thus, we can easily determine the sign of Γ_{231} from these figures.

From (3.111) it follows that

$$\Gamma_{132} = \frac{1}{\gamma_1}\left(\frac{1-(\alpha_1/\alpha_2)^2}{1-\gamma_1^2(\alpha_1/\alpha_2)^2}\right)^{1/2}\Gamma_{231}.$$

Using this formula, we can calculate the values of Γ_{132} from the values of Γ_{231} given in figure 3.12. As the multiplication factor in the above formula does not depend on γ_2, Γ_{132} is again very sensitive to the value of γ_2 and usually increases with increasing γ_2. Γ_{132} may be quite large when α_1/α_2 is small ($\alpha_1/\alpha_2 = 0.2$–0.5), especially when γ_1 is small and γ_2 large. It is always real; it is positive if Γ_{231} is positive, and negative if Γ_{231} is negative.

Numerical values of Γ_{232} are given in figure 3.13. The dependence on γ_2 is even greater than in previous cases. The values of Γ_{232} practically vanish for $\gamma_2 = 0.3$, but are rather large for $\gamma_2 = 0.6$. For small velocity contrast (large α_1/α_2), they depend considerably on γ_1 also. Γ_{232} is always real and positive. Under usual circumstances its values are considerably smaller than those of Γ_{131} for small velocity contrast ($0.6 < \alpha_1/\alpha_2 < 1$).

Head waves of the type 141, 142, 241, 242, and 414 are not as important as the above-discussed waves. (They propagate as S-waves along the interface, and therefore they always come in as later arrivals.) Therefore, we shall not present detailed figures for their H-coefficients, except for Γ_{141}. This wave is very closely related to the reflected compressional wave 11, and can therefore be of some importance. Numerical values of $|\Gamma_{141}|$ and arg Γ_{141} are given in figure 3.14. The numbers near the curves again denote the value of γ_2. The shape of $|\Gamma_{141}|$ is very regular: it initially grows with increasing α_1/α_2, attains a maximum at $\alpha_1/\alpha_2 \sim \frac{1}{2}\gamma_2$, and then decreases. The maximum values are between 1.3 and 2.3, and the values are higher for larger γ_2.

Note that numerical values of $|\Gamma_{142}|$ (which are not presented here) are very similar to those of $|\Gamma_{141}|$, except that the maximum is shifted to about $\alpha_1/\alpha_2 \sim \frac{2}{3}\gamma_2$. The maximum values are about the same as those of $|\Gamma_{141}|$.

(2) *Head waves of the second kind* (figure 3.15) The moduli and arguments of head wave coefficients Γ_{134}, Γ_{234}, and Γ_{412} for $\gamma_1 = \gamma_2 = 1/\sqrt{3}$ are given in figure 3.15. The head wave 412 exists only for $\alpha_1/\alpha_2 > \gamma_2\,(= 0.577)$. (Note that the wave 312 does not exist for $\alpha_1/\alpha_2 < 1$. For $\alpha_1/\alpha_2 > 1$, the numerical values of Γ_{312} and other existing H-coefficients may be found easily from figure 3.15 using (3.111').) The moduli of the H-coefficients under consideration are rather small. $|\Gamma_{234}|$ and $|\Gamma_{134}|$ have their maximum values for very small α_1/α_2, and for large α_1/α_2, they are very small. $|\Gamma_{134}|$ depends considerably on the density ratio ρ_1/ρ_2. The coefficients Γ_{134} and Γ_{234} are real. For small α_1/α_2, they are always positive, but for large α_1/α_2, they may be negative. The value of $|\arg \Gamma_{412}|$ is less than 10°, and we can approximately accept Γ_{412} to be real and positive.

(3) *Head waves of the third kind* (figure 3.16) The moduli of head wave coefficients Γ_{431}, Γ_{432}, and Γ_{214} for $\gamma_1 = \gamma_2 = 1/\sqrt{3}$ are given in figure

3.16. The head wave 214 exists only for $\alpha_1/\alpha_2 > \gamma_2 = 0.577$. The head wave of the third kind, 213, does not exist for $\alpha_1/\alpha_2 < 1$. For $\alpha_1/\alpha_2 > 1$, numerical values of Γ_{213} (and other existing head wave coefficients) may be found from figure 3.16 using (3.111′). Note that the coefficients given in figure 3.16 are connected with those in figure 3.15 by the relation (3.111). Therefore, the arguments of Γ_{431}, Γ_{432}, and Γ_{214} are the same as the arguments of Γ_{134}, Γ_{234}, and Γ_{412} (i.e. $\arg \Gamma_{431} = \arg \Gamma_{134}$, $\arg \Gamma_{432} = \arg \Gamma_{234}$, $\arg \Gamma_{412} = \arg \Gamma_{214}$; see figure 3.15). The moduli of all these coefficients again depend significantly on the density ratio ρ_1/ρ_2. They are usually rather small for large α_1/α_2, and may be large for very small α_1/α_2.

(4) *Head waves of the fourth kind* (figure 3.17) The moduli and arguments of head wave coefficients Γ_{212} and Γ_{434} for $\gamma_1 = \gamma_2 = 1/\sqrt{3}$ are given in figure 3.17. $|\Gamma_{212}|$ and $|\Gamma_{434}|$ are very small for large α_1/α_2, but for small α_1/α_2, they may be rather large, especially $|\Gamma_{212}|$. Note that both these head waves exist also for $\alpha_1/\alpha_2 > 1$ and the coefficients may be found from figure 3.17, using (3.111′) (Γ_{434} from Γ_{212} and Γ_{212} from Γ_{434}). For $\gamma_1 = \gamma_2$, $\arg \Gamma_{434} = 0$ and $\arg \Gamma_{212} \neq 0$ for $\alpha_1/\alpha_2 < 1$.

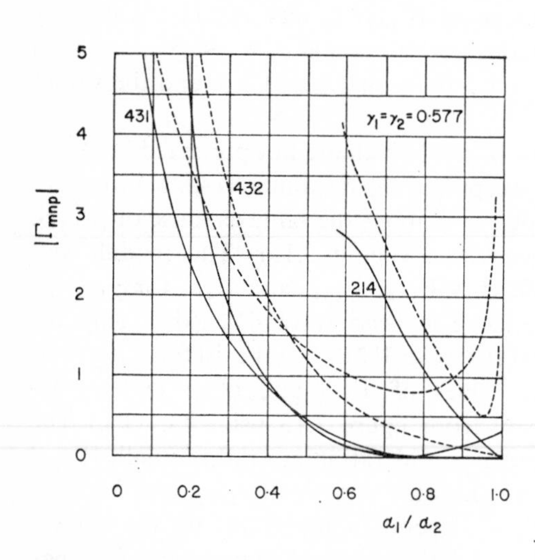

FIGURE 3.16 Moduli of coefficients of head waves of the third kind versus α_1/α_2. (See the text and figure 3.15 for the arguments of these coefficients.) The solid lines are for $\rho_1/\rho_2 = 1.0$ and the dashed for $\rho_1/\rho_2 = 0.7$. $\gamma_1 = \gamma_2 = 0.577$.

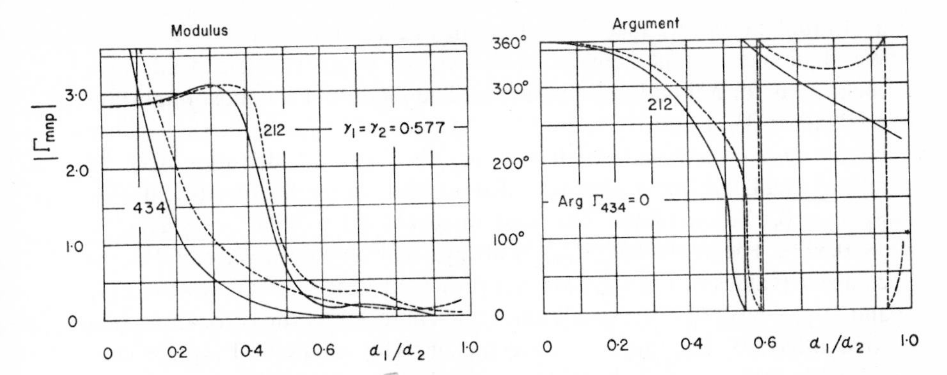

FIGURE 3.17 Coefficients of head waves of the fourth kind versus α_1/α_2. The solid lines are for $\rho_1/\rho_2 = 1.0$ and the dashed for $\rho_1/\rho_2 = 0.7$. $\gamma_1 = \gamma_2 = 0.577$.

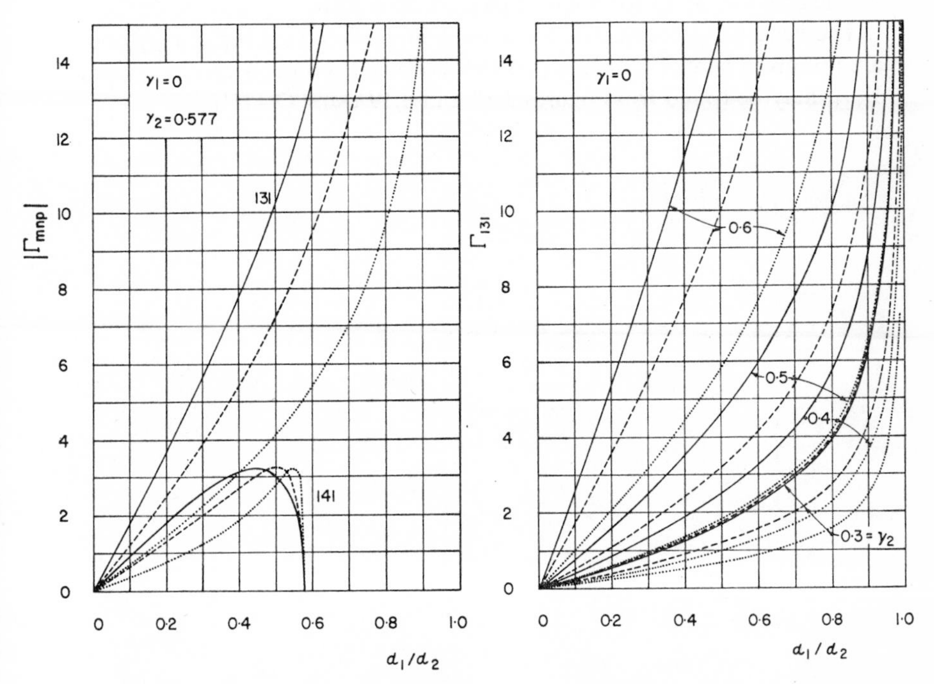

FIGURE 3.18 Moduli of the H-coefficients Γ_{131} and Γ_{141} for a liquid-solid interface versus α_1/α_2 for $\gamma_1 = 0$, $\gamma_2 = 0.577$. The solid lines are for $\rho_1/\rho_2 = 1.0$, the dashed for $\rho_1/\rho_2 = 0.7$, and the dotted for $\rho_1/\rho_2 = 0.4$.

FIGURE 3.19 Γ_{131} for a liquid–solid interface as a function of α_1/α_2, for different values of γ_2. The solid lines are for $\rho_1/\rho_2 = 1.0$, the dashed for $\rho_1/\rho_2 = 0.7$, and the dotted for $\rho_1/\rho_2 = 0.4$.

B Liquid–Solid Interface ($\gamma_1 = 0$) (figures 3.18 and 3.19)

We shall not present here numerical values of all possible head wave coefficients for a liquid–solid interface. The moduli of the most important coefficients, Γ_{131} and Γ_{141}, are given in figure 3.18 for $\gamma_2 = 1/\sqrt{3}$. Since in the case of a liquid–solid interface the ratio of densities may be rather small, we have also included in this figure the coefficients for $\rho_1/\rho_2 = 0.4$ (dotted line). The head wave 141 exists only for $\alpha_1/\alpha_2 < \gamma_2$ ($= 0.577$). The coefficient Γ_{131} is always real and positive, and Γ_{141} is complex (values of arg Γ_{141} are not presented here). The curves for Γ_{131} and $|\Gamma_{141}|$ have approximately the same character as those for the solid–solid interface, but the values are higher. The values of Γ_{131} are considerably higher, especially at large α_1/α_2. Values of Γ_{131} are given in more detail in figure 3.19 for $\gamma_2 = 0.3$, 0.4, 0.5, and 0.6 (in each case for $\rho_1/\rho_2 = 0.4$ dotted line, $\rho_1/\rho_2 = 0.7$ dashed line, $\rho_1/\rho_2 = 1.0$ solid line). These coefficients are larger for higher α_1/α_2, γ_2, and ρ_1/ρ_2.

The head wave coefficients for a solid–liquid interface (i.e. for $\gamma_2 = 0$) are not given here. The most important coefficients for this case, Γ_{313} and Γ_{323}, may be found from figures 3.18 and 3.19 using (3.111′).

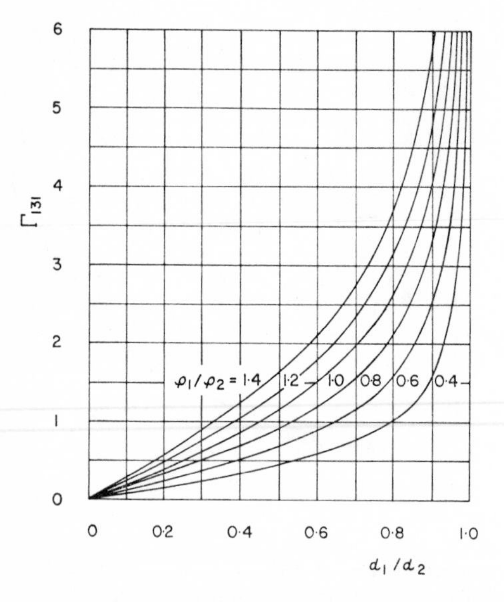

FIGURE 3.20 Γ_{131} for a liquid–liquid interface ($\gamma_1 = \gamma_2 = 0$) as a function of α_1/α_2 for different density ratios.

C Liquid–Liquid Interface ($\gamma_1 = \gamma_2 = 0$) (figure 3.20)

The values of the head wave coefficient Γ_{131} are given in figure 3.20 for six different density ratios: $\rho_1/\rho_2 = 0.4, 0.6, 0.8, 1.0, 1.2, 1.4$. Γ_{131} is higher for higher values of α_1/α_2 and ρ_1/ρ_2. The values are smaller than the corresponding Γ_{131} for solid–solid and liquid–solid interfaces. Note that Γ_{131} is always real and positive.

D Free Interface (figure 3.21)

If the first medium is a vacuum, only the head wave of the type 434 exists. Γ_{434} depends only on one parameter, viz., γ_2. The values of Γ_{434} are always real and positive and increase with γ_2. For small γ_2, they are small, but for $\gamma_2 > 0.5$ they are very high. Therefore, values of $\Gamma_{434}/10$ (instead of Γ_{434}) are given in figure 3.21 for high γ_2. For $\gamma_2 > 0.66$, the values are over 100.

E SH Head Waves (figure 3.22)

The values of the SH head wave coefficient Γ^{SH}_{242} are given in figure 3.22 for six different density ratios: $\rho_1/\rho_2 = 0.4, 0.6, 0.8, 1.0, 1.2, 1.4$. This coefficient is always real and positive and does not depend on γ_1 and γ_2. For small β_1/β_2, it decreases with increasing β_1/β_2, reaches a minimum (near 0.7–0.8), and then increases. Its value is rather high, especially for small and large β_1/β_2, but is smaller for higher ρ_1/ρ_2.

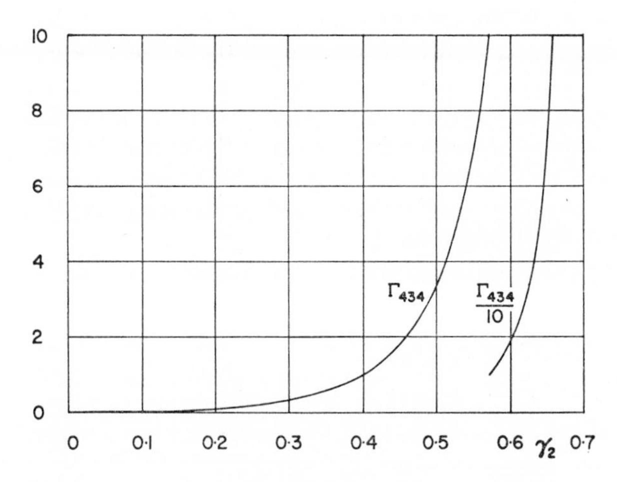

FIGURE 3.21 Γ_{434} as a function of γ_2 for the situation when the first medium is a vacuum.

F

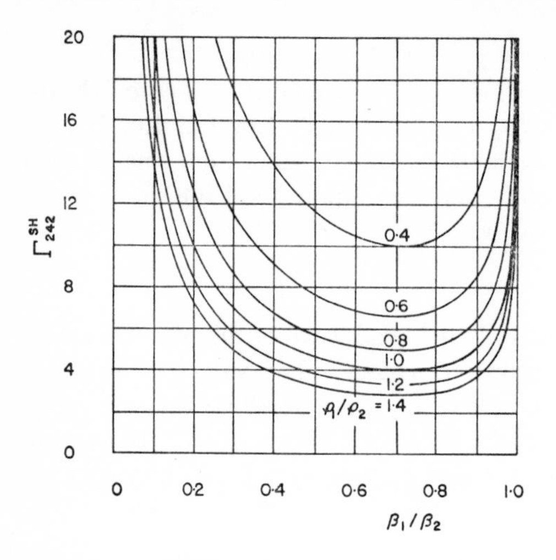

FIGURE 3.22 Γ_{242}^{SH} as a function of β_1/β_2 for different density ratios.

3.6 AMPLITUDES OF HEAD WAVES

The amplitudes of the horizontal and vertical components of the displacement vector of head waves, A_r^{mnp} and A_z^{mnp}, are given by formula (3.93). They generally depend on the following ten parameters: h, H, r, ω, α_1, α_2, β_1, β_2, ρ_1, and ρ_2, or on seven ratios, e.g., r/H, h/H, h/Λ_m, α_1/α_2, γ_1, γ_2, and ρ_1/ρ_2 (where Λ_m is the wavelength of the incident wave, $\Lambda_m = V_m/f = 2\pi V_m/\omega$). If $m = p$, the amplitudes depend only on six ratios: $r/(h+H)$, $(h+H)/\Lambda_m$, α_1/α_2, γ_1, γ_2, ρ_1/ρ_2. The dependence of the amplitudes of head waves on the above parameters or their ratios is often rather complicated, and only in some cases is it possible to reach general conclusions (see, for example, Yepinat'yeva (1959), Červený (1957b), Berry and West (1966), and Werth (1967)).

We can write, from (3.93), (3.90), and table 3.1, the following expressions for A_r^{mnp} and A_z^{mnp}:

$$A_r^{mnp} = c_r\Lambda_m I(r), \qquad A_z^{mnp} = c_z\Lambda_m I(r),$$

where $I(r) = r^{-1/2}(r-r_{mnp}^*)^{-3/2}$. c_r and c_z are multiplication constants which do not depend on the coordinates of the source and the receiver, and are given by

$$\left.\begin{aligned} c_r &= (2\pi)^{-1}c_1|\Gamma_{mnp}| \\ c_z &= (2\pi)^{-1}c_2|\Gamma_{mnp}| \end{aligned}\right\} \quad \text{for } p = 1 \text{ or } 3,$$

$$\left.\begin{aligned} c_r &= (2\pi)^{-1}c_2|\Gamma_{mnp}| \\ c_z &= (2\pi)^{-1}c_1|\Gamma_{mnp}| \end{aligned}\right\} \quad \text{for } p = 2 \text{ or } 4,$$

where

$$c_1 = \frac{V_p}{(V_n^2 - V_m^2)^{1/2}} = \frac{V_p}{V_n}[1-(V_m/V_n)^2]^{-1/2},$$

$$c_2 = \left(\frac{V_n^2 - V_p^2}{V_n^2 - V_m^2}\right)^{1/2} = \left(\frac{1-(V_p/V_n)^2}{1-(V_m/V_n)^2}\right)^{1/2}.$$

Now we shall investigate successively the dependence of the amplitudes of head waves on the individual parameters.

(a) *Dependence on wavelength* The amplitudes of head waves are proportional to the wavelength Λ_m of the incident wave.

(b) *Dependence on frequency* The amplitudes are inversely proportional to the frequency. The higher the frequency of the incident wave, the smaller are the amplitudes. When the amplitude spectrum of the incident wave is $|S(\omega)|$, the amplitude spectrum of the head wave is $|S(\omega)|/\omega$.

(c) *Dependence on epicentral distance* The dependence of the amplitudes on the epicentral distance r is given by the function $I(r)$. The amplitude–distance curve is a smooth, monotonically decreasing curve. Just at the critical distance r^*_{mnp}, the amplitudes are infinite. The above formulae are, however, inexact in the neighbourhood of the critical point and we can accept them as correct only at some distance beyond this point (see chapter 7). The slope of the amplitude curve at a given epicentral distance r is higher for larger r^*_{mnp} (i.e. for larger h, H, V_m/V_n, and V_p/V_n). In seismic prospecting, the slope of the amplitude curve at a given epicentral distance r is often characterized by the *exponent coefficient* ε, which is determined from the assumption that the slope of the amplitude curve at a given epicentral distance is the same as the slope of the curve $1/r^\varepsilon$ at the same distance. From this it follows that $\varepsilon = -\mathrm{d}(\ln A)/\mathrm{d}(\ln r)$ for given r. For head waves, we easily obtain

$$\varepsilon = 2 + \frac{3}{2}\frac{1}{(r/r^*_{mnp})-1}.$$

For $r \to \infty$, $\varepsilon = 2$. Thus, at large epicentral distances r, the amplitudes decrease with increasing r as $1/r^2$. At finite r, however, ε is always larger (see figure 3.23).

(d) *Dependence on H* The dependence of the amplitudes of head waves on the distance H of the receiver from the interface is given by the function

$I(r)$. Since the critical distance r^*_{mnp} increases with increasing H, so do the amplitudes.

(e) *Dependence on h* As for H, the amplitudes increase with increasing distance h of the source from the interface.

(f) *Dependence on elastic parameters* The dependence of the amplitudes on the elastic parameters of the media, α_1, α_2, β_1, β_2, ρ_1, and ρ_2, is very complicated. The function $I(r)$ depends on three velocities (V_m, V_n, and V_p), and the functions c_r and c_z on all six parameters. From this it follows that changes in V_m, V_n, and V_p can cause changes in the form and slope of the amplitude–distance curve. The other parameters, however, do not influence the shape of this curve; they only influence the multiplication constants. If $m = p$, the function $I(r)$ depends, of course, on only two velocities, V_n and V_m $(=V_p)$.

Suppose that all the velocities α_1, β_1, α_2, and β_2 are different from zero and that $m \neq p$; then V_m, V_n, and V_p are three of these velocities. We denote the remaining velocity by V_+. We can construct the ratios $V_{mn} = V_m/V_n$, $V_{pn} = V_p/V_n$, $\rho = \rho_1/\rho_2$, $V_{+n} = V_+/V_n$. (Instead of V_{+n} we can use the ratio of V_+ with any other velocity, e.g., V_{+m}, V_{+p}, V_{n+}). We can see from the above formulae that expressions for $I(r)$, c_1, and c_2 do not contain ρ and V_{+n}. From this it follows that the dependence of the amplitudes of head waves on ρ and V_{+n} is given only by the dependence of the corresponding head wave coefficients on these parameters. The

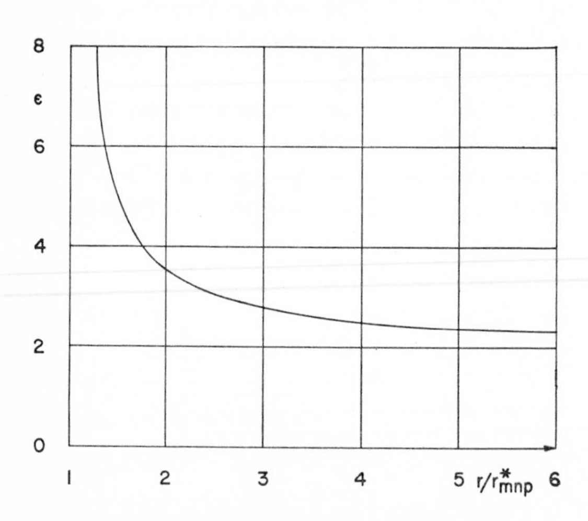

FIGURE 3.23 The exponent coefficient (ε) as a function of epicentral distance.

influence of ρ and V_{+n} on the amplitudes of head waves may thus be appreciated directly from the graphs for head wave coefficients. As an example, let us take $m = 1$, $n = 3$, $p = 2$. Then $V_m = \alpha_1$, $V_n = \alpha_2$, $V_p = \beta_1$, $V_+ = \beta_2$, $V_{mn} = \alpha_1/\alpha_2$, $V_{pn} = \beta_1/\alpha_2$, $V_{+n} = \beta_2/\alpha_2 = \gamma_2$. The amplitude–distance curves of the head wave 132 do not change their shape if ρ and γ_2 change; they differ only if α_1/α_2 or β_1/α_2 changes. The influence of γ_2 and ρ on the amplitudes of head waves can be seen directly from the graphs of the head wave coefficient Γ_{132}.

If $m = p$, only the ratio V_m/V_n has an influence on the function $I(r)$. The other two independent velocity ratios have no influence on the shape of the amplitude-distance curve; only the H-coefficients depend on them.

Now we shall discuss the most important head wave, viz., the compressional head wave 131. Then $V_m = \alpha_1$, $V_n = \alpha_2$, $V_p = \alpha_1$, $V_{mn} = \alpha_1/\alpha_2$. The other independent velocity ratios are $\gamma_1 = \beta_1/\alpha_1$ and $\gamma_2 = \beta_2/\alpha_2$. For given V_{mn}, if γ_1, γ_2, and ρ change, only the values of Γ_{131} change in the formulae for the amplitudes A_r^{131} and A_z^{131}, all other quantities remaining unchanged. Therefore, we can derive a conclusion about the dependence of the amplitude of the head wave 131 on γ_1, γ_2, and ρ directly from the graphs for Γ_{131}. We can see from figure 3.11 that A_r^{131} and A_z^{131} depend only slightly on the density ratio ρ, so that we can put $\rho_1 = \rho_2$ for practical calculations. The dependence of A_r^{131} and A_z^{131} on γ_1 and γ_2 is larger. For the range of γ_1 and γ_2 studied ($0.3 \leqslant \gamma_1 \leqslant 0.6; 0.3 \leqslant \gamma_2 \leqslant 0.6$), the amplitudes are larger for higher γ_2 (when $\gamma_1 =$ const.) and for smaller γ_1 (when $\gamma_2 =$ const.). The value of γ_2 has a particularly strong influence upon A_r^{131} and A_z^{131}. If $\gamma_1 = \gamma_2$, the amplitudes are greater for higher γ_1 $(= \gamma_2)$.

The dependence of A_r^{131} and A_z^{131} on α_1/α_2 is more complicated. If α_1/α_2 changes, so too do the region of existence of the wave $\mathbf{W}^{131}$ and the form and slope of its amplitude-distance curve. All the functions $I(r)$, c_r, and c_z depend on α_1/α_2. The head wave coefficient Γ_{131} is usually an increasing function of α_1/α_2, as are c_r, c_z, and $I(r)$. It follows that the amplitudes of the head wave $\mathbf{W}^{131}$ are larger for higher α_1/α_2 ($\alpha_1/\alpha_2 < 1$, of course), the amplitudes tending to infinity for $\alpha_1/\alpha_2 \to 1$. It seems paradoxical that the amplitudes of head waves grow to infinity when the interface is vanishing. As $\alpha_1/\alpha_2 \to 1$, for given r, the critical distance r^*_{131} increases and for some $\alpha_1/\alpha_2 < 1$ it approaches r. At the critical distance and in its neighbourhood, however, the formulae for the head waves are incorrect. (Moreover, the head wave does not exist for $r^*_{131} > r$, i.e. for $\alpha_1/\alpha_2 > r/[r^2+(h+H)^2]^{1/2}$.) Therefore, when r is given, the case of $\alpha_1/\alpha_2 \to 1$ is not permitted. If we want to investigate the behaviour of the amplitudes A_r^{131} and A_z^{131} for $\alpha_1/\alpha_2 \to 1$, we must change the epicentral distance so as to be always beyond the critical point. We can, for example, put $r = ar^*_{131}$, where $a > 1$ (a = const.). Then we find easily that

$$I(r) = a^{-1/2}(a-1)^{-3/2}(h+H)^{-2}(\alpha_1/\alpha_2)^{-2}[1-(\alpha_1/\alpha_2)^2].$$

We know that for $\alpha_1/\alpha_2 \to 1$, $\Gamma_{131} \sim [1-(\alpha_1/\alpha_2)^2]^{-1/2}$, and from this we have $c_r \sim [1-(\alpha_1/\alpha_2)^2]^{-1}$, $c_z \sim [1-(\alpha_1/\alpha_2)^2]^{-1/2}$. (Here we have supposed also that $\rho_1/\rho_2 \to 1$ and $\gamma_1/\gamma_2 \to 1$ when $\alpha_1/\alpha_2 \to 1$.) If we insert these values into the formulae for A_r^{131} and A_z^{131}, we find that for $\alpha_1/\alpha_2 \to 1$ (at the epicentral distance $r = ar^*_{131}$)

$$A_z^{131} \sim [1-(\alpha_1/\alpha_2)^2]^{1/2}, \qquad A_r^{131} \sim \text{const.}$$

It follows that in the limit of $\alpha_1/\alpha_2 \to 1$, the amplitudes of head waves do not increase but remain constant (the horizontal component), or decrease (the vertical component). This in no way contradicts the former statement that the amplitudes of the head wave 131 are larger for increasing α_1/α_2, at a given epicentral distance r.

(g) *Curves of constant head wave amplitudes* We have found that the amplitudes decrease with increasing epicentral distance r, but increase with increasing H. Therefore, there must be some curve $H = H(r)$ along which the amplitudes remain constant. This curve is determined by the formula $r^{1/2}(r-r^*_{mnp})^{3/2} = c$, where c is a real, positive constant, $0 < c < \infty$. From this we find easily that

$$H = \frac{(V_n^2-V_p^2)^{1/2}}{V_p}\left\{r - \frac{hV_m}{(V_n^2-V_m^2)^{1/2}} - \left(\frac{c^2}{r}\right)^{1/3}\right\}.$$

The amplitudes of head waves $\mathbf{W}^{mnp}$ remain the same along the whole curve.

3.6.1 *Comparison of Amplitudes of Different Types of H and R Waves*

A detailed discussion of this matter would be very lengthy. We shall therefore present here the results of computations for only two characteristic models of the interface: those with a small and a large velocity contrast. It should be stressed again that the following discussion is based only on the study of the leading terms of the ray series for the R and H waves.

Model A: Interfaces with Small Velocity Contrast (figures 3.24 to 3.30)
Interfaces with small velocity contrast are of great importance in seismic investigations of the earth's crust. We shall assume here that the refractive index α_1/α_2 has the value 0.8 (e.g., α_1 = 6.4 km/sec, α_2 = 8 km/sec). The

other parameters are assumed to be $\gamma_1 = \gamma_2 = 1/\sqrt{3}$, $\rho_1/\rho_2 = 1$, $h = H = 30$ km, $f = 10$ cps). These parameters are close to those for the classical one-layer model of the earth's crust with the Mohorovičić discontinuity lying at a depth of 30 km. The amplitudes of the vertical displacement components of the various waves propagating in the two media are given in figures 3.24 and 3.26 to 3.30. Figure 3.25 shows the amplitudes of the *horizontal* components for comparison with figure 3.24. The types of the waves are given schematically above the corresponding figure (solid line for P-wave and dashed line for SV-wave). The effect of the free interface (surface of the earth) is not taken into consideration, and the two media in

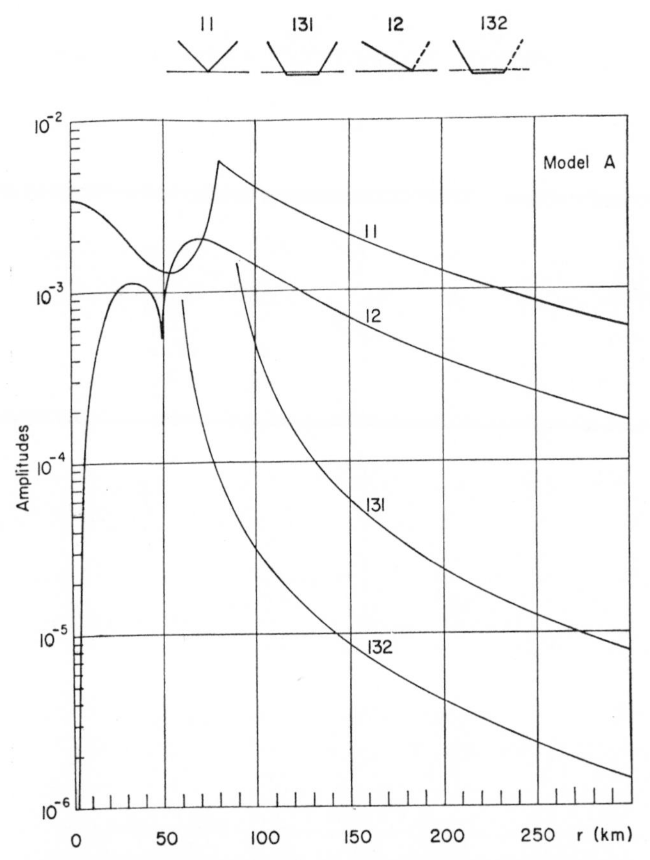

FIGURE 3.24 Amplitudes of the vertical displacement components versus distance, for reflected and head waves for model A for a P-wave source in the first medium.

contact are assumed to be half-spaces. It was shown earlier that the formulae for the amplitudes of head waves are not exact in the neighbourhood of the critical point. Therefore, the amplitude–distance curves of head waves are not drawn right from the critical distance, but from a slightly greater epicentral distance. The critical distances for our model are as follows: $r^*_{232} = 31.2$ km, $r^*_{234} = r^*_{432} = 36.8$ km, $r^*_{212} = r^*_{434} = 42.4$ km, $r^*_{214} = r^*_{412} = 52.5$ km, $r^*_{132} = r^*_{231} = 55.6$ km, $r^*_{134} = r^*_{431} = 61.2$ km, $r^*_{414} = 62.6$ km, $r^*_{131} = r^*_{242} = 80$ km. The source is assumed to radiate P-waves in figures 3.24, 3.25, and 3.27, and SV-waves in figures 3.26, 3.28, 3.29, and 3.30. In figures 3.24, 3.25, and 3.26, both the source and the receiver lie in the first medium; in figures 3.27 and 3.28 the source lies in

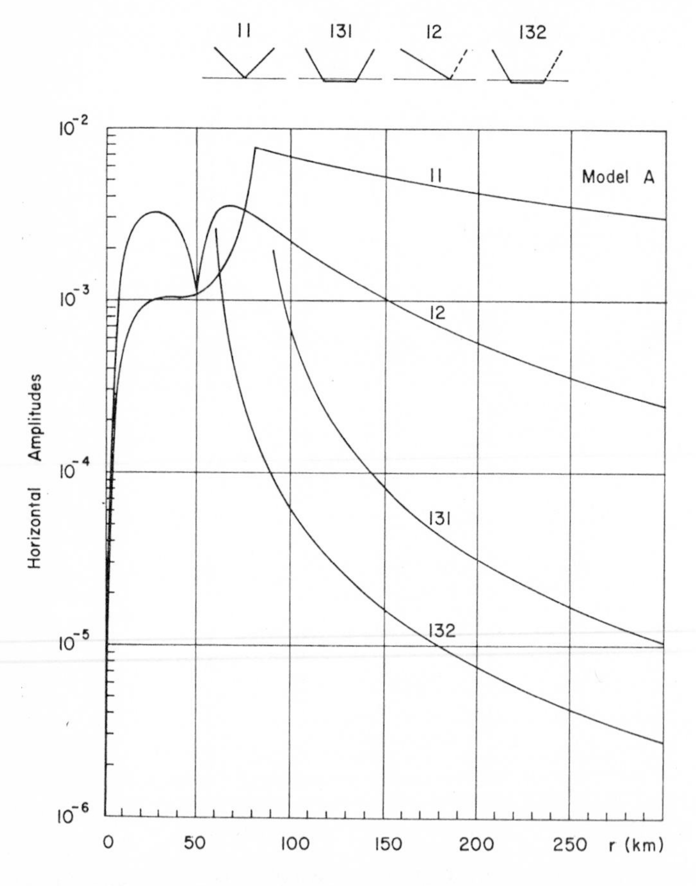

FIGURE 3.25 Amplitudes of the *horizontal* displacement components versus distance, for the same situation as in figure 3.24.

the first medium and the receiver in the second medium; in figure 3.29 the source lies in the second medium and the receiver in the first medium; and in figure 3.30 both the source and the receiver lie in the second medium. All possible types of R-waves and H-waves have been considered. The figures for the source of P-waves lying in the second medium are not given here, as no head wave exists in this case, only reflected and refracted waves being generated. We shall not discuss these figures individually here, as

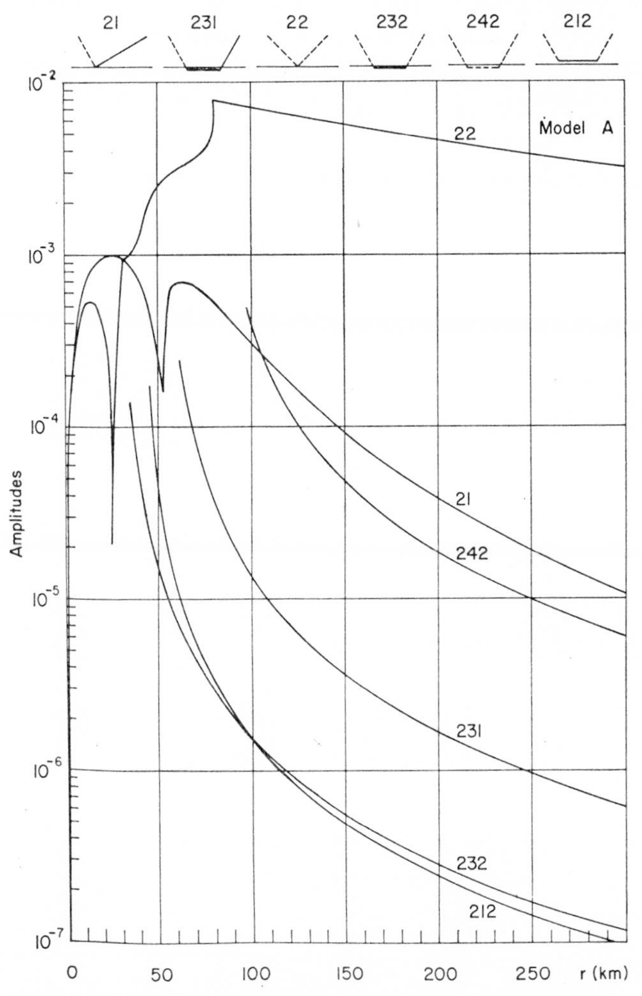

FIGURE 3.26 Amplitudes of the vertical displacement components versus distance, for reflected and head waves for model A for an SV-wave source in the first medium.

they are self-explanatory. We shall only emphasize a few general features of head waves. The neighbourhood of the critical point will be left out of the discussion since the results are incorrect there and, moreover, the head wave interferes with the corresponding R-wave there (see section 3.8). The conclusions given below are valid only from about 15–20 km beyond the critical point and for frequencies near 10 cps. Note that by *amplitude* below we understand the amplitude of the *vertical* component of the displacement vector.

(1) The amplitudes of head waves are always smaller than the amplitudes of reflected and refracted waves, in almost all cases by more than a

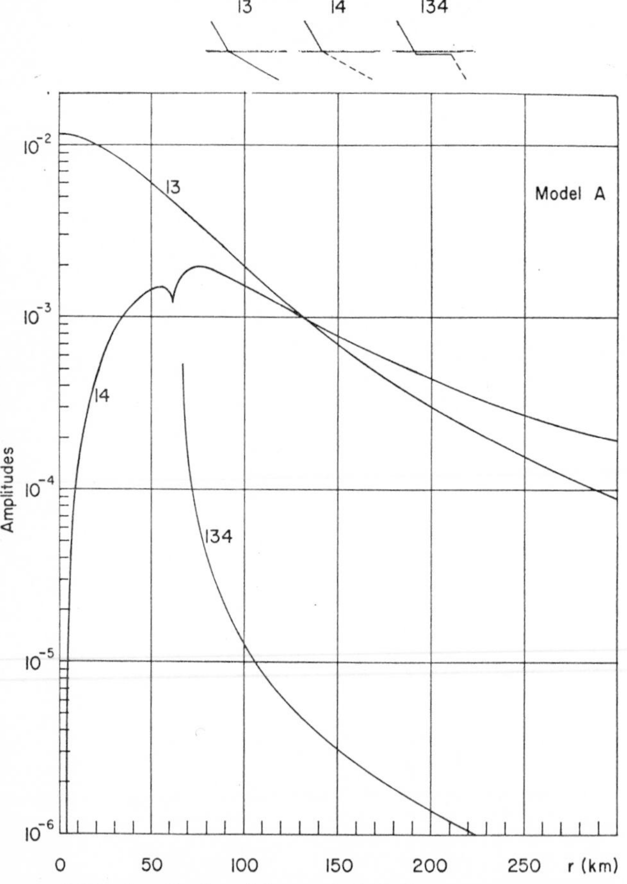

FIGURE 3.27 Amplitudes of the vertical displacement components vs. distance, for refracted and head waves for model A for a P-wave source in the first medium.

factor of ten. The only exception is the reflected wave 21 and the head wave 242, whose amplitudes differ from each other only slightly. Usually, at epicentral distances of about 300 km, differences between the amplitudes of R-waves and H-waves amount to $1\frac{1}{2}$–2 orders of magnitude or more, and in some cases to 4 orders of magnitude (see, for example, the reflected wave 22 and H-waves 232 and 212 in figure 3.26).

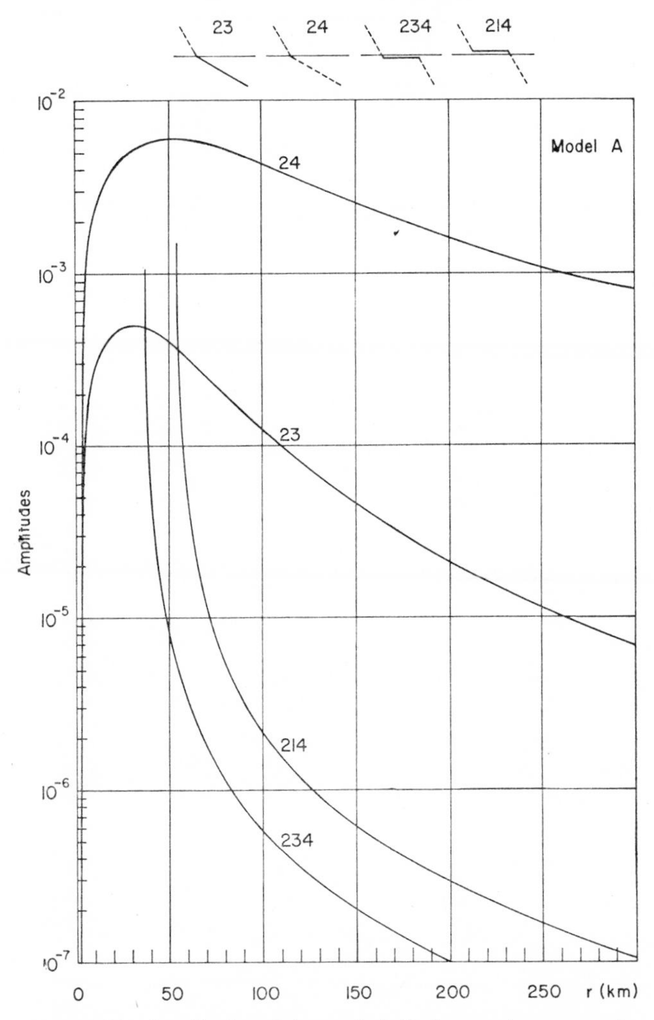

FIGURE 3.28 Amplitudes of the vertical displacement components vs. distance, for refracted and head waves for model A for an SV-wave source in the first medium.

(2) At small epicentral distances, the amplitudes of H-waves decrease with increasing epicentral distance more rapidly than those of R-waves. At large distances, however, only the R-waves 22 and 44 decrease less rapidly, the slope of the amplitude–distance curves of other R-waves being nearly the same as that of head waves. For some R-waves, the slope is even higher, e.g., the transformed R-waves 21, 23, 41, 43.

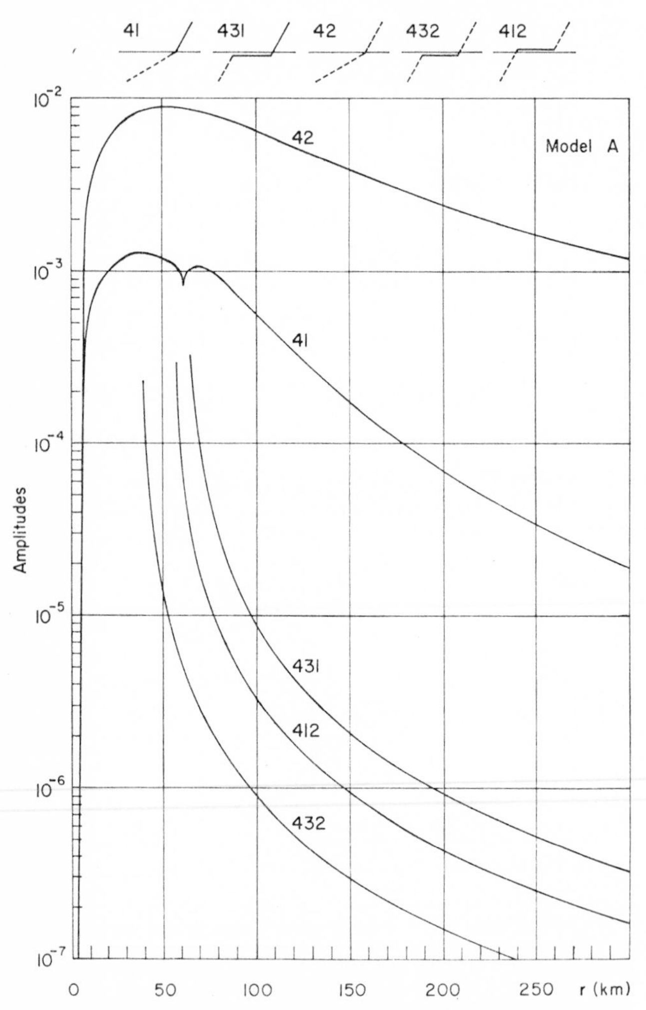

FIGURE 3.29 Amplitudes of the vertical displacement components vs. distance, for refracted and head waves for model A for an SV-wave source in the second medium.

(3) The ratio of the vertical amplitude of an R-wave (A_z^{mp}) to the amplitude of an arbitrary H-wave depends on the epicentral distance r in three different ways according to the values of m and p:

(α) It increases with increasing r, at small epicentral distances very rapidly and at large r approximately as r^1, for $mp = 22$ or 44.

(β) It increases with increasing r at small epicentral distances; it remains

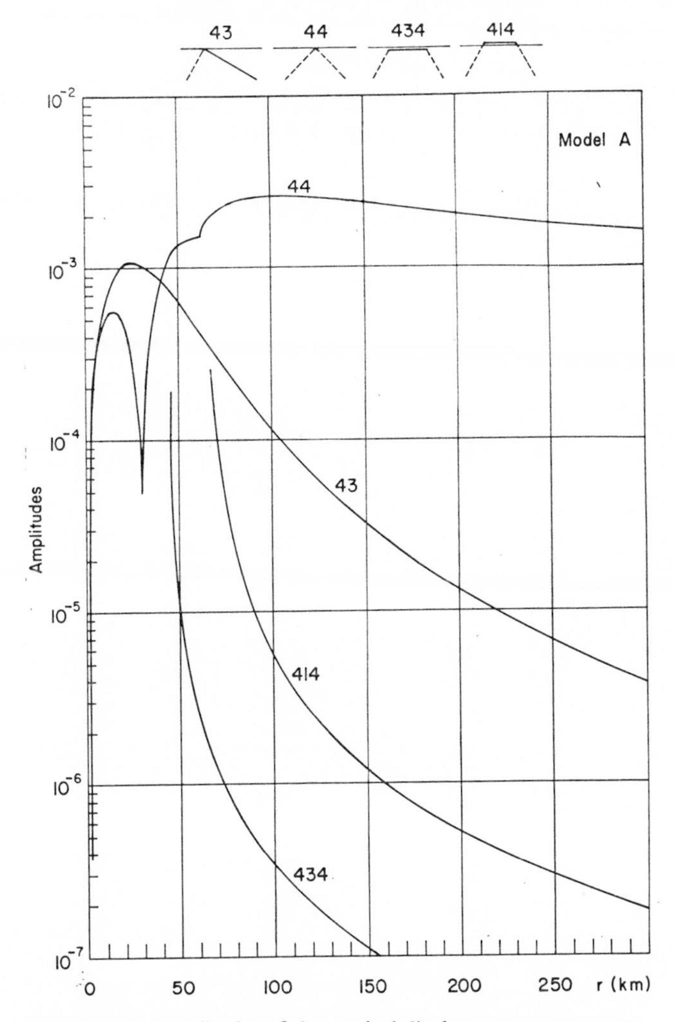

FIGURE 3.30 Amplitudes of the vertical displacement components vs. distance, for reflected and head waves for model A for an SV-wave source in the second medium.

approximately constant at large r for $mp = 11$ or 33. If $m \neq p$, the ratio remains constant when $V_p > V_m$ and $p = 2$ or 4, or when $V_p < V_m$.

(γ) It increases with increasing r at small epicentral distances, and decreases approximately as r^{-1} at large epicentral distances if $V_p > V_m$, $p = 1$ or 3, and $m \neq p$. At some epicentral distance the ratio reaches a maximum value.

Note that the ratio of amplitudes of the important compressional waves 11 and 131 is of the β type.

(4) The amplitudes of different types of head waves are very different. The larger amplitudes correspond to the waves 131 and 242, the other head waves being about one order of magnitude (or more) weaker. The trans-

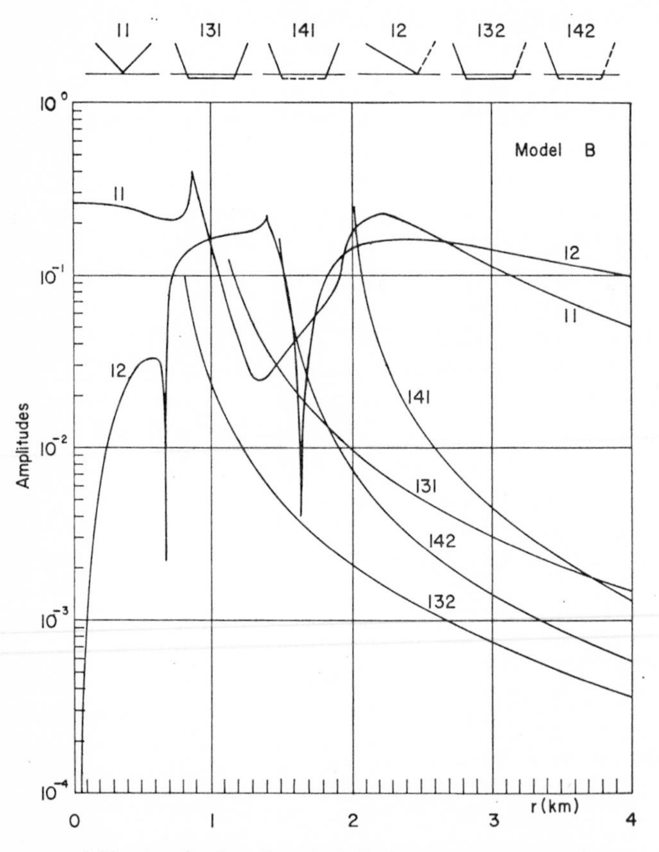

FIGURE 3.31 Amplitudes of vertical displacement components vs. distance, for reflected and head waves for model B for a P wave source in the first medium.

formed head waves 132 and 231 have relatively large amplitudes. H-waves of the fourth kind (212 and 434) are very weak, as are some of the second and third kind (134, 234, 214, 432). Among the basic head waves, wave 232 has the smallest amplitude.

(5) The amplitudes of the various types of R-waves are also very different. Usually the amplitudes of the transformed waves (12, 21, 23, 41, 43) are smaller than those of pure compressional or shear waves (11, 22, 24, 42, 44). The differences in the amplitudes of different types of R-waves sometimes amount to a few orders of magnitude (see, for example, the R-waves 22 and 21). The only exceptions are the waves 13 and 14.

Model B: Interfaces with Large Velocity Contrast (figures 3.31 to 3.37)

Interfaces with large velocity contrast are often of great importance in seismic prospecting. We shall assume here that the refractive index α_1/α_2 has the value 0.4 (e.g., $\alpha_1 = 2$ km/sec, $\alpha_2 = 5$ km/sec). The other parameters are assumed to be $\gamma_1 = \gamma_2 = 1/\sqrt{3}$, $\rho_1/\rho_2 = 0.75$, $h = H = 1$ km, $f = 50$ cps. The amplitudes of the vertical displacement components of different waves propagating in the two media are given in figures 3.31 and 3.33 to 3.37. Figure 3.32 shows the *horizontal* displacement component for comparison with figure 3.31. The system of figures is the same as for model A with small velocity contrast, and all the remarks about the figures in that case are applicable here also. The critical distances for the present model are $r^*_{232} = 0.475$, $r^*_{132} = r^*_{231} = 0.674$, $r^*_{131} = r^*_{242} = 0.873$, $r^*_{234} = r^*_{432} = 0.944$, $r^*_{134} = r^*_{431} = 1.143$, $r^*_{142} = r^*_{241} = 1.397$, $r^*_{212} = r^*_{434} = 1.414$, $r^*_{141} = 1.921$. We shall again discuss the figures only briefly and, as in the previous case, we shall not consider the neighbourhood of the critical points. The situation is now more complicated than for small velocity contrast, mainly because of the more complicated shape of the amplitude–distance curves of R-waves at short epicentral distances (up to 2 km). These curves have local maxima and deep minima in this region. At larger distances (from about 2.5 km), the situation becomes simpler. When we speak of large and small epicentral distances here, we mean distances larger and smaller than 2.5 km ($r/h = 2.5$).

At large epicentral distances, the head wave amplitudes are again smaller than the amplitudes of reflected and refracted waves, the differences being greater than one order of magnitude. The amplitudes of the various head waves, however, do not differ as much as in the case of small velocity contrast, the maximum difference at a given epicentral distance being not more than $1\frac{1}{4}$ orders of magnitude (see figure 3.33). The relative amplitudes of the various kinds of head waves are different than for small velocity contrast. For example, the large amplitudes of the head waves of the fourth

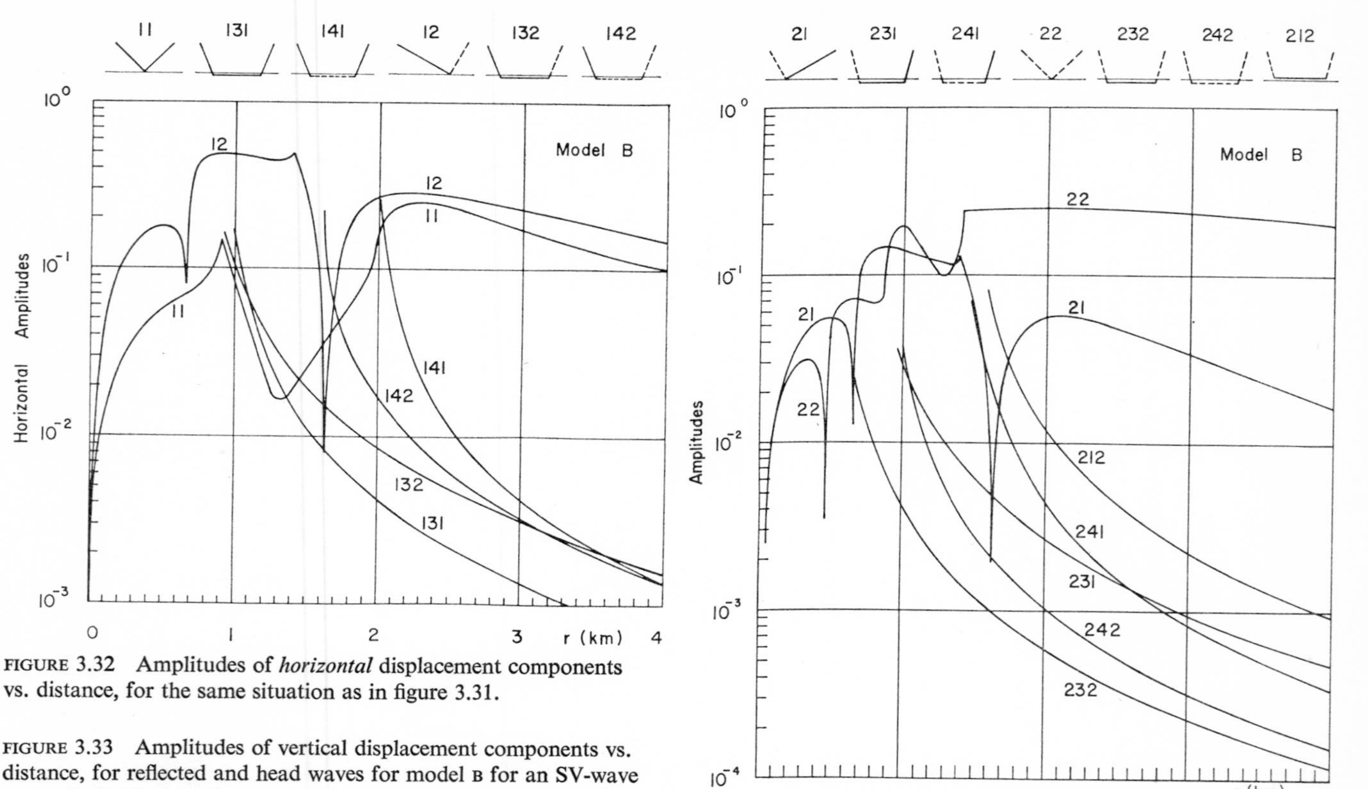

FIGURE 3.32 Amplitudes of *horizontal* displacement components vs. distance, for the same situation as in figure 3.31.

FIGURE 3.33 Amplitudes of vertical displacement components vs. distance, for reflected and head waves for model B for an SV-wave source in the first medium.

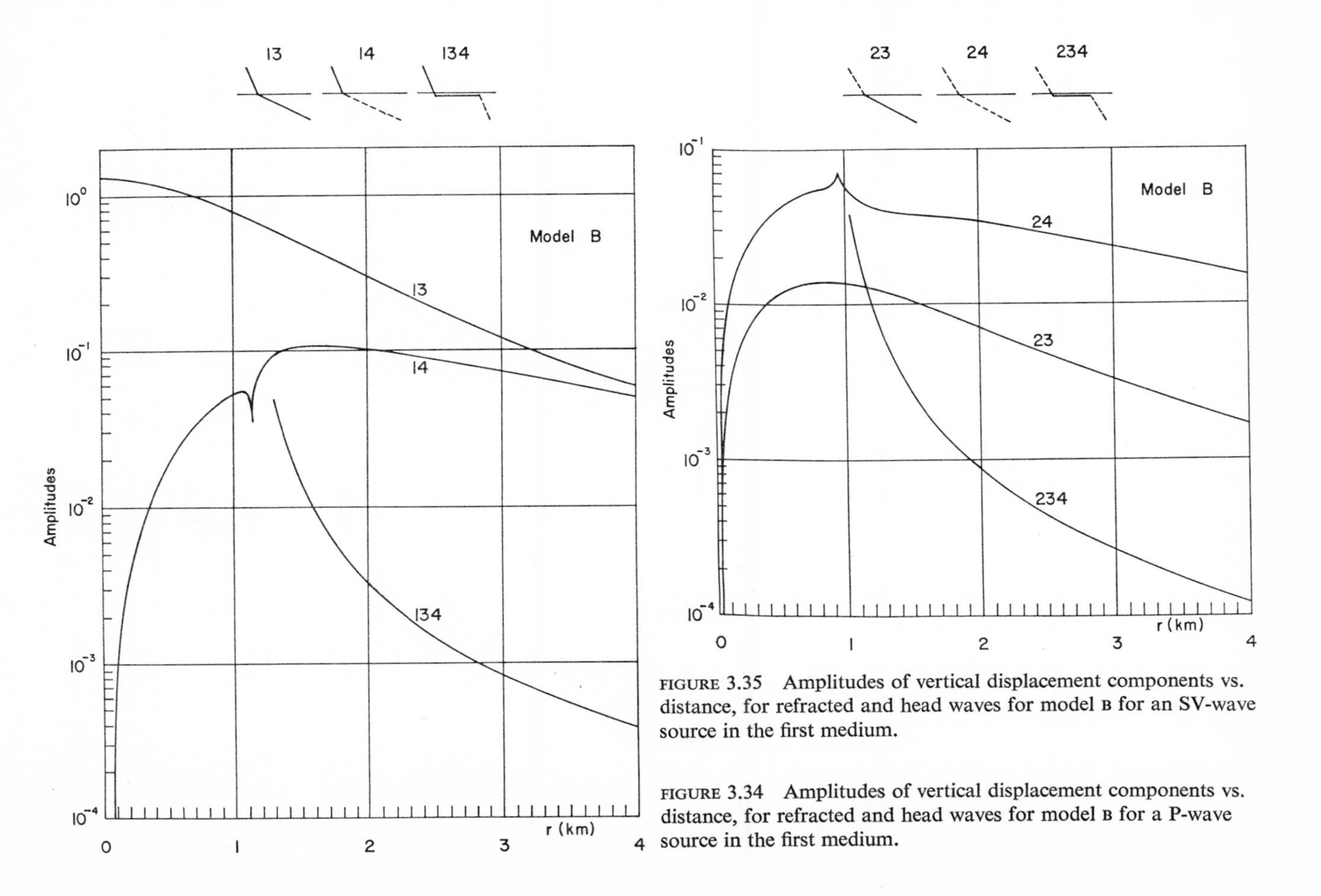

FIGURE 3.35 Amplitudes of vertical displacement components vs. distance, for refracted and head waves for model B for an SV-wave source in the first medium.

FIGURE 3.34 Amplitudes of vertical displacement components vs. distance, for refracted and head waves for model B for a P-wave source in the first medium.

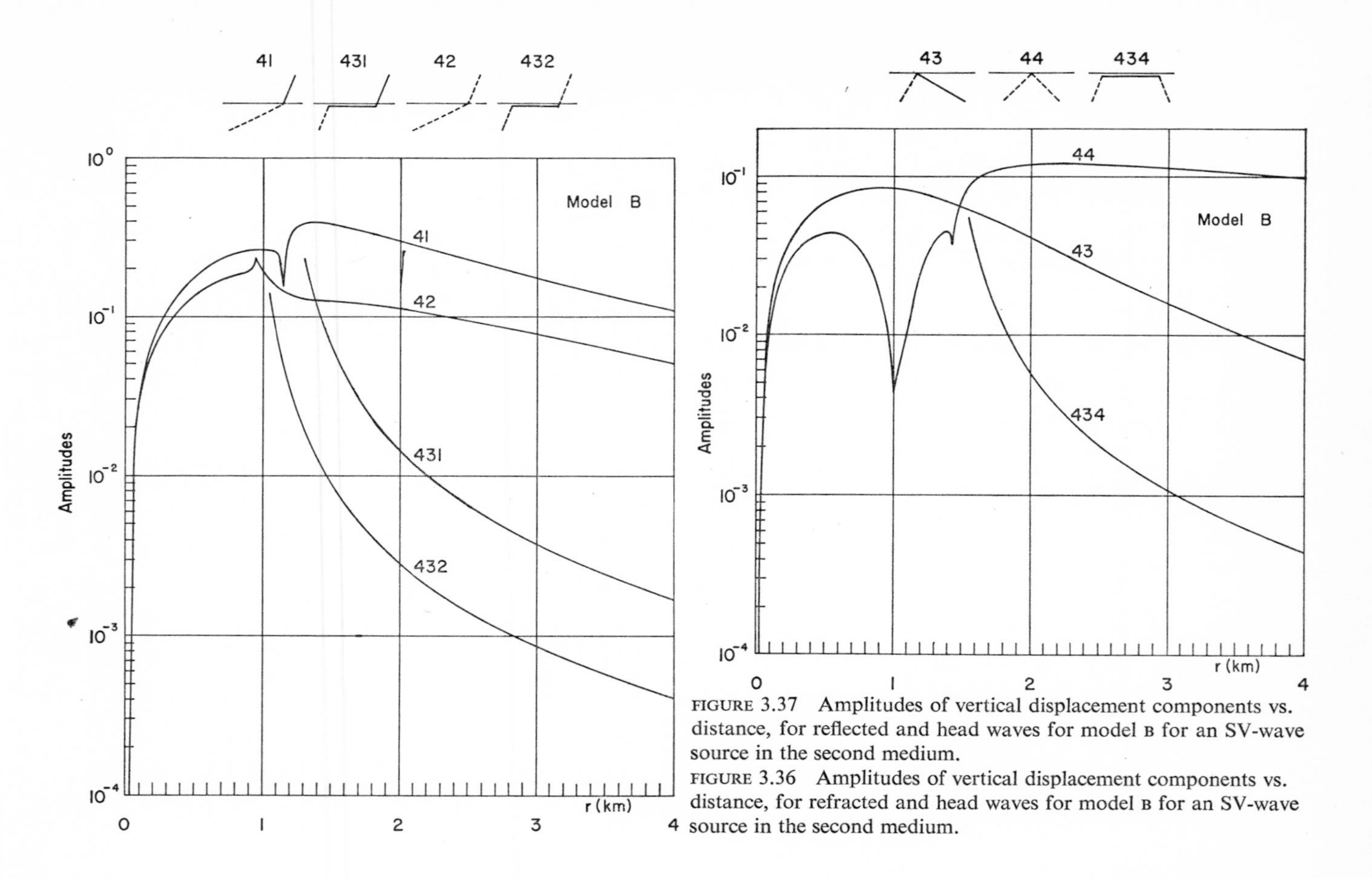

FIGURE 3.37 Amplitudes of vertical displacement components vs. distance, for reflected and head waves for model B for an SV-wave source in the second medium.

FIGURE 3.36 Amplitudes of vertical displacement components vs. distance, for refracted and head waves for model B for an SV-wave source in the second medium.

kind (the wave 212 is the strongest head wave in figure 3.33) are very surprising. On the other hand, the wave 242, which was the strongest for small velocity contrast, is very weak now.

The situation is more complicated at small epicentral distances, especially when both receiver and source lie in the first medium (figures 3.31 and 3.33). We can see that the amplitudes of head waves may even be greater than those of reflected waves at these epicentral distances (see, for example, the head wave 131 and the R-wave 11 in figure 3.31). However, the regions where the head wave amplitudes may be greater are usually very small, and are caused by the deep minima in the amplitude–distance curves of R-waves. Usually, of course, the R-wave amplitudes are again greater than the H-wave. The relation between the amplitudes of different types of R-waves is also not as simple as in the previous case (see figures 3.31 and 3.33).

It is clear from the above figures for both the large and small velocity contrasts that (according to theoretical calculations) the head wave amplitudes are, as a rule, more than ten times smaller than the R-wave amplitudes (with some exceptions at short epicentral distances for a large velocity contrast, caused by the deep minima of the amplitude–distance curves of R-waves). It should be extremely difficult, therefore, to register head waves at all. In practice, however, head waves are observed every day. The amplitude ratio of head waves and reflected waves determined experimentally is usually larger than the ratio determined from the above figures. This increase in the amplitude ratio may be ascribed to two groups of factors. The first group result in an increase in the head wave amplitudes while the reflected wave amplitudes remain approximately the same. Such factors are, for example, a positive velocity gradient below the interface, or curvature of the interface. The second group result in a decrease in the reflected wave amplitudes while the head wave amplitudes remain approximately the same. Such factors are mainly connected with the inhomogeneity of the overburden (layering of the overburden, existence of high-velocity layers in it, existence of a vertical velocity gradient inside it). Any of these factors may change the H-wave or R-wave amplitudes to the extent of some orders of magnitude. Most of these possibilities will be considered in the following chapters.

3.7 IMPULSE SOURCE AND CONSTRUCTION OF THEORETICAL SEISMOGRAMS

In the previous sections we assumed that the source was harmonic; these results will now be generalized to the case of a pulse source. The displacement vector of the impulse wave will again be denoted by $\mathscr{W}$, instead of $\mathbf{W}$ which has been used for harmonic waves. The displacement vector of the impulse wave radiating from the source is given by (2.176). For the displacement vector of an arbitrary type of head wave, $\mathscr{W}^{mnp}$, we can then write

$$\mathscr{W}^{mnp} = \frac{1}{\pi} \operatorname{Re} \int_0^\infty S(\omega)\mathbf{W}^{mnp}\, d\omega, \tag{3.113}$$

where $\mathbf{W}^{mnp}$ is the displacement vector of the harmonic H-wave. The expressions for $\mathbf{W}^{mnp}$ are known from previous sections (see (3.58)). Inserting into (3.113) yields

$$\mathscr{W}^{mnp} = \frac{V_n \tan \theta_m^*}{r^{1/2}l^{3/2}}\mathbf{n}_p^* \frac{1}{\pi} \operatorname{Re}\left(\Gamma_{mnp} \int_0^\infty \frac{S(\omega)}{i\omega} \exp[i\omega(t-\tau_{mnp})] d\omega\right), \tag{3.114}$$

where l is defined by (3.56) and Γ_{mnp} is the H-coefficient. If Γ_{mnp} is real, we obtain very simply from (3.114)

$$\mathscr{W}^{mnp} = \frac{V_n \cdot \Gamma_{mnp} \tan \theta_m^*}{r^{1/2}l^{3/2}} F(t-\tau_{mnp})\mathbf{n}_p^*, \tag{3.115}$$

where

$$F(t) = \frac{1}{\pi} \operatorname{Re} \int_0^\infty \frac{S(\omega)}{i\omega} e^{i\omega t}\, d\omega = \int_{-\infty}^{t} f(u)\, du. \tag{3.116}$$

Thus the form of the pulse of head waves differs from that of incident waves. It turns out to be an integral of the incident pulse function in the time domain when the H-coefficient Γ_{mnp} is real. Note that many H-coefficients are always real (for example, Γ_{131}, Γ_{132}, Γ_{134}, Γ_{231}, Γ_{213}, Γ_{313}, Γ_{312}; all SH H-coefficients; H-coefficients for a liquid–liquid interface). For these waves, (3.115) is the final formula for their pulse displacement vector.

If Γ_{mnp} is complex, the expression for $\mathscr{W}^{mnp}$ is a little more complicated. We find from (3.114) that

$$\mathscr{W}^{mnp} = \frac{V_n|\Gamma_{mnp}| \tan \theta_m^*}{r^{1/2}l^{3/2}}\mathbf{n}_p^*\left(\cos(\arg \Gamma_{mnp})\frac{1}{\pi} \operatorname{Re}\int_0^\infty \frac{S(\omega)}{i\omega} \exp[i\omega(t-\tau_{mnp})] d\omega \right.$$
$$\left. -\sin(\arg \Gamma_{mnp})\frac{1}{\pi} \operatorname{Im} \int_0^\infty \frac{S(\omega)}{i\omega} \exp[i\omega(t-\tau_{mnp})]\, d\omega\right).$$

We introduce

$$G(t) = -\frac{1}{\pi} \operatorname{Im} \int_0^\infty \frac{S(\omega)}{i\omega} e^{i\omega t} d\omega = \int_{-\infty}^t g(u)\, du, \tag{3.117}$$

where $g(u)$ is given by (2.184). Then we can write

$$\mathscr{W}^{mnp} = \frac{V_n|\Gamma_{mnp}| \tan \theta_m^*}{r^{1/2}l^{3/2}} \mathbf{n}_p^*[\cos(\arg \Gamma_{mnp})F(t-\tau_{mnp}) + \sin(\arg \Gamma_{mnp})G(t-\tau_{mnp})]. \tag{3.118}$$

We now introduce the notation

$$F(t; \chi) = F(t) \cos \chi + G(t) \sin \chi, \tag{3.119}$$

where $F(t)$ and $G(t)$ are given by (3.116) and (3.117). Then we can write from (3.118).

$$\mathscr{W}^{mnp} = \frac{V_n|\Gamma_{mnp}| \tan \theta_m^*}{r^{1/2}l^{3/2}} F(t-\tau_{mnp}; \arg \Gamma_{mnp})\mathbf{n}_p^*. \tag{3.120}$$

Note that the function $F(t; \chi)$ may be expressed by means of $f(t; \chi)$ (given by (2.186)) as follows:

$$F(t; \chi) = \int_{-\infty}^t f(u; \chi)\, du.$$

The form of the head wave signal $F(t-\tau_{mnp}; \arg \Gamma_{mnp})$ will always be different from that of the incident wave. For a given type of wave, it does not depend on the position of the source or the receiver, as $\arg \Gamma_{mnp}$ does not depend on it. The form of this signal will, of course, change if the elastic parameters of the interface change.

For the r and z components of the displacement vector $\mathscr{W}^{mnp}$ we can write from (3.120)

$$\begin{bmatrix} \mathscr{W}_r^{mnp} \\ \mathscr{W}_z^{mnp} \end{bmatrix} = \frac{|D_{mnp}|}{r^{1/2}l^{3/2}} F(t-\tau_{mnp}; \arg \Gamma_{mnp}) \begin{bmatrix} \mathbf{n}_{pr}^* \\ \mathbf{n}_{pz}^* \end{bmatrix}. \tag{3.121}$$

The values of n_{pr}^* and n_{pz}^* are given in table 3.1, and the formulae for D_{mnp}, l, τ_{mnp}, and Γ_{mnp} in sections 3.2 and 3.3.

The *theoretical seismogram* for the horizontal or vertical component of displacement, $\mathscr{W}_r(t)$ or $\mathscr{W}_z(t)$, is obtained simply by the algebraic addition of the pulses of all the waves that are received at a given point. If the receiver lies in the same medium as the source, we get

$$\mathscr{W}_r(t) = \mathscr{W}_r^{\,m} + \sum_{(n)} \mathscr{W}_r^{\,mn} + \sum_{(n,p)} \mathscr{W}_r^{\,mnp},$$
$$\mathscr{W}_z(t) = \mathscr{W}_z^{\,m} + \sum_{(n)} \mathscr{W}_z^{\,mn} + \sum_{(n,p)} \mathscr{W}_z^{\,mnp}, \tag{3.122}$$

where the summations over (n) and (n, p) are understood to include all possible values of n and p ($n = 1, 2, 3$, or 4; $p = 1, 2, 3$, or 4) which will produce a wave that arrives at the receiver. If the receiver does not lie in the same medium as the source, the formulae (3.122) remain valid, but we must eliminate the first terms, $\mathscr{W}_r^{\,m}$ and $\mathscr{W}_z^{\,m}$, characterizing the direct wave.

3.8
INTERFERENCE WAVES

In the previous section we derived the formulae for theoretical seismograms as a superposition of pulses of all types of waves arriving at the receiver. In some cases some of these waves interfere, whereas in others they exist separately. The question whether any given waves interfere or not depends on their travel times and phase shifts, and on the length of the pulse of the incident wave. There is, however, a region in which the H-wave $\mathscr{W}^{mnp}$ will *always* interfere with the corresponding R-wave $\mathscr{W}^{mp}$. This is the so-called *critical region* lying immediately beyond the critical point $r = r^*_{mnp}$ (see Kosminskaya, 1946).

The formulae for $\mathscr{W}^{mp}$ and $\mathscr{W}^{mnp}$ obtained using the ray theory are not quite exact in the critical region (see section 3.3.3). Therefore, the formulae for the interference wave, $\mathscr{W}^{mp} + \mathscr{W}^{mnp}$ (obtained as a superposition of these two waves), are also inaccurate in this region. The more exact formulae for this interference wave will be given in chapter 7. The region in which the formulae for the individual waves lack precision, however, is usually smaller than the region of interference of the waves. Only in very narrow regions close to the critical point are the ray formulae quite inapplicable, so that we can find some properties of the interference wave from the ray formulae. The accuracy of these formulae will be discussed in chapter 7.

We shall now discuss the length of the region of interference of the waves $\mathscr{W}^{mp}$ and $\mathscr{W}^{mnp}$. As a first approximation, we shall not consider the phase shifts of the individual waves and we shall assume that the duration of the pulses of both waves equals δt (the same as that of the pulse of the incident wave, $\mathscr{W}^{m}$).

The travel times of the waves $\mathscr{W}^{mnp}$ and $\mathscr{W}^{mp}$ are as follows:

$$\tau_{mnp} = \frac{r}{V_n}+\frac{h}{V_m}\left[1-\left(\frac{V_m}{V_n}\right)^2\right]^{1/2}+\frac{H}{V_p}\left[1-\left(\frac{V_p}{V_n}\right)^2\right]^{1/2}, \tag{3.123}$$

$$\tau_{mp} = \frac{r}{V_m}\sin\theta_m+\frac{h}{V_m}\cos\theta_m+\frac{H}{V_p}\cos\theta_p.$$

The second of these equations is parametric, with the parameter θ_m, so we must add another equation for the determination of the dependence of r on θ_m, viz.,

$$r = h\tan\theta_m+H\tan\theta_p. \tag{3.124}$$

Here, of course, $\sin\theta_p = (V_p/V_m)\sin\theta_m$.

It is very well known, and we can easily see from (3.123), that right at the critical point $\tau_{mnp} = \tau_{mp}$ (at that point $\sin\theta_m = V_m/V_n$, $\sin\theta_p = V_p/V_n$). The difference $\tau_{mp}-\tau_{mnp}$ is always positive and increases with increasing distance from the critical point (see figure 3.5). If the duration δt of the pulse is larger than the difference $\tau_{mp}-\tau_{mnp}$, the two waves interfere; if it is smaller, both waves exist independently. Hence in the interference zone of the two waves,

$$\tau_{mp}-\tau_{mnp} \leqslant \delta t. \tag{3.125}$$

The end of the interference zone is determined by the expression

$$\tau_{mp}-\tau_{mnp} = \delta t. \tag{3.126}$$

If we insert (3.123) and (3.124) into this expression, we can determine the horizontal distance from the source (r^{I}_{mnp}) at which the interference zone ends.

In the general case of $m \neq p$, it is not possible to find the explicit formula for r^{I}_{mnp}; (3.126) must be solved numerically. For $m = p$ we obtain

$$r^{\mathrm{I}}_{mnm} = r^*_{mnm}+\frac{V_m\delta t}{1-(V_m/V_n)^2}\left\{\frac{V_m}{V_n}+\left[1+\frac{2(h+H)}{V_m\delta t}\sqrt{1-\left(\frac{V_m}{V_n}\right)^2}\right]^{1/2}\right\}. \tag{3.127}$$

Since usually

$$\frac{2(h+H)}{V_m\delta t}\left[1-\left(\frac{V_m}{V_n}\right)^2\right]^{1/2} \gg 1, \tag{3.128}$$

we can simplify equation (3.127) to

$$r^{\mathrm{I}}_{mnm} = r^*_{mnm}+\frac{V_m\delta t}{1-(V_m/V_n)^2}\left\{\frac{V_m}{V_n}+\left[\frac{2(h+H)}{V_m\delta t}\right]^{1/2}\left[1-\left(\frac{V_m}{V_n}\right)^2\right]^{1/4}\right\}. \tag{3.129}$$

This formula was given earlier by Yepinat'yeva (1960) and Červený (1962a, 1966a) for the special case of longitudinal waves.

For the length of the interference zone Δr_{mnp}, we can write

$$\Delta r_{mnp} = r^{\mathrm{I}}_{mnp} - r^{*}_{mnp}. \tag{3.130}$$

For $p = m$, we obtain from (3.129)

$$\Delta r_{mnm} = \frac{V_m \delta t}{1-(V_m/V_n)^2}\left\{\frac{V_m}{V_n}+\left[\frac{2(h+H)}{V_m \delta t}\right]^{1/2}\left[1-\left(\frac{V_m}{V_n}\right)^2\right]^{1/4}\right\}, \tag{3.131}$$

from which it is seen that the interference zone may be often very long,

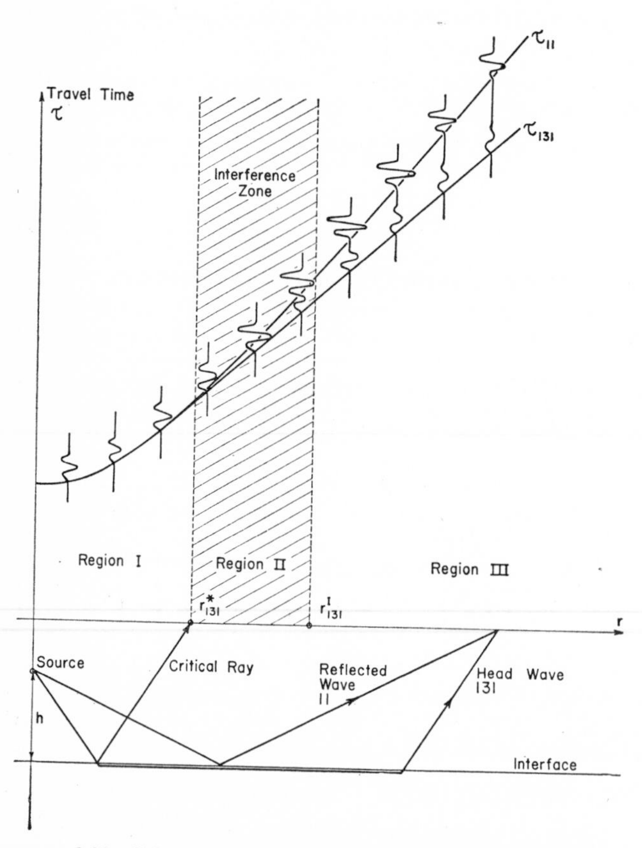

FIGURE 3.38 Schematic representation of the interference zone of the reflected wave 11 and the head wave 131.

especially when there is small velocity contrast at the interface ($V_m \lesssim V_n$) and, of course, for large $V_m \delta t$. In investigations of, for example, the structure of the earth's crust by seismic methods, the length of the interference zone is often tens of kilometres. Then the interference zone of the compressional waves $\mathscr{W}^{11}$ and $\mathscr{W}^{131}$ is very important. For the length of this zone we obtain, from (3.131), the formula

$$\Delta r_{131} = \frac{\alpha_1 \delta t}{1-(\alpha_1/\alpha_2)^2}\left\{\frac{\alpha_1}{\alpha_2}+\left[\frac{2(h+H)}{\alpha_1 \delta t}\right]^{1/2}\left[1-\left(\frac{\alpha_1}{\alpha_2}\right)^2\right]^{1/4}\right\}. \tag{3.132}$$

The values of Δr_{131} for different values of $\frac{1}{2}(h+H)$, $\alpha_1 \delta t$, and α_1/α_2 are given in table 3.3.

If we are interested in properties of the waves $\mathscr{W}^{mp}$ and $\mathscr{W}^{mnp}$, we can divide the profile where measurements are made into three regions:

Region I: $0 \leqslant r \leqslant r^*_{mnp}$. Here only the wave $\mathscr{W}^{mp}$ exists.
Region II: $r^*_{mnp} < r < r^{\mathrm{I}}_{mnp}$. The waves $\mathscr{W}^{mp}$ and $\mathscr{W}^{mnp}$ interfere with each other.
Region III: $r > r^{\mathrm{I}}_{mnp}$. The waves $\mathscr{W}^{mp}$ and $\mathscr{W}^{mnp}$ exist independently of each other.

A schematic picture of these three regions is given in figure 3.38 for compressional waves $\mathscr{W}^{11}$ and $\mathscr{W}^{131}$. Thus we can record the pure H-waves $\mathscr{W}^{mnp}$ only at distances $r > r^{\mathrm{I}}_{mnp}$, and not from the critical point r^*_{mnp}. The interference wave $\mathscr{W}^{mp}+\mathscr{W}^{mnp}$ existing in the interference zone (region II) has properties which are quite different from those of the pure head wave. Its travel time is, of course, given by the travel time of the head wave (τ_{mnp}) but its amplitude properties are quite different, its amplitudes being usually higher than those of head waves. Since the travel time of the interference wave is the same as that of the head wave, the former wave is often taken as a head wave in seismic field measurements. It is, therefore, necessary to derive some properties of this interference wave.

We introduce, in region II, the displacement vector

$$\overline{\mathscr{W}}^{mnp} = \mathscr{W}^{mp}+\mathscr{W}^{mnp}. \tag{3.133}$$

The r and z components of this vector are obtained from (2.187) and (3.121) as

$$\overline{\mathscr{W}}_r^{mnp} = \frac{|R_{mp}|}{L} f(t-\tau_{mp};\ \arg R_{mp}) n_{pr} + \frac{V_n|\Gamma_{mnp}|\tan\theta_m^*}{r^{1/2} l^{3/2}} F(t-\tau_{mnp};\ \arg \Gamma_{mnp}) n^*_{pr}, \tag{3.134a}$$

$$\overline{\mathscr{W}}_z^{mnp} = \frac{|R_{mp}|}{L} f(t-\tau_{mp};\ \arg R_{mp}) n_{pz} + \frac{V_n|\Gamma_{mnp}|\tan\theta_m^*}{r^{1/2}l^{3/2}} F(t-\tau_{mnp};\ \arg \Gamma_{mnp}) n_{pz}^*, \tag{3.134b}$$

TABLE 3.3

Length of the interference zone of compressional waves 11 and 131 (α_1 and α_2 are the compressional wave velocities in the first and the second medium; δt is the duration of the pulse)

$(h+H)/2$ (km)	α_1/α_2	Values of Δr_{131} (km)		
		$\alpha_1\delta t = 1$ km	$\alpha_1\delta t = 2$ km	$\alpha_1\delta t = 3$ km
10	0.75	13	20	25
	0.80	16	24	30
	0.85	20	29	38
	0.90	27	41	52
	0.95	46	71	92
	0.98	96	150	198
20	0.75	18	27	34
	0.80	21	32	40
	0.85	26	39	50
	0.90	36	53	68
	0.95	61	92	118
	0.98	125	192	249
30	0.75	22	32	40
	0.80	26	38	47
	0.85	32	47	59
	0.90	43	63	80
	0.95	72	108	138
	0.98	148	224	288
40	0.75	25	37	46
	0.80	29	43	54
	0.85	36	53	66
	0.90	49	72	90
	0.95	82	122	155
	0.98	167	251	321
50	0.75	28	41	51
	0.80	33	47	59
	0.85	40	58	73
	0.90	54	79	99
	0.95	91	134	170
	0.98	184	275	350

If we know the form of the incident pulse $f(t)$, we can construct, using (3.134), the theoretical form of the interference wave $\overline{\mathcal{W}}^{mnp}$. In figure 3.39, theoretical seismograms of the vertical component of the compressional wave $\overline{\mathcal{W}}^{131} = \mathcal{W}^{11} + \mathcal{W}^{131}$ for the simple model of a one-layer earth's crust ($\alpha_1 = 6.4$ km/sec, $\alpha_2 = 8$ km/sec, $\gamma_1 = \gamma_2 = 1/\sqrt{3}, \rho_1/\rho_2 = 1$, $h = H = 30$ km) are given. The critical distance is 80 km for this model.

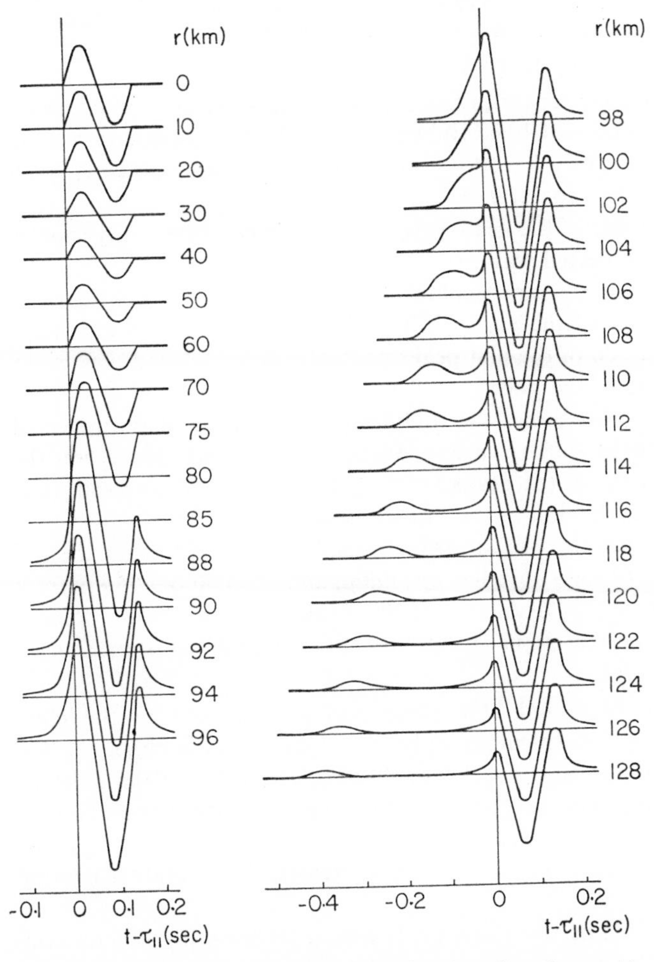

FIGURE 3.39 Theoretical seismogram showing the reflected wave 11 before the critical point and the reflected wave 11 and the head wave 131 beyond the critical point (at a distance of 80 km). The waves 11 and 131 interfere between the critical point and the epicentral distance 108 km (approximately), and after that they are separated.

Before the critical point (region I), the theoretical seismograms of the reflected wave $\mathscr{W}^{11}$ are presented. The source function is sinusoidal with a duration of one period, i.e., $f(t) = \sin \Omega_0 t$ for $0 \leqslant t \leqslant 2\pi/\Omega_0$ and $f(t) = 0$ for $t < 0$ and $t > 2\pi/\Omega_0$ ($\Omega_0 = 2\pi f_0$, $f_0 = 6.4$ cps). Note that $\alpha_1 \delta t = 1$ km in our case. The theoretical seismograms in the close neighbourhood of the critical point (at epicentral distances $r = 80$ and 85 km) are not given in the figure because of increased inaccuracy (see section 3.3.3).

The main conclusions to be drawn from figure 3.39 are:

(1) The head wave separates from the reflected wave at approximately the epicentral distance of 105–110 km. The length of the interference zone is therefore about 25–30 km (this corresponds to the value of $\Delta r^{\mathrm{I}}_{131} = 26$ km given in table 3.3 for $\alpha_1 \delta t = 1$ km, $\alpha_1/\alpha_2 = 0.8$.

(2) The amplitude of the head wave in region III is substantially smaller than that of the reflected wave.

(3) The character of the pulse of the reflected wave changes considerably under the influence of the phase shift in the manner described in section 2.6.5. The amplitude of the first peaks gradually decreases and new peaks appear at the end of the pulse.

(4) The figure also shows the time lead of the peaks of the reflected pulse, i.e. the peaks shift towards the onset of the pulse. Therefore, the time-distance curves of the peaks are not parallel to the theoretical time-distance curves of the first arrivals of the reflected wave.

(5) By measuring the peak-to-trough amplitudes we find that the impulse of the interference wave has maximum amplitude about 95 km from the source. Before this distance, the amplitude increases monotonically and beyond it it decreases. The properties of the amplitude-distance curves will be discussed later.

We have not presented here theoretical seismograms in the close neighbourhood of the critical point, but it will be shown in chapter 7 that there are no peculiarities in behaviour there. The form of the reflected pulse changes only very little on passing through the critical point and the amplitudes increase monotonically there. The maximum of the amplitude curve at about 95 km from the source is, therefore, the absolute maximum of this curve.

Equations (3.134) hardly allow us, however, to draw any general conclusions about the dynamic properties (amplitudes, spectra) of the interference wave. We shall therefore investigate the properties of the interference wave $\overline{\mathscr{W}}^{mnp}$ in the frequency domain. We introduce

$$\overline{\mathbf{W}}^{mnp} = \mathbf{W}^{mp} + \mathbf{W}^{mnp} \tag{3.135}$$

For the r and z components we can write

$$\bar{W}_r^{mnp} = A_r^{mp} \exp[i\omega(t-\tau_{mp})+i\chi_r^{mp}]+A_r^{mnp} \exp[i\omega(t-\tau_{mnp})+i\chi_r^{mnp}],$$
$$\bar{W}_z^{mnp} = A_z^{mp} \exp[i\omega(t-\tau_{mp})+i\chi_z^{mp}]+A_z^{mnp} \exp[i\omega(t-\tau_{mnp})+i\chi_z^{mnp}], \quad (3.136)$$

which may be written in the form

$$\bar{W}_r^{mnp} = \bar{A}_r^{mnp} \exp[i\omega(t-\tau_{mnp})+i\bar{\chi}_r^{mnp}],$$
$$\bar{W}_z^{mnp} = \bar{A}_z^{mnp} \exp[i\omega(t-\tau_{mnp})+i\bar{\chi}_z^{mnp}], \quad (3.137)$$

where $\bar{A}_r^{mnp}$ and $\bar{A}_z^{mnp}$ are the amplitudes of the horizontal and vertical components of the displacement vector of the wave and $\bar{\chi}_r^{mnp}$ and $\bar{\chi}_z^{mnp}$ are their phase shifts.

From (3.136) and (3.137) we obtain the expressions for the amplitudes of the components of the interference wave:

$$\bar{A}_r^{mnp} = \{(A_r^{mp})^2+(A_r^{mnp})^2+2A_r^{mp}A_r^{mnp} \cos \Theta^{mnp}\}^{1/2},$$
$$\bar{A}_z^{mnp} = \{(A_z^{mp})^2+(A_z^{mnp})^2+2A_z^{mp}A_z^{mnp} \cos \Theta^{mnp}\}^{1/2}, \quad (3.138)$$

where

$$\Theta^{mnp} = \omega(\tau_{mp}-\tau_{mnp})+\arg \Gamma_{mnp}-\arg R_{mp}-\tfrac{1}{2}\pi. \quad (3.139)$$

In this expression $\arg \Gamma_{mnp}$ is always independent of the epicentral distance. The function $\arg R_{mp}$ depends on the distance, but always lies in the range $0 \leqslant \arg R_{mp} < 2\pi$. Therefore, the dependence of Θ^{mnp} on the epicentral distance will be given mainly by the first term, $\omega(\tau_{mp}-\tau_{mnp})$, except very close to the critical point where $\tau_{mp} \simeq \tau_{mnp}$. As $\tau_{mp}-\tau_{mnp}$ increases with increasing distance, Θ^{mnp} is usually a monotonically increasing function of the epicentral distance. The amplitude–distance curve of the interference wave will, therefore, be an oscillating function since (3.138) contains the factor $\cos \Theta^{mnp}$. When the functions A_r^{mp}, A_z^{mp}, A_r^{mnp}, and A_z^{mnp} do not change very rapidly with the epicentral distance, the maxima of the interference amplitude-distance curve lie at the points where

$$\Theta^{mnp} = 2\pi k \qquad (k = 0, 1, \ldots). \quad (3.140)$$

Using (3.139) and (3.140), we can write

$$\tau_{mp}-\tau_{mnp} = (1/2\pi)\{(2k+\tfrac{1}{2})\pi+\arg R_{mp}-\arg \Gamma_{mnp}\}T, \quad (3.141)$$

where T is the period of the *harmonic* wave ($T = 1/f = 2\pi/\omega$).

In (3.141), $\arg R_{mp}$ is a function of the distance from the source, but it plays the role of a correction term only. Therefore, we can use the method of successive approximations for the solution of this equation. First we insert arg $R_{mp} = 0$ and calculate the first approximations for the epicentral distances corresponding to the maxima of the amplitude curve. There we calculate the values of arg R_{mp} and, inserting into (3.141), we find the second approximations for the positions of the maxima. If the value of arg R_{mp} changes only slowly at the distances where the maxima are expected to lie, we can take arg R_{mp} to be constant. Then we can write

$$\tau_{mp} - \tau_{mnp} = \eta_k T, \tag{3.142}$$

where η_k is a constant given by

$$\eta_k = k + \tfrac{1}{4} + (1/2\pi)(\arg R_{mp} - \arg \Gamma_{mnp}). \tag{3.143}$$

Equation (3.142) has the same nature as the condition for the end of the interference zone, (3.126). We can solve it and find the position of the maxima of the amplitude curve.

We shall not be interested here in the position of all maxima, but only in the position of the first maximum (for $k = 0$), which is of the greatest importance. For example, in investigating the earth's crust using compressional reflected and head waves, this maximum is also the absolute maximum for the reflected wave. Thus, the best conditions for the reception of waves generated on the interface exist at the distance mentioned above. Note that this maximum was mentioned in the discussion of figure 3.39.

In figure 3.40, the amplitude curve for $\bar{A}_z^{131}$ is given for the same model

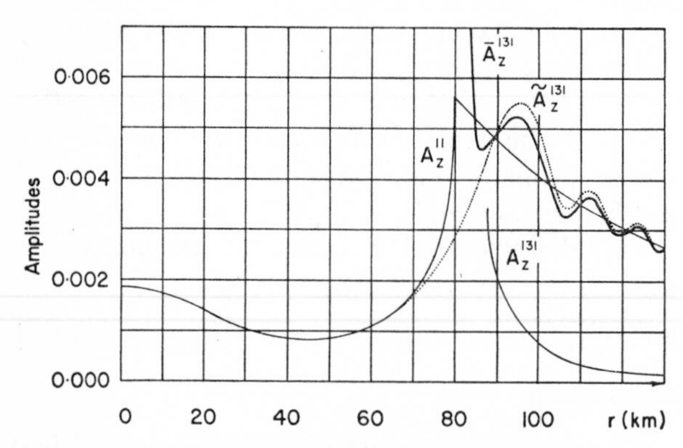

FIGURE 3.40 Comparison of the amplitude–distance curves of the reflected wave (A_z^{11}), head wave (A_z^{131}), and interference wave ($\bar{A}_z^{131}$) – all calculated by the ray method – and the interference wave ($\tilde{A}_z^{131}$) calculated by more exact methods. $\alpha_1/\alpha_2 = 0.8$, $\gamma_1 = \gamma_2 = 0.577$, $\rho_1/\rho_2 = 1$, $h = H = 30$ km, $f = 6.4$ cps.

of the earth's crust as in figure 3.39 (the frequency is assumed to be 6.4 cps). For comparison, the amplitude curve of the interference wave calculated by exact methods ($\tilde{A}_z^{131}$; see chapter 7) and the amplitude curves of the reflected wave A_z^{11} and the head wave A_z^{131}, calculated by the ray method, are also given. We can see from this figure that the curve $\bar{A}_z^{131}$ differs only slightly in character from that obtained by exact methods, except in the close neighbourhood of the critical point. The epicentral distance of the maximum of the amplitude curve $\bar{A}_z^{131}$ practically coincides with the position of the maximum of the exact amplitude curve. We can, therefore, seek the distance where the amplitude is maximum from (3.142).

We substitute $k = 0$ in (3.142) and denote the epicentral distance at which the maximum lies by r^M_{mnp}. As in the determination of r^I_{mnp}, we are not able to find an explicit formula for r^M_{mnp} if $m \neq p$ (and the equation (3.142) must be solved numerically).

For $m = p$, we obtain, as in (3.129),

$$r^M_{mnm} = r^*_{mnm} + \frac{V_m \eta_0}{f(1-(V_m/V_n)^2)}\left\{\frac{V_m}{V_n} + \left[\frac{2(h+H)f}{V_m \eta_0}\right]^{1/2}\left[1-\left(\frac{V_m}{V_n}\right)^2\right]^{1/4}\right\}. \tag{3.144}$$

For the compressional wave $\overline{\mathbf{W}}^{131}$, we find from (3.144) that

$$r^M_{131} = r^*_{131} + \xi\left(\frac{r^*_{131}}{f}\right)^{1/2} + \frac{1}{2f}\xi^2\left(\frac{\alpha_1}{\alpha_2}\right)^2, \tag{3.145}$$

or

$$r^M_{131} = r^*_{131} + \tilde{\xi}(\Lambda_1 r^*_{131})^{1/2} + \frac{\Lambda_1}{2}\tilde{\xi}^2\left(\frac{\alpha_1}{\alpha_2}\right)^2, \tag{3.146}$$

where

$$\xi = (2\eta_0\alpha_2)^{1/2}\left[1-\left(\frac{\alpha_1}{\alpha_2}\right)^2\right]^{-1/2}, \qquad \tilde{\xi} = (2\eta_0)^{1/2}\left(\frac{\alpha_1}{\alpha_2}\right)^{-1/2}\left[1-\left(\frac{\alpha_1}{\alpha_2}\right)^2\right]^{-1/2} \tag{3.147}$$

and Λ_1 is the wavelength of the incident wave ($\Lambda_1 = \alpha_1/f$).

The above formulae allow us to calculate r^M_{131} if η_0 is known. From (3.143) we find that $\eta_0 = \frac{1}{4}+(1/2\pi)\arg R_{11}$. For $\arg R_{11}$ we can take some average value, as was mentioned earlier. It was shown by Červený (1963b) that for the broad range of frequencies and parameters of the interfaces representative of the earth's crust, we can take (approximately) $\arg R_{11} = 3/4$ in the above formula for η_0. We then obtain (approximately) $\eta_0 = 0.37$. Therefore, $(2\eta_0)^{1/2} = 0.86$ and

$$\xi \sim 0.86\,\alpha_2^{1/2}\left[1-\left(\frac{\alpha_1}{\alpha_2}\right)^2\right]^{-1/2}, \qquad \tilde{\xi} \sim 0.86\left(\frac{\alpha_1}{\alpha_2}\right)^{-1/2}\left[1-\left(\frac{\alpha_1}{\alpha_2}\right)^2\right]^{-1/2}. \tag{3.147'}$$

Table 3.4 gives the values of r_{131}^{M} (found by using (3.146) and (3.147′)) for several models of interface and several wavelengths. We see from this table that the distance of the maximum beyond the critical point increases with increasing depth of the interface, increasing wavelength (or decreasing frequency), and decreasing departure of α_1/α_2 from unity. These conclusions can, of course, be derived directly from (3.144) through (3.147′).

Since all the calculations are only approximate (i.e. valid only for high frequencies), the second and third terms in (3.145) have the nature of correction terms. We can neglect the third term and insert $r_{131}^{*} \sim r_{131}^{M}$ in the second term. Then we have

$$r_{131}^{M} = r_{131}^{*} + \xi(r_{131}^{M}/f)^{1/2}. \tag{3.148}$$

Let us consider the significance of this relation. Assume that we have made seismic measurements on some profile and obtained the pulses of the interference wave. If the spectrum of this wave is so wide that the amplitude–distance curves can be found for a wide enough range of frequencies, we can obtain different r_{131}^{M} for different frequencies f. We can now plot the dependence of r_{131}^{M} on $(r_{131}^{M}/f)^{1/2}$. According to (3.148), this dependence will be linear, and the line will intersect the vertical axis at a distance r_{131}^{*} from the origin. It is thus possible to determine the position of the critical point (critical distance r_{131}^{*}) without knowing anything about the parameters of the interface.

The above formulae can be invalid when the functions A_r^{mp}, A_z^{mp}, A_r^{mnp}, and A_z^{mnp} change very rapidly in the neighbourhood of the critical point (see formulae (3.138)). This applies especially to the case of large velocity contrast at the interface, liquid–solid interface, etc.

Note that formulae (3.144) to (3.146) give the position of the maximum of the amplitude–distance curve for *harmonic* waves. In reality, however, we use not harmonic but impulse waves. The relation between the position of the maximum of the amplitude curve of *harmonic* and *impulse* waves is not obvious, as the impulse signal is given by the continuous spectrum of frequencies, and the position of the maximum for each frequency is different. However, when the amplitude spectrum of the incident impulse wave is narrow, with the prevailing frequency f_M, then the maximum of the amplitude–distance curve of the impulse interference reflected-head wave lies at about the same epicentral distance as the maximum of the harmonic amplitude–distance curve constructed for the frequency f_M. In the first approximation in this case, the formulae for the position of the maximum of the harmonic amplitude–distance curve may therefore also be used for the impulse wave. For example, we can see from figures 3.39

and 3.40 that the maximum amplitudes are observed at about 95 km in both cases.

In section 3.6, the amplitudes of reflected and head waves and their ratios were studied. We found that the amplitudes of head waves are usually smaller than those of reflected waves. The ratio of these amplitudes is quite dependent on the epicentral distance. To calculate the ratio at a given

TABLE 3.4

Values of r_{131}^{M} for different $(h+H)/2$, α_1/α_2, and λ (km)

$(h+H)/2$ (km)	α_1/α_2	Values of r_{131}^{M} (km)			
		$\Lambda = 0$	$\Lambda = 1$	$\Lambda = 2$	$\Lambda = 3$
10	0.75	22	31	34	38
	0.80	27	36	40	44
	0.85	32	44	49	54
	0.90	41	57	64	70
	0.95	61	87	100	110
	0.98	98	151	178	202
20	0.75	45	56	61	65
	0.80	53	66	72	77
	0.85	65	80	87	93
	0.90	83	103	113	121
	0.95	122	157	173	187
	0.98	197	268	302	331
30	0.75	68	81	87	92
	0.80	80	95	102	108
	0.85	97	116	124	131
	0.90	124	149	161	170
	0.95	183	225	244	260
	0.98	295	380	420	453
40	0.75	91	106	113	118
	0.80	107	124	132	138
	0.85	129	151	160	167
	0.90	165	194	207	217
	0.95	243	291	313	331
	0.98	394	490	535	572
50	0.75	113	130	138	144
	0.80	133	153	161	168
	0.85	161	185	196	204
	0.90	206	238	253	264
	0.95	304	357	381	401
	0.98	492	599	648	688

distance, we must know the amplitudes of both waves. However, there are some points where the formula for this ratio becomes very simple. One of these points is the end of the interference zone. We shall again consider only the waves 11 and 131 and denote the ratio of the amplitudes A_z^{131} and A_z^{11} at the end of the interference zone by $(A_z^{131}/A_z^{11})_{\mathrm{I}}$. We shall assume that the amplitude spectrum of the impulse is narrow and that it has only one maximum at the frequency f_M. Using (2.150), (3.63), and (3.132), we find for $f = f_M$, after some simplification, that

$$(A_z^{131}/A_z^{11})_{\mathrm{I}} = F_{\mathrm{I}}\left(\frac{h+H}{2\Lambda_1}\right)^{-1/4}\left(\frac{\alpha_1\delta t}{\Lambda_1}\right)^{-3/4}\frac{1}{|R_{11}|},$$

where $F_{\mathrm{I}} = \Gamma_{131}[1-(\alpha_1/\alpha_2)^2]^{3/8}/[4\pi(\alpha_1/\alpha_2)^{1/2}]$, R_{11} is the R-coefficient, δt the duration of the impulse, and $\Lambda_1 = \alpha_1/f_M$. We shall now consider interfaces with a small velocity contrast. Then we can put $|R_{11}| \sim 1$. If the pulse is nearly sinusoidal with the duration of one period, we have $\alpha_1\delta t \sim \Lambda_1$. Therefore,

$$(A_z^{131}/A_z^{11})_{\mathrm{I}} = F_{\mathrm{I}}\left(\frac{h+H}{2\Lambda_1}\right)^{-1/4}.$$

When the duration of the pulse is s periods, we must multiply the above formula by $s^{-3/4}$.

The function F_{I} varies only slightly with the parameters of the media. For example, for $\gamma_1 = \gamma_2 = 1/\sqrt{3}$ and $0.7 < \alpha_1/\alpha_2 < 0.98$, it lies between 0.2 and 0.25. Thus, we can write approximately

$$(A_z^{131}/A_z^{11})_{\mathrm{I}} \sim 0.22\left(\frac{h+H}{2\Lambda_1}\right)^{-1/4}.$$

From this we have $0.07 < (A_z^{131}/A_z^{11})_{\mathrm{I}} < 0.12$ for $10 < (h+H)/2\Lambda_1 < 100$. The general result is, therefore, that the head wave 131 has, at the end of the interference zone, an amplitude about one order of magnitude smaller than the reflected wave 11 (if the velocity contrast at the interface is small) and that $(A_z^{131}/A_z^{11})_{\mathrm{I}}$ depends only slightly on the parameters of the media and on the frequency. If the duration of the pulse is longer than one period, $(A_z^{131}/A_z^{11})_{\mathrm{I}}$ is even smaller. This conclusion may not hold if there is a large velocity contrast (since the function $|R_{11}|$ can be very different from 1 in this case) or if the interference zone is very long (small frequency and/or α_1/α_2 near 1). Note that similar formulae and conclusions are applicable for $(A_r^{131}/A_r^{11})_{\mathrm{I}}$ also.

The other distance at which the ratio of the amplitudes of the head and reflected waves may be easily determined is that for which the amplitude–

distance curve of the interference reflected-head wave has its maximum value, viz., r^{M}_{131}. We denote this ratio by $(A_z^{131}/A_z^{11})_M$. At this point the two waves are not yet separated from each other, and the ratio determines their relative contribution to the resulting wave. Using (2.150), (3.63), (3.146), and (3.147′), we find approximately

$$(A_z^{131}/A_z^{11})_M = F_M\left(\frac{h+H}{2\Lambda_1}\right)^{-1/4}\frac{1}{|R_{11}|},$$

where $F_M = 2.1\,F_{\mathrm{I}}$. It follows that the ratio of the amplitudes of the H and R waves at the epicentral distance r^{M}_{131} is about twice as great as at the distance r^{I}_{131} (when $\delta t = 1/f_M$). For $\gamma_1 = \gamma_2 = 1/\sqrt{3}$, $0.7 < \alpha_1/\alpha_2 < 0.98$, and $10 < (h+H)/2\Lambda_1 < 100$, we have $0.15 < (A_z^{131}/A_z^{11})_M < 0.25$.

3.9
GENERAL EXPRESSIONS FOR HEAD WAVES

In the previous sections we have derived expressions for head waves generated at a plane interface between two homogeneous media when a spherical wave is incident on it. In reality, however, we meet this situation very rarely. For example, if we study the propagation of head waves in layered media, the incident wave responsible for the H-wave is usually some type of multiply reflected and refracted wave with an arbitrarily shaped wave front, the media are very often inhomogeneous, and the interface curved. We cannot use the ray theory to study head waves in all these situations, although it does enable us to find considerably more general expressions for head waves than have been derived in this chapter.

In this section we shall find general expressions connecting head waves with some zero-order waves (e.g., the wave incident on the interface or some types of reflected or refracted waves). The formulae for the zero-order waves were found in chapter 2 for many types of media, and these general expressions will make it possible to investigate more complicated situations than the incidence of a spherical wave on a plane interface between two homogeneous media (see following chapters).

We shall consider here mainly head waves satisfying the following conditions:

(1) The interface at which the head wave is generated is of first order.

(2) The wave incident on the interface is a simple, non-interfering wave.

(3) The R-wave which propagates along the interface, and is responsible for generation of the head wave, is a simple, non-interfering wave.

(4) The ray of the R-wave which corresponds to the critical angle of incidence is parallel to the interface (see O^*A in figure 3.2).

We shall call head waves satisfying the above conditions *pure head waves.* Apart from a few exceptions, the ray theory cannot be used to study head waves which are not pure. The physical meaning of the first condition is quite clear. The second and the third conditions guarantee that we can write a ray series for the incident wave and for the R-wave responsible for generation of the head wave, except near some singular points. If the fourth condition is not fulfilled and no ray of the R-wave is parallel to the interface, head waves in a pure form cannot arise.

What do we understand by *non-interfering* waves? In the *time domain*, when we study propagation of impulses of short duration, the meaning of the term is clear. In the *frequency domain*, we call the wave $\mathbf{W}(M)$ with the phase function $\tau(M)$ a non-interfering wave at M if the phase functions

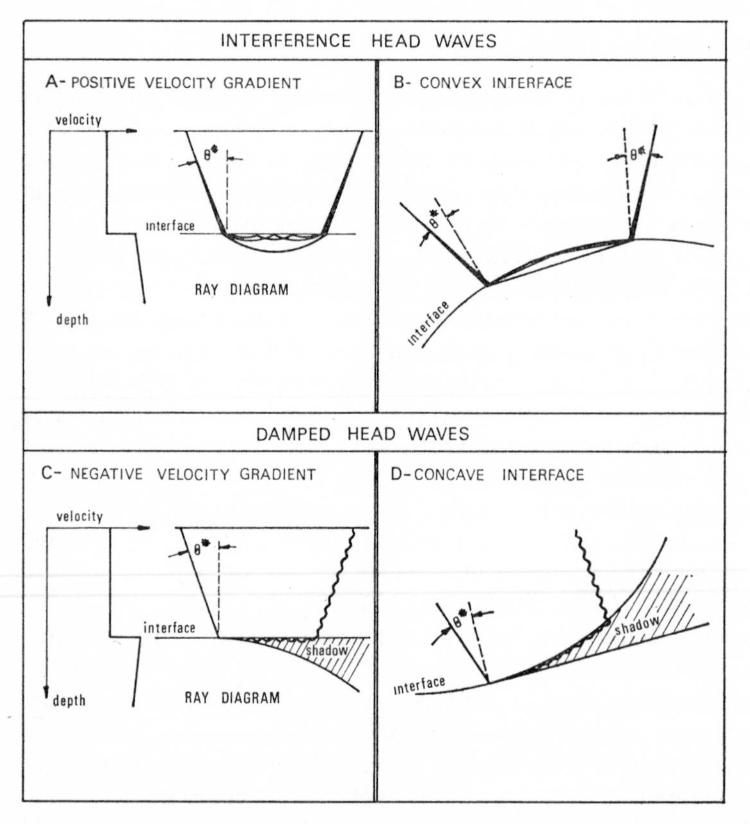

FIGURE 3.41 Schematic representation of interference and damped head waves.

$\tau_s(M)$ of all other waves arriving at M satisfy the condition $|\tau(M) - \tau_s(M)| > \delta$, where δ is some constant. We can, for example, put $\delta = T$, where T is the period of the harmonic process ($T = 2\pi/\omega$). We do not need to specify δ in more detail for the considerations that follow.

Many types of head waves exist which do not fulfil all of the above conditions and therefore cannot be investigated by the ray method. For example, head waves from a thin layer do not satisfy condition 3, and head waves from a transition layer do not fulfil conditions 1, 2, and 3. Some other waves which do not satisfy these conditions have travel time–distance curves very similar to those of pure head waves, but not exactly the same. These are also very often called head waves. To investigate waves which do not fulfil the above conditions, we must use either wave methods (see sections 1.1 and 7.3) or modifications of the ray method, such as the parabolic equation method (see Yanovskaya, n.d.). We shall present here, for illustrative purposes, some examples of waves which are kinematically very similar to head waves but are not pure head waves since they do not fulfil condition 3 or 4. These waves can often be very important in seismological applications (see figure 3.41).

In figure 3.41A, we consider a situation with a positive velocity gradient below the plane interface. When the angle of incidence is a little less than the critical angle, the ray of the R-wave has a turning point in the second medium, and later strikes the interface. It can be reflected from the interface (from below) and strike it again, and so on. Thus, the wave propagating along the interface which is responsible for the generation of head waves has a clear interference character, as it is composed of waves reflected from the interface, from below, 0, 1, 2, ... times. These waves gradually separate out from the interference complex as the distance increases. First to separate is the wave without any reflection from below the interface (this is sometimes called the refracted or diving wave), next the wave with one reflection, then the wave with two reflections, and so on. For individual waves which become separated from the interference complex, the standard zero-order formulae of the ray theory can usually be used (see chapter 2). However, the ray theory cannot be applied to the interference wave which propagates along the interface and is responsible for the generation of head waves. If the velocity gradient is small and the epicentral distance is not large, the wave radiated into the first medium by the interference wave propagating along the interface will be kinematically very similar to the pure head wave. We shall call this wave the *interference head wave*. The amplitude and spectral properties of an interference head wave are very different from those of a pure head wave (which arises when the velocity gradient vanishes) even for very small gradients. Generally, amplitudes of an interference head wave are considerably higher than those of the pure head wave. Properties of interference head waves in situations involving small velocity gradients have been investigated in detail by Chekin (1965, 1966), Buldyrev and Lanin (1965, 1966), and Lanin (1966, 1968). We shall deal with this problem in chapter 6.

A similar interference head wave can arise from a curved interface between two homogeneous media when the interface is convex for an observer situated on the side of the incident wave (for basic head waves, see figure 3.41B). This situation has also been investigated by Buldyrev and Lanin (1965, 1966).

Quite a different situation arises if there is a negative velocity gradient below the interface (see figure 3.41C). Then the ray of the R-wave corresponding to the critical angle of incidence is not parallel to the interface, but deviates into the second medium, and a shadow zone is formed along the interface beyond the critical point. Thus, condition 4 is not fulfilled. However, we know from wave theory that some diffracted waves exist in the shadow zone, too. These waves, propagating along the interface, will radiate some weak waves into the first medium (see wavy line in figure 3.41C). The kinematic properties of these waves are very similar to those of pure head waves (which would exist as the velocity gradient vanishes), but their amplitudes will be damped with increasing distance more rapidly than those of pure head waves. These waves, which may be called *damped head waves*, have been investigated by Chekin (1964). Similar waves are generated at a curved interface between two homogeneous media when the interface is concave for an observer situated on the side of the incident wave (for basic head waves, see figure 3.41D).

The interference and damped head waves are of considerable importance in the interpretation of seismic refraction measurements, especially in crustal explosion seismology, for we could hardly expect the material in the earth's crust to be perfectly homogeneous (i.e. the velocity does not depend, even slightly, on depth) and all the interfaces to be perfect planes. Even very small departures from the ideal case of plane interfaces and a homogeneous substratum cause significant changes in head wave amplitudes. In particular, interference head waves are very important, as their amplitudes can be rather large (see detailed discussion in chapter 6).

The interference and damped head waves cannot arise when the interface is *plane* and when the medium in which the ray of the R-wave responsible for the generation of head waves) runs along the interface (i.e. is parallel to it) is *homogeneous*. In this section, we shall consider only those situations where interference and damped head waves do not arise. It should be noted, however, that these restrictions (such as homogeneity of the substratum) are not quite necessary for the generation of pure head waves; they can exist in some other situations also (e.g., horizontal gradients of velocity in the substratum).

We shall call the medium in which the incident wave propagates the M-medium, the medium in which the ray of the head wave runs along the interface the N-medium, and the medium in which the head wave propagates away from the interface the P-medium. Two of these media can, of course, coincide (for head waves of the fourth kind all the media are the same). In seismic refraction experiments, the M and P media almost always correspond to the overburden and the N-medium to the substratum. As was mentioned earlier, we assume that the *N-medium is homogeneous.* The M and P media may be inhomogeneous (with gradients of velocity small in comparison with the frequency), and the wave front of the incident wave may be of arbitrary shape. Also, we shall assume here that all four conditions formulated above are fulfilled.

Note that the assumption that the N-medium is homogeneous makes the P-medium also homogeneous for head waves of the second kind (see figure 3.6), the M-medium homogeneous for head waves of the third kind, and both the M and P media homogeneous for head waves of the fourth kind. However, for basic head waves, which are most important in seismic measurements, both M and P media may be inhomogeneous.

We shall now derive two expressions connecting the H-wave $\mathbf{W}^{mnp}$ with the R-wave $\mathbf{W}^{mn}$. We shall obtain the formulae for the head wave only at the point A, lying on the interface (see figure 3.2), for we can then use the methods for continuation of the displacement vector along a ray (see chapter 2) to determine the H-wave at an arbitrary point of the ray in the P-medium.

In fact, we have derived one general expression earlier. The procedure used in the beginning of this chapter, especially the application of interface conditions, did not depend on the shape of the wave front of the incident wave or on possible inhomogeneities of the M and P media. The expression (3.46) is quite general. Only later in this chapter did we use the assumption that the incident wave is spherical and is given by (3.1).

We shall now assume that the displacement vector of the incident wave, $\mathbf{W}^m$, is given by

$$\mathbf{W}^m = U^m \exp[i\omega(t-\tau_m)]\mathbf{n}_m, \qquad (3.149)$$

where U^m, τ_m, and $\mathbf{n}_m$ may be arbitrary, keeping in mind the general limitations imposed by the ray theory. (In the special cases when $U^m = 1/R_0$ and $\tau_m = R_0/V_m$, (3.149) yields the spherical wave (3.1).)

If we know $\mathbf{W}^m$, we can determine the displacement vector $\mathbf{W}^{mn}$ of the

R-wave (responsible for the generation of the head wave) by using the methods outlined in chapter 2, i.e.,

$$\mathbf{W}^{mn} = U^{mn} \exp[i\omega(t-\tau_{mn})]\mathbf{n}_n. \tag{3.150}$$

The displacement vector of the head wave at the point A lying on the interface is then given by

$$\mathbf{W}^{mnp}(A) = U^{mnp}(A) \exp\{i\omega[t-\tau_{mnp}(A)]\}\mathbf{n}_p(A). \tag{3.151}$$

We now want to determine $U^{mnp}(A)$, $\tau_{mnp}(A)$, and $\mathbf{n}_p(A)$ from the knowledge of U^m, τ_m, and $\mathbf{n}_m$ or U^{mn}, τ_{mn}, and $\mathbf{n}_n$. As the unit vector $\mathbf{n}_p(A)$ can be determined simply, we shall be interested here only in the functions $U^{mnp}(A)$ and $\tau_{mnp}(A)$.

As mentioned above, a general formula connecting $\mathbf{W}^{mnp}$ with $\mathbf{W}^{mn}$ is given by (3.46). From (3.46) and (3.19′), we obtain

$$\begin{aligned} U^{mnp}(A) &= \frac{1}{i\omega} V_n M_{np}(A) \frac{\partial U^{mn}(A)}{\partial H}, \\ \tau_{mnp}(A) &= \tau_{mn}(A). \end{aligned} \tag{3.152}$$

This is our first general expression for head waves. The derivative $\partial U^{mn}(A)/\partial H$, denoting $\partial U^{mn}/\partial H$ at the point A, can be found if we know the analytic expressions for U^{mn} (see chapter 2). The function $M_{np}(A)$ is given by (3.43) or as a solution of the set of equations (3.39). Formula (3.152) can be used for an arbitrarily shaped incident wave front and for inhomogeneous M and P media. Note that $M_{np}(A)$ depends on the position of the point A when the P-medium is horizontally inhomogeneous. When we calculate $M_{np}(A)$ from (3.43) or (3.39), we must insert *local* velocities and densities at A. When the P-medium is only vertically inhomogeneous, $M_{np}(A)$ is constant and does not depend on the position of the point A.

We can easily derive another simple expression for head waves from (3.152). We shall define the function $\tilde{U}^{mn}$ by the relation

$$U^{mn} = \tilde{U}^{mn} \cos \theta_n. \tag{3.153}$$

Then, from (3.47),

$$\frac{\partial U^{mn}(A)}{\partial H} = \frac{1}{l}(\tilde{U}^{mn})_{\theta_n=\pi/2}, \tag{3.154}$$

where l is the distance between O^* and A. Inserting this into (3.152) yields

$$\begin{aligned} U^{mnp}(A) &= \frac{1}{i\omega} V_n M_{np}(A)(\tilde{U}^{mn})_{\theta_n=\pi/2}, \\ \tau_{mnp}(A) &= \tau_{mn}(A). \end{aligned} \tag{3.155}$$

This is another general expression for head waves. It can be used under the same circumstances as (3.152). Note that the expression for $\tilde{U}^{mn}$ can be easily found from known U^{mn}, as the expression for U^{mn} usually contains the multiplication factor $\cos\theta_n$.

The formulae (3.152) and (3.155) connect the head wave $\mathbf{W}^{mnp}$ with the R-wave $\mathbf{W}^{mn}$, which is responsible for its generation. Now we shall derive two very important relations between the *head wave* $\mathbf{W}^{mnp}$ and the *incident wave* $\mathbf{W}^{m}$. For this purpose, we first find the relation between U^{mn} and U^{m} and then use (3.152) or (3.155). This relation is given by (2.95). If we suppose that the incident wave is known at the point O, lying just at the interface, and that the medium in which the R-wave propagates is homogeneous, we can write

$$U^{mn}(\bar{M}) = U^{m}(O)R_{mn}G(\theta_n), \tag{3.156}$$

where

$$G(\theta_n) = \left(\frac{\mathrm{d}\sigma'(O)}{\mathrm{d}\sigma(\bar{M})}\right)^{1/2} \tag{3.156'}$$

(see (2.95)). $\mathrm{d}\sigma'(O)$ and $\mathrm{d}\sigma(\bar{M})$ are the cross-sectional areas of an elementary ray tube of the R-wave at the point O (on the side of the generated wave) and at the point $\bar{M}$, respectively. The function G does not depend only on θ_n, of course, but also on other parameters. One possible form of the function $G(\theta_n)$ is given by (2.62), where θ_n corresponds to θ_ν.

Since $U^{m}(O) \to U^{m}(O^*)$ and $R_{mn} \to R^*_{mn}$ (see (3.13) for R^*_{mn}) for $\theta_n \to \frac{1}{2}\pi$, we obtain from (3.153)

$$\frac{\partial U^{mn}(A)}{\partial H} = \frac{U^{m}(O^*)R^*_{mn}}{l}\left(\frac{G(\theta_n)}{\cos\theta_n}\right)_{\theta_n=\pi/2}.$$

Inserting this into (3.152), we obtain a third general expression for head waves:

$$U^{mnp}(A) = U^{m}(O^*)\frac{V_n M_{np}(A)R^*_{mn}}{i\omega l}\left(\frac{G(\theta_n)}{\cos\theta_n}\right)_{\theta_n=\pi/2},$$
$$\tau_{mnp}(A) = \tau_m(O^*)+l/V_n. \tag{3.157}$$

These formulae give the relation between a head wave generated on the interface and the wave incident on it. The function $U^{m}(O^*)$ and the R-coefficient R^*_{mn} are to be determined at the point O^* where the critical ray strikes the interface.

When the M and the P media are inhomogeneous only in the vertical direction, then the values R^*_{mn} and $M_{np}(A)$ do not depend on the position of the points O^* and A. We recall that the H-coefficient Γ_{mnp} was introduced for homogeneous media as a product $R^*_{mn} M_{np}$ (see (3.50)). Thus, if the M and P media are vertically inhomogeneous, we obtain from (3.157)

$$U^{mnp}(A) = U^{m}(O^*)\frac{V_n\Gamma_{mnp}}{i\omega l}\left(\frac{G(\theta_n)}{\cos\theta_n}\right)_{\theta_n=\pi/2}. \tag{3.157'}$$

Now we shall use (2.62) and express $G(\theta_n)/\cos\theta_n$ in terms of the principal radii of curvature of the incident wave at O^*. We obtain

$$\left(\frac{G(\theta_n)}{\cos\theta_n}\right)_{\theta_n=\pi/2} = \{\cos^2\theta_n+(A+B)l\cos^2\theta_n+(AB\cos^2\theta_n-C^2)l^2\}^{-1/2}_{\theta_n=\pi/2},$$

where A, B, and C are given by (2.60). (We must remember that our notation in this section is different from that in section 2.3.2, and insert in (2.60) $\theta_\nu = \theta_n$, $\theta_0 = \theta_m$, $V_\nu = V_n$, $V_0 = V_m$.) Since $B\cos^2\theta_n \to 0$ for $\theta_n \to \frac{1}{2}\pi$, we can write

$$\left(\frac{G(\theta_n)}{\cos\theta_n}\right)_{\theta_n=\pi/2} = l^{-1/2}\{A\cos^2\theta_n(1+Bl)-C^2l\}^{-1/2}_{\theta_n=\pi/2}. \tag{3.158}$$

We shall give below the final formulae only for the case when the M-medium is homogeneous. Since $1/R_{\|} = 1/R_{\perp} = 0$ (the interface is plane), we obtain from (2.60), using Euler's formulae,

$$\begin{aligned}\{A\cos^2\theta_n(1+Bl)\}_{\theta_n=\pi/2} &= \frac{V_n\cos^2\theta_m^*}{V_m r_{\|}(O^*)}\left(1+\frac{lV_n}{r_{\perp}(O^*)V_m}\right),\\ \{C^2l\}_{\theta_n=\pi/2} &= \frac{V_n^2\cos^2\theta_m^*\tan^2 2\varphi\, l}{4V_m^2}\left(\frac{1}{r_{\|}(O^*)}-\frac{1}{r_{\perp}(O^*)}\right)^2,\end{aligned} \tag{3.159}$$

where $r_{\|}(O^*)$ and $r_{\perp}(O^*)$ are the radii of curvature of the wave front of the incident wave at O^* associated with the normal sections of the wave front with $E_{\|}$ and with the plane perpendicular to it, respectively. φ is the angle between $E_{\|}$ and a principal normal section of the wave front (it is immaterial which principal normal section we consider).

From (3.158) and (3.159) it follows that

$$\left(\frac{G(\theta_n)}{\cos\theta_n}\right)_{\theta_n=\pi/2} = \frac{\tan\theta_m^*[r_{\|}(O^*)r_{\perp}(O^*)]^{1/2}}{l^{1/2}\left[r_{\perp}(O^*)\frac{V_m}{V_n}+l\left\{1-\frac{1}{4}r_{\|}(O^*)r_{\perp}(O^*)\tan^2 2\varphi\left(\frac{1}{r_{\|}(O^*)}-\frac{1}{r_{\perp}(O^*)}\right)^2\right\}\right]^{1/2}}. \tag{3.160}$$

Inserting this expression into (3.157) yields

$$U^{mnp}(A) = \frac{V_n U^m(O^*) M_{np}(A) R^*_{mn} \tan\theta_m^* [r_{\|}(O^*) r_{\perp}(O^*)]^{1/2}}{i\omega l^{3/2} \left[r_{\perp}(O^*) \frac{V_m}{V_n} + l \left\{ 1 - \frac{1}{4} r_{\|}(O^*) r_{\perp}(O^*) \tan^2 2\varphi \left(\frac{1}{r_{\|}(O^*)} - \frac{1}{r_{\perp}(O^*)} \right)^2 \right\} \right]^{1/2}} \tag{3.161}$$

which is yet another expression for head waves. $\tau_{mnp}(A)$ is again given by (3.157).

In the following chapters we shall use (3.161) for a simpler situation in which both M and P media are homogeneous and $\varphi(O^*) = 0$. Then (3.161) reduces to

$$U^{mnp}(A) = U^m(O^*) \frac{V_n \Gamma_{mnp} \tan\theta_m^* [r_{\|}(O^*) r_{\perp}(O^*)]^{1/2}}{i\omega l^{3/2} [l + r_{\perp}(O^*) V_m / V_n]^{1/2}}. \tag{3.162}$$

All the parameters corresponding to the incident wave in (3.161) and (3.162) are to be determined at the point O^*, where the critical ray strikes the interface. Formulae similar to (3.162) were first given by Alekseyev and Gel'chinskiy (1958, 1959) and Petrashen (1959), and similar to (3.161) by Alekseyev and Gel'chinskiy (1961).

We can see from (3.161) and (3.162) that properties of head waves are quite dependent on the radii of curvature of the wave front of the incident wave at the point of incidence O^*. It was shown in chapter 2 that the ratio of the amplitudes of reflected (refracted) waves and the incident wave, determined just at the interface, does not depend on the curvature of the wave front of the incident wave (in zero-order approximation). For head waves, however, the situation is different; their properties change with the curvature of the wave front of the incident wave, even if the incident wave remains unaltered in other characteristics.

When a spherical wave is incident on the interface, we obtain from (3.151) and (3.162) the formula (3.51). In this case $U^m(O^*) = 1/R_0$, $r_{\|}(O^*) = r_{\perp}(O^*) = R_0$, $[r_{\perp}(O^*) V_m / V_n + l]^{1/2} = r^{1/2}$, where R_0 is the distance between the point O^* and the source and r is the epicentral distance of the point A. From this it follows that, for a spherical wave,

$$U^m(O^*) \left(\frac{r_{\|}(O^*) r_{\perp}(O^*)}{l + r_{\perp}(O^*) V_m / V_n} \right)^{1/2} = \frac{1}{r^{1/2}}.$$

Inserting this relation into (3.162) and (3.151) yields (3.51).

It should be noted that the wave (3.151) is a head wave only if $m \neq n$, $n \neq p$ and if $V_m < V_n$ and $V_p < V_n$ along the whole interface.

4

HEAD WAVES IN MULTILAYERED MEDIA

In this chapter we shall investigate properties of head waves propagating in multilayered media. The layers are assumed to be thick in comparison with the wavelength. Interference effects connected with thin layers are not considered. We shall assume that the velocity depends only on the z-coordinate and that the interfaces separating the layers are *plane* and *parallel to each other*. The boundary at which the head wave is generated will be called the *B-interface* (see figure 4.1). We shall make a significant assumption that the velocity of propagation is constant in the layer in which the head wave propagates along the B-interface; in all other layers, the velocity may change with the z-coordinate. When there are rapid changes of velocity within short distances (comparable with the wavelengths), however, the formulae will be less exact, or inapplicable. We shall also assume that the velocities are discontinuous across the B-interface

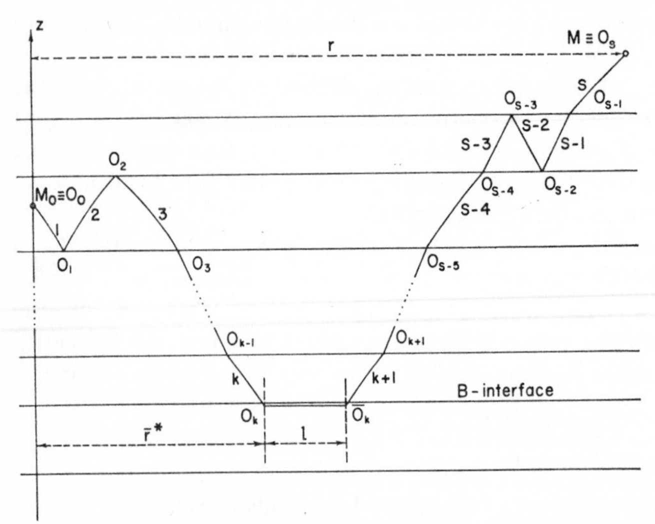

FIGURE 4.1 Various symbols connected with the ray of a head wave in a vertically inhomogeneous multilayered medium with plane, parallel interfaces.

and that the ray of the head wave has no turning point inside any layer. The source and the receiver may be (individually) located in any of the layers and the ray of the head wave may be reflected or refracted any number of times along its path.

4.1 HARMONIC SOURCE

We shall assume that a harmonic, symmetric source of elastic waves lies on the z-axis at the point M_0 (see figure 4.1). The displacement vector of the spherical wave emanating from the source is given by (2.110). This wave may be of P, SV, or SH type. We seek the displacement vector of any head wave that arrives at the receiver, M, situated at a horizontal distance r from the z-axis (see figure 4.1). We shall denote this displacement vector by $\mathbf{W}^*(M)$.

The ray of the H-wave is composed of $s+1$ elements: the first, second, ..., kth elements from the source to the B-interface, the $(k+1)$th, $(k+2)$th, ..., sth elements from the B-interface to the receiver, and one element along the B-interface. Within a given layer the H-wave may propagate either as a P or SV wave (in the case of a P or SV source) or as an SH-wave (SH-source). The propagation velocity along the jth element of the ray (between the points O_{j-1} and O_j) is denoted by $v_j(z)$ $(j = 1, 2, \ldots, s)$. It may be either compressional or shear velocity, depending on the type of

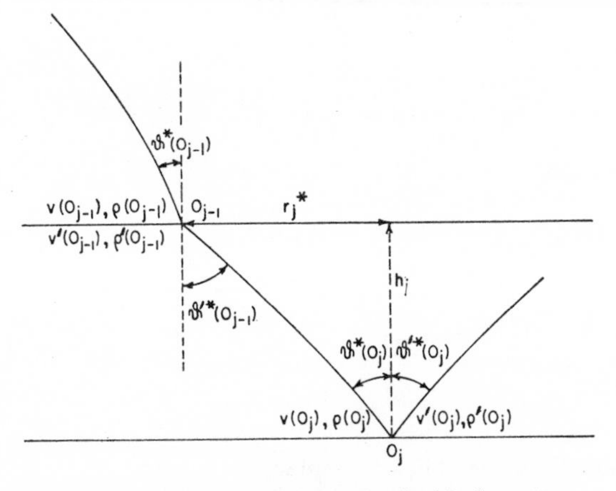

FIGURE 4.2 More symbols associated with the propagation of a ray in a vertically inhomogeneous multilayered medium.

wave along the element. The velocity with which the H-wave propagates along the B-interface will be denoted by $\bar{v}$ ($\bar{v}$ = const.).

Other symbols are clear from figures 4.1 and 4.2. They are the same as in section 2.6.1 (see figures 2.14 and 2.15) with only one difference: the angle $\vartheta_j(z)$ in that section depended on the ray under consideration but the angle $\vartheta_j^*(z)$ introduced here corresponds to the critical angle and is given by

$$\sin \vartheta_j^*(z) = v_j(z)/\bar{v}. \tag{4.1}$$

Similarly, the angles $\vartheta^*(O_j)$ and $\vartheta'^*(O_j)$ are given by

$$\sin \vartheta^*(O_j) = v(O_j)/\bar{v}, \qquad \sin \vartheta'^*(O_j) = v'(O_j)/\bar{v}. \tag{4.2}$$

It follows from (4.1) that the parameter Θ, defined by

$$\Theta = \frac{\sin \vartheta_j^*(z)}{v_j(z)} = \frac{1}{\bar{v}} \qquad (j = 1, 2, \ldots, s), \tag{4.3}$$

remains constant along the whole ray of the head wave.

Also, it follows from (4.1) that the head wave can exist only if

$$v_j(z)/\bar{v} < 1, \tag{4.4}$$

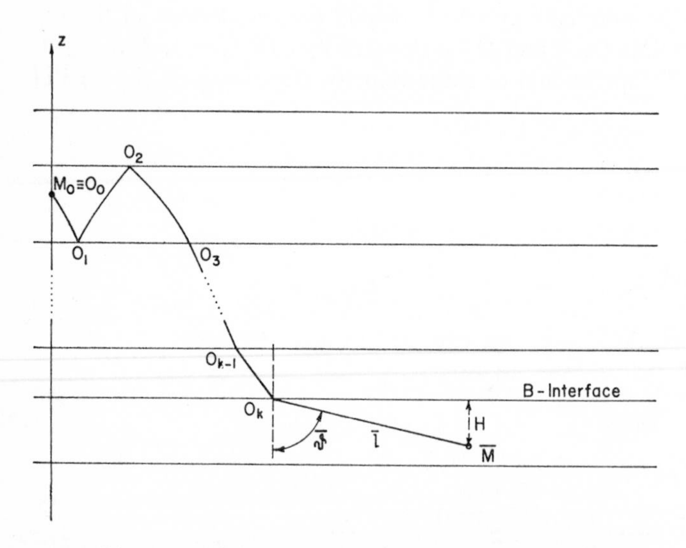

FIGURE 4.3 A ray of the R-wave (which generates the head waves presented in figure 4.1) in a vertically inhomogeneous multilayered medium, arriving at an arbitrary point $\overline{M}$ inside a homogeneous layer surrounding the B-interface.

for $j = 1, 2, \ldots, s$ and arbitrary z. Thus, the velocity of propagation along any segment of the ray of the head wave must be smaller than the velocity $\bar{v}$ with which this wave propagates along the B-interface.

There is an infinite number of multiply reflected waves (R-waves) propagating in the medium under consideration. The R-wave from which the head wave is generated will be called the B-wave. The B-wave has the same succession of ray elements between the source and the B-interface ($j = 1, 2, \ldots, k$) as the H-wave, and the $(k+1)$th element of the ray lies in the layer surrounding the B-interface. For $\vartheta_j(z) = \vartheta_j^*(z)$, the $(k+1)$th element of the ray of the B-wave is parallel to the interface. Generally, for $\vartheta_j(z) < \vartheta_j^*(z)$, the $(k+1)$th element of the ray of the B-wave deviates from the B-interface (see figure 4.3). The velocity with which the B-wave propagates along the B-interface is, of course, $\bar{v}$.

In order to determine the displacement vector of the head wave, $\mathbf{W}^*(M)$, at an arbitrary point M we must first know the displacement vector of the head wave at the point $\bar{O}_k$ lying on the B-interface (see figure 4.1). For this we must know the properties of the B-wave (see figure 4.3). Expressions for an arbitrary type of R-wave were given in section 2.6.1 (see (2.126) to (2.128) and (2.121)).

We shall now find the displacement vector $\mathbf{W}^{\mathrm{B}}$ of the B-wave at the point $\bar{M}$ lying in the homogeneous layer bounded by the B-interface (see figure 4.3). We introduce the following notation: H is the distance of the point $\bar{M}$ from the B-interface, $\bar{l}$ the length of the element of the ray of the B-wave between the point O_k and $\bar{M}$, $\bar{\vartheta}$ the acute angle between this element of the ray and the z-axis. As this element of the ray lies in a homogeneous layer, $v'(O_k) = v(\bar{M}) = \bar{v}$, $\vartheta'(O_k) = \vartheta(\bar{M}) = \bar{\vartheta}$, $\rho'(O_k) = \rho(\bar{M})$. Inserting into (2.126) to (2.128) yields

$$\mathbf{W}^{\mathrm{B}}(\bar{M}) = U^{\mathrm{B}}(\bar{M}) \exp\{i\omega[t-\tau^{\mathrm{B}}(\bar{M})]\}\mathbf{n}^{\mathrm{B}}(\bar{M}), \tag{4.5}$$

where

$$U^{\mathrm{B}}(\bar{M}) = \frac{R_k}{L}\left(\frac{v_0\rho_0}{v(O_k)\rho(O_k)}\right)^{1/2} \prod_{j=1}^{k-1}\left(\frac{v'(O_j)\rho'(O_j)}{v(O_j)\rho(O_j)}\right)^{1/2} R_j, \tag{4.6}$$

$$\tau^{\mathrm{B}}(\bar{M}) = \frac{\bar{l}}{\bar{v}} + \sum_{j=1}^{k} \tau_j. \tag{4.7}$$

In the above expressions R_j ($j = 1, 2, \ldots, k$) are the appropriate R-coefficients at the points O_j. Depending on the type of wave, each R_j can be any one of the sixteen possible coefficients of reflection or refraction (R_{mn}), formulae for which were given in chapter 2, (2.74). Note that the value of ϵ is zero in the present case. The values of τ_j are given by (2.125). For L, we find, from (2.121) and (2.141),

$$L = \left\{ r\left(\sum_{j=1}^{k} \frac{\partial r_j}{\partial \vartheta_0} + \frac{\bar{v} H \cos \vartheta_0}{v_0 \cos^3 \bar{\vartheta}} \right) \frac{\cos \vartheta(O_k)}{\sin \vartheta_0} \right\}^{1/2} \prod_{j=1}^{k-1} \left(\frac{\cos \vartheta(O_j)}{\cos \vartheta'(O_j)} \right)^{1/2}, \tag{4.8}$$

where r_j is given by (2.119).

Inserting $\bar{l} = H/\cos \bar{\vartheta}$ and $v_0/\bar{v} = \sin \vartheta_0/\sin \bar{\vartheta}$ into (4.8) yields

$$L = \frac{\bar{v}}{v_0 \cos \bar{\vartheta}} \left\{ \frac{r \cos \vartheta_0 \cos \vartheta(O_k)}{\sin \bar{\vartheta}} \left[\bar{l} + \frac{v_0 \cos^2 \bar{\vartheta}}{\bar{v} \cos \vartheta_0} \sum_{j=1}^{k} \frac{\partial r_j}{\partial \vartheta_0} \right] \prod_{j=1}^{k-1} \left(\frac{\cos \vartheta(O_j)}{\cos \vartheta'(O_j)} \right) \right\}^{1/2}. \tag{4.9}$$

From (4.6) and (4.9) it follows that

$$U^{\mathrm{B}}(\bar{M}) = \tilde{U}^{\mathrm{B}}(\bar{M}) \cos \bar{\vartheta}, \tag{4.10}$$

where

$$\tilde{U}^{\mathrm{B}}(\bar{M}) = \frac{R_k v_0 \left(\dfrac{v_0 \rho_0}{v(O_k)\rho(O_k)} \right)^{1/2} \displaystyle\prod_{j=1}^{k-1} \left(\frac{v'(O_j)\rho'(O_j)}{v(O_j)\rho(O_j)} \right)^{1/2} R_j}{\bar{v} \left\{ \dfrac{r \cos \vartheta_0 \cos \vartheta(O_k)}{\sin \bar{\vartheta}} \left[\bar{l} + \dfrac{v_0 \cos^2 \bar{\vartheta}}{\bar{v} \cos \vartheta_0} \displaystyle\sum_{j=1}^{k} \frac{\partial r_j}{\partial \vartheta_0} \right] \displaystyle\prod_{j=1}^{k-1} \left(\frac{\cos \vartheta(O_j)}{\cos \vartheta'(O_j)} \right) \right\}^{1/2}} \tag{4.11}$$

If we know $U^{\mathrm{B}}(\bar{M})$, we can determine $\mathbf{W}^*(\bar{O}_k)$ by the same method as in section 3.9. The general expressions (3.151) and (3.155) given there are quite applicable in the present case. Using the notation of this chapter, we can write

$$\mathbf{W}^*(\bar{O}_k) = \frac{\bar{v}}{i\omega l} M(\bar{O}_k)(\tilde{U}^{\mathrm{B}}(\bar{M}))_{\bar{\vartheta}=\pi/2} \exp\{i\omega[t - \tau^*(\bar{O}_k)]\} \mathbf{n}^*(\bar{O}_k), \tag{4.12}$$

where $\tau^*(\bar{O}_k) = \tau^{\mathrm{B}}(\bar{O}_k)$, i.e., from (4.7),

$$\tau^*(\bar{O}_k) = \frac{l}{\bar{v}} + \sum_{j=1}^{k} \tau_j^*. \tag{4.13}$$

Values of τ_j^* are given by (2.125), where $\vartheta_0 = \vartheta_0^*$ is substituted. l is the distance between the points $\bar{O}_k$ and O_k (see figure 4.1) so that

$$l = r(\bar{O}_k) - r(O_k) \tag{4.14}$$

(i.e. $\bar{l} \to l$ when $\bar{\vartheta} \to \frac{1}{2}\pi$), where $r(O_k)$ is the interface critical distance, $\bar{r}^*$, given by the relation

$$r(O_k) = \bar{r}^* = \sum_{j=1}^{k} r_j^*. \tag{4.15}$$

Values of r_j^* are given by (2.119), where $\vartheta_0 = \vartheta_0^*$ is substituted. The function $M(\bar{O}_k)$ in (4.12) can be any one of the ten possible functions M_{np} given by (3.43). The choice of the particular M_{np} depends on the types of waves under consideration and on which side of the B-interface they propagate. In any discussion involving the B-interface, the upper layer is taken to be the first medium and the lower the second medium.

To find the final expression for $\mathbf{W}^*(\bar{O}_k)$, we must determine $(\tilde{U}^{\text{B}}(\bar{M}))_{\bar{\vartheta}=\pi/2}$. For $\bar{\vartheta} = \frac{1}{2}\pi$, we have $\cos\bar{\vartheta} = 0$, $\bar{l} = l$, $\vartheta(O_j) = \vartheta^*(O_j)$, $\vartheta'(O_j) = \vartheta'^*(O_j)$ (where $\vartheta^*(O_j)$ and $\vartheta'^*(O_j)$ are given by (4.2)). For $(\tilde{U}^{\text{B}}(\bar{M}))_{\bar{\vartheta}=\pi/2}$, we have, from (4.11),

$$(\tilde{U}^{\text{B}}(\bar{M}))_{\bar{\vartheta}=\pi/2} = \frac{R_k^* \tan\vartheta_0^* \delta_1^*}{l^{1/2}[r(\bar{O}_k)]^{1/2}} \prod_{j=1}^{k-1} R_j^*, \tag{4.16}$$

where

$$\delta_1^* = \left(\frac{v_0\rho_0 \cos\vartheta_0^*}{v(O_k)\rho(O_k)\cos\vartheta^*(O_k)}\right)^{1/2} \prod_{j=1}^{k-1}\left(\frac{v'(O_j)\rho'(O_j)\cos\vartheta'^*(O_j)}{v(O_j)\rho(O_j)\cos\vartheta^*(O_j)}\right)^{1/2}, \tag{4.17}$$

and R_j^* are the coefficients of reflection or refraction, R_j, where we have inserted $\vartheta_0 = \vartheta_0^*$, or $\Theta = 1/\bar{v}$; i.e.,

$$R_j^* = (R_j)_{\Theta=1/\bar{v}}. \tag{4.18}$$

Substituting (4.16) into (4.12) yields

$$\mathbf{W}^*(\bar{O}_k) = \frac{\bar{v}\Gamma_k \tan\vartheta_0^*\delta_1^*}{i\omega l^{3/2}[r(\bar{O}_k)]^{1/2}} \prod_{j=1}^{k-1} R_j^* \exp\{i\omega[t-\tau^*(\bar{O}_k)]\}\mathbf{n}^*(\bar{O}_k), \tag{4.19}$$

where Γ_k is the head wave coefficient given by

$$\Gamma_k = R_k^* M(\bar{O}_k). \tag{4.20}$$

Γ_k is one of the twenty-six possible head wave coefficients, Γ_{mnp}, formulae for which were given in chapter 3 (see (3.99) for P and SV waves, and (3.106) for SH-waves). The indices m, n, p are to be determined by the procedure described in chapter 3. m characterizes the wave incident on the B-interface, n the wave propagating along the B-interface, and p the head wave generated on the B-interface. If the wave in question lies above the B-interface, the index is 1 for a P-wave and 2 for an SV or SH wave; if the wave lies below the interface, the index is 3 for a P-wave and 4 for an SV or SH wave. The velocities and densities, α_1, β_1, ρ_1 and α_2, β_2, ρ_2, are to be taken just above and below the B-interface.

The equation (4.19) is the final expression for the displacement vector of the head wave under consideration at the point $\bar{O}_k$. Using the formulae for continuation of the displacement vector along a ray, we can find the displacement vector of this wave at an arbitrary point O_s $(\equiv M)$. Note that the zeroth term of the ray series vanishes identically for head waves, and therefore the first term becomes the leading term of this series. We can easily see that the zero-order formulae for the continuation of the displacement vector along a ray can be used for the leading term of the ray series (in spite of the fact that this term contains $(i\omega)^{-1}$); see section 2.1.

The formulae for the displacement vectors of the head wave at the points $\bar{O}_k$ and M may be written in the form

$$\mathbf{W}^*(\bar{O}_k) = U^*(\bar{O}_k)\exp\{i\omega[t-\tau^*(\bar{O}_k)]\}\mathbf{n}^*(\bar{O}_k), \tag{4.21}$$

$$\mathbf{W}^*(M) = U^*(M)\exp\{i\omega[t-\tau^*(M)]\}\mathbf{n}^*(M), \tag{4.22}$$

where $U^*(M)$ and $\tau^*(M)$ are to be determined; $U^*(\bar{O}_k)$ is given, from (4.19), by

$$U^*(\bar{O}_k) = \frac{\bar{v}\Gamma_k \tan\vartheta_0^*}{i\omega[r(\bar{O}_k)]^{1/2}l^{3/2}}\left(\frac{v_0\rho_0\cos\vartheta_0^*}{v(O_k)\rho(O_k)\cos\vartheta^*(O_k)}\right)^{1/2} \times\prod_{j=1}^{k-1}\left(\frac{v'(O_j)\rho'(O_j)\cos\vartheta'^*(O_j)}{v(O_j)\rho(O_j)\cos\vartheta^*(O_j)}\right)^{1/2} R_j^*. \tag{4.23}$$

Using (2.98) and (2.99), we find for $U^*(M)$ and $\tau^*(M)$:

$$U^*(M) = \frac{U^*(\bar{O}_k)}{L(M,\bar{O}_k)}\left[\frac{v'(\bar{O}_k)\rho'(\bar{O}_k)}{v(M)\rho(M)}\right]^{1/2}\prod_{j=k+1}^{s-1}\left[\frac{v'(O_j)\rho'(O_j)}{v(O_j)\rho(O_j)}\right]^{1/2} R_j^*, \tag{4.24}$$

$$\tau^*(M) = \tau^*(\bar{O}_k) + \sum_{j=k+1}^{s}\tau_j^*, \tag{4.25}$$

where τ_j^* are given by (2.125) (when $\vartheta_0 = \vartheta_0^*$ is substituted) and $L(M,\bar{O}_k)$ is given by (see (2.100))

$$L(M,\bar{O}_k) = \left(\frac{d\sigma(M)}{d\sigma(\bar{O}_k)}\right)^{1/2}\prod_{j=k+1}^{s-1}\left(\frac{d\sigma(O_j)}{d\sigma'(O_j)}\right)^{1/2}, \tag{4.26}$$

$d\sigma$ being the cross-sectional area of the ray tube of the head wave under consideration. We find easily that

$$\left(\frac{d\sigma(M)}{d\sigma(\bar{O}_k)}\right)^{1/2} = \left(\frac{r(M)}{r(\bar{O}_k)}\right)^{1/2}\left(\frac{\cos\vartheta^*(M)}{\cos\vartheta'^*(\bar{O}_k)}\right)^{1/2}, \tag{4.27}$$

where $\vartheta'^*(\bar{O}_k)$ is the acute angle between the ray and the z-axis at the point $\bar{O}_k$. Using (2.115), we obtain

$$L(M,\bar{O}_k) = \left(\frac{r(M)}{r(\bar{O}_k)}\right)^{1/2}\left(\frac{\cos\vartheta^*(M)}{\cos\vartheta'^*(\bar{O}_k)}\right)^{1/2}\prod_{j=k+1}^{s-1}\left(\frac{\cos\vartheta^*(O_j)}{\cos\vartheta'^*(O_j)}\right)^{1/2}. \tag{4.28}$$

Inserting (4.28) and (4.23) into (4.24) yields

$$U^*(M) = \frac{\bar{v}\Gamma_k\tan\vartheta_0^*\,\delta^*}{i\omega[r(M)]^{1/2}l^{3/2}}\prod_{\substack{j=1\\ j\neq k}}^{s-1} R_j^*, \tag{4.29}$$

where

$$\delta^* = \left(\frac{v_0\rho_0 \cos\vartheta_0^*}{v(M)\rho(M)\cos\vartheta^*(M)}\right)^{1/2} \left(\frac{v'(\bar{O}_k)\rho'(\bar{O}_k)\cos\vartheta'^*(\bar{O}_k)}{v(O_k)\rho(O_k)\cos\vartheta^*(O_k)}\right)^{1/2}$$
$$\times \prod_{\substack{j=1\\ j\neq k}}^{s-1} \left(\frac{v'(O_j)\rho'(O_j)\cos\vartheta'^*(O_j)}{v(O_j)\rho(O_j)\cos\vartheta^*(O_j)}\right)^{1/2}. \quad (4.30)$$

We can see from the above equation that δ^* is a real and positive constant. It depends only on the parameters of the media and on the z-coordinates of the source and the receiver. It does not depend on the epicentral distance of the receiver as long as its z-coordinate does not change. The constant δ^* is composed of three terms: the first term depends only on the z-coordinates of the source and the receiver, the second on the properties of the B-interface, and the third on the properties of the intermediate interfaces.

The parameters $v'(\bar{O}_k)$, $\rho'(\bar{O}_k)$, and $\vartheta'^*(\bar{O}_k)$ do not depend on the position of the point $\bar{O}_k$ on the B-interface. We can therefore put $v'(\bar{O}_k) = v'(O_k)$, $\rho'(\bar{O}_k) = \rho'(O_k)$, and $\vartheta'^*(\bar{O}_k) = \vartheta'^*(O_k)$. We must remember, however, that these parameters are to be taken at the B-interface on the side of the generated head wave. Then we can write for δ^*

$$\delta^* = \left(\frac{v_0\rho_0 \cos\vartheta_0^*}{v(M)\rho(M)\cos\vartheta^*(M)}\right)^{1/2} \prod_{j=1}^{s-1} \left(\frac{v'(O_j)\rho'(O_j)\cos\vartheta'^*(O_j)}{v(O_j)\rho(O_j)\cos\vartheta^*(O_j)}\right)^{1/2}. \quad (4.31)$$

If any of the interfaces, say the lth interface, represents a second-order discontinuity (i.e. velocity and density do not undergo a discontinuous change across this interface, but their gradients do), we insert into (4.31)

$$\left(\frac{v'(O_l)\rho'(O_l)\cos\vartheta'^*(O_l)}{v(O_l)\rho(O_l)\cos\vartheta^*(O_l)}\right)^{1/2} = 1. \quad (4.31')$$

As all angles in the expression (4.31) are given by (4.2), we can write

$$\delta^* = \left\{\frac{v_0\rho_0}{v(M)\rho(M)} \prod_{j=1}^{s-1} \frac{v'(O_j)\rho'(O_j)}{v(O_j)\rho(O_j)}\right\}^{1/2} \left\{\frac{\bar{v}^2 - v_0^{\,2}}{\bar{v}^2 - v^2(M)} \prod_{j=1}^{s-1} \frac{\bar{v}^2 - v'^2(O_j)}{\bar{v}^2 - v^2(O_j)}\right\}^{1/4}. \quad (4.32)$$

It is clear from figure 4.1 that

$$r(M) = r(\bar{O}_k) + \sum_{j=k+1}^{s} r_j^*, \quad (4.33)$$

where r_j^* is given by (2.119) (when ϑ_0^* is substituted for ϑ_0). Inserting (4.33) into (4.14) yields

$$l = r(M) - r^*, \tag{4.34}$$

where

$$r^* = \sum_{j=1}^{s} r_j^*. \tag{4.35}$$

r^* is called the *critical distance*. It is the distance of the critical ray from the z-axis, measured at the z-coordinate of the receiver. It is obvious that the head wave under consideration exists only at distances larger than the critical distance, i.e.

$$r(M) > r^*. \tag{4.36}$$

Final Expressions (Summary)

The displacement vector of the head wave at the point M is given by

$$\mathbf{W}^*(M) = U^*(M) \exp\{i\omega[t - \tau^*(M)]\}\mathbf{n}^*(M), \tag{4.37}$$

where values of $\mathbf{n}^*(M)$, $\tau^*(M)$, and $U^*(M)$ are as follows:

(1) $\mathbf{n}^*(M)$ is a unit vector at the point M. The direction of this vector depends on the character of the wave at the point M: if this wave is a P-wave, $\mathbf{n}^*(M)$ lies along the ray (positive in the direction of propagation of the wave); if the wave is an SV-wave, $\mathbf{n}^*(M)$ lies in the plane of the ray and is perpendicular to the ray (it is positive if its projection on the r-axis is positive); and for an SH-wave, $\mathbf{n}^*(M)$ is perpendicular to the ray plane.

(2) The travel time, $\tau^*(M)$, is given by

$$\tau^*(M) = \frac{r - r^*}{\bar{v}} + \sum_{j=1}^{s} \tau_j^* \tag{4.38}$$

(which follows from (4.25), (4.13), and (4.34)). Here, r^* is the critical distance,

$$r^* = \sum_{j=1}^{s} r_j^*$$

(see (4.35)); r_j^* and τ_j^* are given by the expressions

$$r_j^* = \left| \frac{1}{\bar{v}} \int_{z(O_{j-1})}^{z(O_j)} \frac{v_j(z)\mathrm{d}z}{[1 - v_j^2(z)/\bar{v}^2]^{1/2}} \right|, \tag{4.39}$$

$$\tau_j^* = \left| \int_{z(O_{j-1})}^{z(O_j)} \frac{\mathrm{d}z}{v_j(z)[1 - v_j^2(z)/\bar{v}^2]^{1/2}} \right|. \tag{4.40}$$

We can see from (4.38) that for a given z-coordinate of the receiver, the *time-distance* curves for the head waves are straight lines. A simpler formula follows from (4.38) by rearranging the terms:

$$\tau^*(M) = \frac{r}{\bar{v}} + \sum_{j=1}^{s} \bar{\tau}_j^*, \tag{4.41}$$

where

$$\bar{\tau}_j^* = \left| \int_{z(O_{j-1})}^{z(O_j)} \frac{[1-v_j^2(z)/\bar{v}^2]^{1/2}}{v_j(z)} \, dz \right|. \tag{4.42}$$

(3) $U^*(M)$ is given by the expression (see 4.29)

$$U^*(M) = \frac{\bar{v}\Gamma_k \tan \vartheta_0^* \delta^*}{i\omega r^{1/2}(r-r^*)^{3/2}} \prod_{\substack{j=1 \\ j \neq k}}^{s-1} R_j^*, \tag{4.43}$$

where r is the horizontal distance of the receiver, M, from the z-axis, r^* is the critical distance (see (4.35) and (4.39)), Γ_k is the head wave coefficient on the B-interface, R_j^* are the R-coefficients at the intermediate interfaces (taken for $\Theta = 1/\bar{v}$), and δ^* is given by (4.31) or (4.32). Note that, for $\tan \vartheta_0^*$, we can insert into (4.43) the expression $\tan \vartheta_0^* = v_0/(\bar{v}^2 - v_0^2)^{1/2}$. If any of the intermediate interfaces are second-order discontinuities, $R_j^* = 1$ for the coefficient of refraction.

Displacement Components

We can write the displacement vector $\mathbf{W}^*(M)$ in component form. For P or SV waves, we have

$$\begin{aligned} W_r^*(M) &= A_r^* \exp[i\omega(t-\tau^*)+i\chi_r^*], \\ W_z^*(M) &= A_z^* \exp[i\omega(t-\tau^*)+i\chi_z^*], \end{aligned} \tag{4.44}$$

when all the terms are to be taken at the point M, A_r^* and A_z^* are the *amplitudes* of the horizontal and vertical components of displacement, and χ_r^* or χ_z^* are their *phase-shifts*. They are given by

$$\begin{bmatrix} A_r^* \\ A_z^* \end{bmatrix} = \frac{\bar{v}\delta^* \tan \vartheta_0^*}{\omega r^{1/2}(r-r^*)^{3/2}} \left| \Gamma_k \prod_{\substack{j=1 \\ j \neq k}}^{s-1} R_j^* \right| \begin{bmatrix} |n_r^*| \\ |n_z^*| \end{bmatrix}, \tag{4.45}$$

$$\begin{bmatrix} \chi_r^* \\ \chi_z^* \end{bmatrix} = -\tfrac{1}{2}\pi + \arg \Gamma_k + \sum_{\substack{j=1 \\ j \neq k}}^{s-1} \arg R_j^* + \begin{bmatrix} \arg n_r^* \\ \arg n_z^* \end{bmatrix}, \tag{4.46}$$

where n_r^* and n_z^* are the r and z components of the unit vector $\mathbf{n}^*$ at the point M, which can be expressed as

$$n_r^* = v(M)/\bar{v}, \qquad n_z^* = \pm[1-(v(M)/\bar{v})^2]^{1/2}, \tag{4.47}$$

if the head wave is of P-type at the point M, and as

$$n_r^* = [1-(v(M)/\bar{v})^2]^{1/2}, \qquad n_z^* = \mp v(M)/\bar{v}, \tag{4.48}$$

if this wave is of SV-type. The upper sign applies if the wave is propagating upwards, the lower sign if it is propagating downwards at the point M.

For SH-waves, the formulae are simpler, as Γ_k is always real and positive and the unit vector $\mathbf{n}^*$ is in the φ direction. We have

$$W_\varphi^*(M) = A_\varphi^* \exp[i\omega(t-\tau^*)+i\chi_\varphi^*], \tag{4.49}$$

where

$$A_\varphi^* = \frac{\bar{v}\delta^*\Gamma_k^{\mathrm{SH}}\tan\vartheta_0^*}{\omega r^{1/2}(r-r^*)^{3/2}}\left|\prod_{\substack{j=1\\ j\neq k}}^{s-1} R_j^{\mathrm{SH}*}\right|, \tag{4.50}$$

$$\chi_\varphi^* = -\tfrac{1}{2}\pi + \sum_{\substack{j=1\\ j\neq k}}^{s-1} \arg R_j^{\mathrm{SH}*}. \tag{4.51}$$

The above formulae, (4.45) to (4.51), are applicable if the receiver lies *inside* the elastic medium. If, on the other hand, the receiver is located on a free surface (on the surface of the earth, for example), formulae (4.45), (4.46), and (4.50) change. The components of the vector $\mathbf{n}^*(M)$ (i.e. n_r^* and n_z^* for P or SV waves) must be replaced by the corresponding coefficients of conversion; see formulae (2.89) and (2.90). These coefficients of conversion, q_r and q_z, are to be taken for the critical ray, i.e., for $\Theta = 1/\bar{v}$. We shall mark these by an asterisk, i.e.,

$$q_r^* = (q_r)_{\Theta=1/\bar{v}}, \qquad q_z^* = (q_z)_{\Theta=1/\bar{v}}. \tag{4.52}$$

The final expressions for the amplitudes and phase shifts of the H-wave under consideration at the point M lying on the earth's surface are the following:

$$\begin{bmatrix} A_r^* \\ A_z^* \end{bmatrix} = \frac{\bar{v}\delta^*\tan\vartheta_0^*}{\omega r^{1/2}(r-r^*)^{3/2}}\left|\Gamma_k\prod_{\substack{j=1\\ j\neq k}}^{s-1} R_j^*\right|\begin{bmatrix} |q_r^*| \\ |q_z^*| \end{bmatrix}, \tag{4.53}$$

$$\begin{bmatrix} \chi_r^* \\ \chi_z^* \end{bmatrix} = -\tfrac{1}{2}\pi + \arg\Gamma_k + \sum_{\substack{j=1\\ j\neq k}}^{s-1} \arg R_j^* + \begin{bmatrix} \arg q_r^* \\ \arg q_z^* \end{bmatrix}. \tag{4.54}$$

If the wave is incident on the earth's surface as a P-wave, we must take (for q_r^* and q_z^*) q_{3r}^* and q_{3z}^*. If it is incident as an SV-wave, we must substitute q_{4r}^* and q_{4z}^* (see equations (2.89) and (2.90)).

In the case of SH-waves recorded on the earth's surface, A_φ^* given by (4.50) must be multiplied by 2 because of the free surface effects. χ_φ^* remains valid as given by (4.51).

4.2
PULSE SOURCE

If the source is not harmonic, as we have assumed so far in this chapter, we can employ the procedure outlined in section 3.7 to obtain the displacement vector of the transient head wave. We get

$$\mathscr{W}^*(M) = \frac{\bar{v}\delta^* \tan \vartheta_0^*}{r^{1/2}(r-r^*)^{3/2}} \left| \Gamma_k \prod_{\substack{j=1 \\ j\neq k}}^{s-1} R_j^* \right| F(t-\tau^*(M); \chi^*)\mathbf{n}^*(M), \tag{4.55}$$

where

$$\chi^* = \arg \Gamma_k + \sum_{\substack{j=1 \\ j\neq k}}^{s-1} \arg R_j^* \tag{4.56}$$

and the function $F(t; \chi)$ is given by (3.119).

For the horizontal and vertical components of the displacement vector of the transient head wave, we obtain

$$\begin{aligned} \mathscr{W}_r^*(M) &= \mathscr{A}_r^* F(t-\tau^*(M); \bar{\chi}_r^*), \\ \mathscr{W}_z^*(M) &= \mathscr{A}_z^* F(t-\tau^*(M); \bar{\chi}_z^*), \end{aligned} \tag{4.57}$$

where

$$\begin{bmatrix} \mathscr{A}_r^* \\ \mathscr{A}_z^* \end{bmatrix} = \omega \begin{bmatrix} A_r^* \\ A_z^* \end{bmatrix} = \frac{\bar{v}\delta^* \tan \vartheta_0^*}{r^{1/2}(r-r^*)^{3/2}} \left| \Gamma_k \prod_{\substack{j=1 \\ j\neq k}}^{s-1} R_j^* \right| \begin{bmatrix} |n_r^*| \\ |n_z^*| \end{bmatrix}, \tag{4.58}$$

$$\begin{bmatrix} \bar{\chi}_r^* \\ \bar{\chi}_z^* \end{bmatrix} = \tfrac{1}{2}\pi + \begin{bmatrix} \chi_r^* \\ \chi_z^* \end{bmatrix} = \arg \Gamma_k + \sum_{\substack{j=1 \\ j\neq k}}^{s-1} \arg R_j^* + \begin{bmatrix} \arg n_r^* \\ \arg n_z^* \end{bmatrix}. \tag{4.59}$$

If the receiver lies on the earth's surface, the components of the unit vector $\mathbf{n}^*(M)$ must be replaced by the corresponding coefficients of conversion.

For the φ component of the SH head wave, we find

$$\mathscr{W}_\varphi^*(M) = \mathscr{A}_\varphi^* F(t-\tau^*(M), \bar{\chi}_\varphi^*), \tag{4.60}$$

where

$$\mathscr{A}_\varphi{}^* = \omega A_\varphi{}^* = \frac{\bar{v}\delta^*\Gamma_k^{\mathrm{SH}} \tan \vartheta_0{}^*}{r^{1/2}(r-r^*)^{3/2}} \left| \prod_{\substack{j=1 \\ j\neq k}}^{s-1} R_j{}^{\mathrm{SH}*} \right|,$$

$$\bar{\chi}_\varphi{}^* = \tfrac{1}{2}\pi + \chi_\varphi{}^* = \sum_{\substack{j=1 \\ j\neq k}}^{s-1} \arg R_j{}^{\mathrm{SH}*}. \tag{4.61}$$

If the receiver lies on the earth's surface, $\mathscr{A}_\varphi{}^*$ has to be multiplied by 2.

4.3
PARTICULAR TYPES OF HEAD WAVES

4.3.1 *Symmetric Head Waves*

Symmetric head waves are very important in seismic experiments. They have the important characteristic that the branch of the ray between the source and the B-interface is equivalent to the branch between the receiver and the B-interface. Examples of symmetric head waves are given in figure 4.4. For symmetric head waves, s is always an even number, i.e. $s = 2k$. The jth element of the ray ($j = 1, 2, \ldots, k$) corresponds to the $(s+1-j)$th element, i.e.,

$$r_j{}^* = r^*_{s+1-j}, \qquad \tau_j{}^* = \tau^*_{s+1-j}. \tag{4.62}$$

For P and SV waves, both elements must be of the same type (P or SV). The elastic parameters and the angles between the ray and the z-axis at

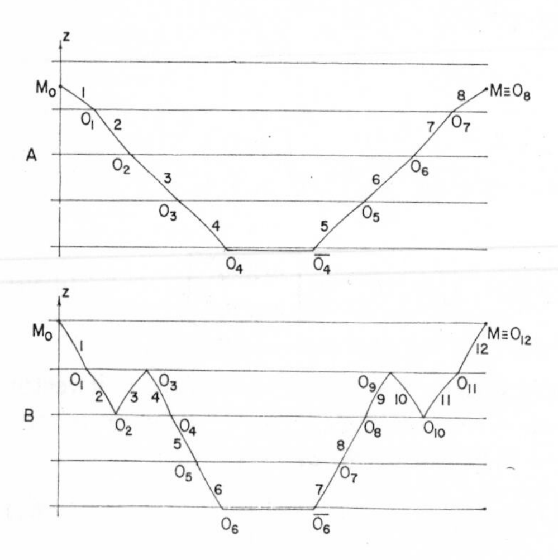

FIGURE 4.4 Examples of symmetric head waves.

the point O_j are the same as at the point O_{s-j}. Since the direction of propagation of the wave is opposite in the two branches of the ray, we can write

$$\begin{aligned} v(O_j) &= v'(O_{s-j}), \quad \rho(O_j) = \rho'(O_{s-j}), \quad \vartheta^*(O_j) = \vartheta'^*(O_{s-j}), \\ v'(O_j) &= v(O_{s-j}), \quad \rho'(O_j) = \rho(O_{s-j}), \quad \vartheta'^*(O_j) = \vartheta^*(O_{s-j}), \end{aligned} \tag{4.63}$$

from which it follows that

$$\left(\frac{v'(O_j)\rho'(O_j)\cos\vartheta'^*(O_j)}{v(O_j)\rho(O_j)\cos\vartheta^*(O_j)}\right)^{1/2} = \left(\frac{v'(O_{s-j})\rho'(O_{s-j})\cos\vartheta'^*(O_{s-j})}{v(O_{s-j})\rho(O_{s-j})\cos\vartheta^*(O_{s-j})}\right)^{-1/2}. \tag{4.64}$$

Inserting this in (4.31) yields

$$\delta^* = 1. \tag{4.65}$$

The formulae for the displacement vector of symmetric head waves, their components, amplitudes, and phase shifts are given in section 4.1, but they are all simplified by the substitution $\delta^* = 1$.

The expressions for the critical distance, r^*, and the travel time, $\tau^*(M)$, also become simpler in the case of symmetric head waves. Using (4.35), (4.38), (4.41), and (4.63), we can write

$$r^* = 2\sum_{j=1}^{k} r_j^*, \tag{4.66}$$

$$\tau^*(M) = \frac{r-r^*}{\bar{v}} + 2\sum_{j=1}^{k} \tau_j^*, \tag{4.67}$$

or

$$\tau^*(M) = \frac{r}{\bar{v}} + 2\sum_{j=1}^{k} \bar{\tau}_j^*. \tag{4.67'}$$

4.3.2 *Partially Symmetric Head Waves*

For partially symmetric head waves, the z-coordinates of the source and the receiver need not be the same. However, the two branches of the ray from the B-interface to the source and to the receiver must be symmetric until we reach the source or the receiver. We shall investigate only those partially symmetric head waves for which the branch of the ray between the source and the B-interface is shorter than that between the receiver and the B-interface. These waves are very important in seismology. (Consider the situation in which the source is inside the medium and the receiver is on the earth's surface.) Examples of these partially symmetric head waves are given in figure 4.5.

The expressions for δ^* are again simpler than those in the general case. If $s = 2k$ (i.e., the source and receiver lie in the same layer), we find from (4.31), using (4.64), that

$$\delta^* = \left(\frac{v_0 \rho_0 \cos \vartheta_0^*}{v(M)\rho(M) \cos \vartheta^*(M)}\right)^{1/2} \tag{4.68}$$

(see figure 4.5, A and C).

If $s > 2k$, δ^* is given by the expression

$$\delta^* = \left(\frac{v_0 \rho_0 \cos \vartheta_0^*}{v(M)\rho(M) \cos \vartheta^*(M)}\right)^{1/2} \prod_{j=2k}^{s-1} \left(\frac{v'(O_j)\rho'(O_j) \cos \vartheta'^*(O_j)}{v(O_j)\rho(O_j) \cos \vartheta^*(O_j)}\right)^{1/2} \tag{4.69}$$

(see figure 4.5B).

We denote the point corresponding to the source on the symmetrical part of the ray by M_0' (see figure 4.5). Then we can write for the critical distance, r^*, and the time of arrival, $\tau^*(M)$, the expressions:

$$r^* = 2\sum_{j=1}^{k} r_j^* + \Delta r_{2k}^* + \sum_{j=2k+1}^{s} r_j^*, \tag{4.70}$$

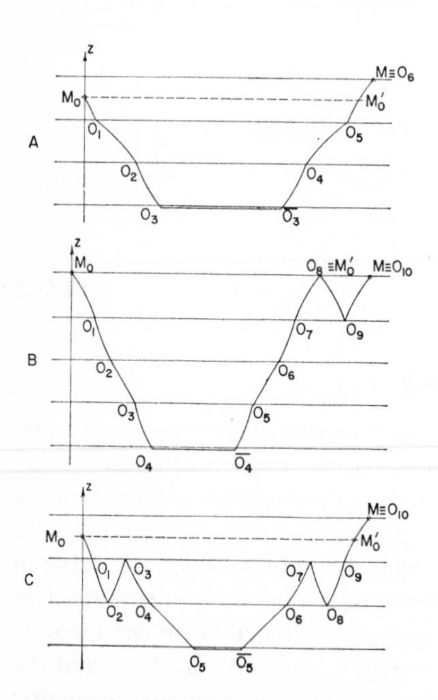

FIGURE 4.5 Examples of partially symmetric head waves.

$$\tau^*(M) = \frac{r-r^*}{\bar{v}} + 2\sum_{j=1}^{k} \tau_j^* + \Delta\tau_{2k}^* + \sum_{j=2k+1}^{s} \tau_j^*, \tag{4.71}$$

or

$$\tau^*(M) = \frac{r}{\bar{v}} + 2\sum_{j=1}^{k} \bar{\tau}_j^* + \Delta\bar{\tau}_{2k}^* + \sum_{j=2k+1}^{s} \bar{\tau}_j^*, \tag{4.71'}$$

where

$$\Delta r_{2k}^* = \left| \int_{z(M_0')}^{z(O_{2k})} \frac{v_{2k}(z)\mathrm{d}z}{\bar{v}[1-v_{2k}^2(z)/\bar{v}^2]^{1/2}} \right|, \tag{4.72}$$

$$\Delta\tau_{2k}^* = \left| \int_{z(M_0')}^{z(O_{2k})} \frac{\mathrm{d}z}{v_{2k}(z)[1-v_{2k}^2(z)/\bar{v}^2]^{1/2}} \right|, \tag{4.73}$$

and

$$\Delta\bar{\tau}_{2k}^* = \left| \int_{z(M_0')}^{z(O_{2k})} \frac{[1-v_{2k}^2(z)/\bar{v}^2]^{1/2}\mathrm{d}z}{v_{2k}(z)} \right|. \tag{4.73'}$$

4.3.3 *Fundamental Head Waves*

A fundamental head wave is a symmetric or a partially symmetric head wave which is *compressional* along the whole ray and does not have any segment formed by reflection at an intermediate interface. The branch of the ray between the source and the B-interface is descending, whereas the other branch (between the B-interface and the receiver) is ascending. Examples of fundamental head waves are given in figures 4.4A and 4.5A. Fundamental head waves are the most important head waves in seismic prospecting and in investigations of the earth's crust by seismic methods. Only these head waves can be registered as first arrivals on seismograms; all other head waves arrive later. (This is, of course, true only if the source is radiating P-waves. Otherwise, by definition, there can be no fundamental head waves.) Therefore, we shall give explicit formulae for this case.

All the R-coefficients (R_j^*) on the descending part of the ray ($j = 1, 2, \ldots, k-1$) are of the type R_{13}^*, whereas on the ascending part of the ray ($j = k+1, \ldots, s-1$), they are of the type R_{31}^*. They are always real and positive. The head wave coefficient Γ_k is of the type Γ_{131}, again always real and positive. Moreover, values of n_r^* and n_z^*, given by (4.47), and the coefficients of conversion, q_{3r}^* and q_{3z}^*, are real and positive.

The amplitudes and phase shifts of the r and z displacement components of fundamental head waves are given by the following expressions, which are derived from (4.45) and (4.46):

$$\begin{bmatrix} A_r^* \\ A_z^* \end{bmatrix} = \frac{\bar{v}\delta^* \tan \vartheta_0{}^*\Gamma_k}{\omega r^{1/2}(r-r^*)^{3/2}} \prod_{\substack{j=1 \\ j\neq k}}^{s} R_j{}^* \begin{bmatrix} n_r{}^* \\ n_z{}^* \end{bmatrix}, \tag{4.74}$$

$$\begin{bmatrix} \chi_r{}^* \\ \chi_z{}^* \end{bmatrix} = -\tfrac{1}{2}\pi. \tag{4.75}$$

If the point M lies on the earth's surface, we substitute q_{3r}^* and q_{3z}^* for $n_r{}^*$ and $n_z{}^*$ in (4.74). Values of δ^*, r^*, and $\tau^*(M)$ are given by the same formulae as for symmetric or partially symmetric head waves. If both the source and the receiver lie on the earth's surface, a fundamental head wave is symmetric and $\delta^* = 1$ (r^* and $\tau^*(M)$ are then given by (4.66) and (4.67)). For the velocities v, we substitute the compressional velocities α in all the formulae for fundamental head waves. For the case of an impulsive source, we find, from (4.59) and (4.75), that

$$\bar{\chi}_r{}^* = \bar{\chi}_z{}^* = 0. \tag{4.76}$$

Inserting this into (4.57) yields

$$\begin{bmatrix} \mathscr{W}_r{}^*(M) \\ \mathscr{W}_z{}^*(M) \end{bmatrix} = \begin{bmatrix} \mathscr{A}_r{}^* \\ \mathscr{A}_z{}^* \end{bmatrix} F(t-\tau^*(M)), \tag{4.77}$$

where $\mathscr{A}_r{}^*$ and $\mathscr{A}_z{}^*$ are given by (4.58) and the function $F(t)$ is an integral obtained from the source function $f(t)$ (see (3.116)).

4.4 HOMOGENEOUS LAYERS

We shall assume in this section that the velocity and density within individual layers are constant, i.e., $v_j(z) = v_j =$ const., $\rho_j(z) = \rho_j =$ const. Then, for $j = 2, 3, \ldots, s-1$,

$$v'(O_{j-1}) = v(O_j), \quad \rho'(O_{j-1}) = \rho(O_j), \quad \vartheta'^*(O_{j-1}) = \vartheta^*(O_j). \tag{4.78}$$

Similarly, for $j = 1$, $v(O_1) = v_0$, $\rho(O_1) = \rho_0$, $\vartheta^*(O_1) = \vartheta_0{}^*$, and for $j = s$, $v'(O_{s-1}) = v(M)$, $\rho'(O_{s-1}) = \rho(M)$, $\vartheta'^*(O_{s-1}) = \vartheta^*(M)$. Inserting this into (4.31) yields

$$\delta^* = 1. \tag{4.79}$$

Note that if the layers are homogeneous, $\delta^* = 1$ for all head waves and not only for symmetric head waves, which were discussed in the previous section.

For r_j^*, τ_j^*, and $\bar{\tau}_j^*$, we find

$$r_j^* = h_j \tan \vartheta_j^* = h_j \frac{v_j}{[\bar{v}^2 - v_j^2]^{1/2}}, \tag{4.80}$$

$$\tau_j^* = \frac{h_j}{v_j \cos \vartheta_j^*} = \frac{h_j}{v_j[1-(v_j^2/\bar{v}^2)]^{1/2}}, \tag{4.81}$$

$$\bar{\tau}_j^* = \frac{h_j \cos \vartheta_j^*}{v_j} = \frac{h_j}{v_j}[1-(v_j/\bar{v})^2]^{1/2}. \tag{4.82}$$

The critical distance, r^*, and the travel time, $\tau^*(M)$, are then given by

$$r^* = \sum_{j=1}^{s} \frac{h_j v_j}{[\bar{v}^2 - v_j^2]^{1/2}}, \tag{4.83}$$

$$\tau^*(M) = \frac{r - r^*}{\bar{v}} + \sum_{j=1}^{s} \frac{h_j \bar{v}}{v_j[\bar{v}^2 - v_j^2]^{1/2}}, \tag{4.84}$$

or

$$\tau^*(M) = \frac{r}{\bar{v}} + \sum_{j=1}^{s} h_j \left(\frac{1}{v_j^2} - \frac{1}{\bar{v}^2}\right)^{1/2}. \tag{4.84'}$$

The displacement vector of the head wave at the point M is given by (4.37) and (4.43), and the r and z components by (4.44) to (4.48). If the receiver lies on the earth's surface, appropriate changes should be made, as discussed in section 4.1. In the case of a pulse source, the expressions in section 4.2 are again applicable.

For symmetric or partially symmetric head waves, we do not find any further substantial simplification of the formula when all the layers are homogeneous. For r^* and $\tau^*(M)$, of course, we need not calculate r_j^* and τ_j^* twice in symmetrical parts of the ray, but can calculate them once and multiply by 2. For a fundamental head wave in the case of a layered overburden (with homogeneous layers), the formulae (4.74) to (4.77) are valid if we insert $\delta^* = 1$ and use (4.83) to (4.84′).

4.5 CONSTANT VELOCITY GRADIENT INSIDE LAYERS

The expressions for head waves in layered media, given in sections 4.1 to 4.3, may be used for arbitrary, smooth velocity functions $v_j(z)$ inside the layers. In this section, we shall discuss a specific velocity variation for which velocity gradients inside the layers are constant. Suppose that the velocity–depth function is given by

$$v_j(d) = v'(O_{j-1})\{1 + b_j(d - d(O_{j-1}))\}, \tag{4.85}$$

where $j = 1, 2, \ldots, s$, $d = -z$, $b_j = \text{const}$. From (2.153) and (2.156) we obtain

$$r_j^* = \frac{\bar{v}h_j}{v(O_j)-v'(O_{j-1})}[\cos\vartheta'^*(O_{j-1})-\cos\vartheta^*(O_j)], \tag{4.86}$$

$$\tau_j^* = \frac{h_j}{v(O_j)-v'(O_{j-1})}\ln\frac{v(O_j)[1+\cos\vartheta'^*(O_{j-1})]}{v'(O_{j-1})[1+\cos\vartheta^*(O_j)]}. \tag{4.87}$$

Substituting (4.2) and (2.152) into (4.86) and (4.87) yields

$$r_j^* = \frac{\bar{v}h_j}{v(O_j)-v'(O_{j-1})}\{[1-(v'(O_{j-1})/\bar{v})^2]^{1/2}-[1-(v(O_j)/\bar{v})^2]^{1/2}\}, \tag{4.88}$$

$$\tau_j^* = \frac{h_j}{v(O_j)-v'(O_{j-1})}\ln\frac{v(O_j)\{1+[1-(v'(O_{j-1})/\bar{v})^2]^{1/2}\}}{v'(O_{j-1})\{1+[1-(v(O_j)/\bar{v})^2]^{1/2}\}}. \tag{4.89}$$

The expressions (4.86) to (4.89) are indefinite for $b_j = 0$ (i.e., for $v(O_j) = v'(O_{j-1})$) and numerically inexact for small b_j. For r_j^*, however, we can write, from (4.88),

$$r_j^* = \frac{h_j(v(O_j)+v'(O_{j-1}))}{[\bar{v}^2-v'^2(O_{j-1})]^{1/2}+[\bar{v}^2-v^2(O_j)]^{1/2}}. \tag{4.90}$$

This expression may be used for an arbitrary value of the gradient parameter b_j. For zero gradient (i.e., for a homogeneous layer), it gives the same result as (4.80).

We are unable to find a similar expression for τ_j^* which is suitable for an arbitrary b_j. For small b_j, we find from (4.89)

$$\tau_j^* = \frac{h_j}{v(O_j)[1-(v(O_j)/\bar{v})^2]^{1/2}}\left(1+\frac{v(O_j)-v'(O_{j-1})}{2v(O_j)}\,\frac{1-2(v(O_j)/\bar{v})^2}{1-(v(O_j)/\bar{v})^2}\right). \tag{4.91}$$

The first term in this expression is the same as for a homogeneous layer ($b_j = 0$), and the other is a correction term.

Inserting r_j^* and τ_j^* into (4.35) and (4.38), we find the critical distance r^* and the time of arrival $\tau^*(M)$. The displacement vector, amplitudes, phase shifts, etc. of the head wave are then given by (4.37) and (4.43) to (4.61). For symmetric, partially symmetric, and fundamental head waves, we can use the formulae of section 4.3.

4.6 ARBITRARY VELOCITY–DEPTH FUNCTIONS

The same procedure as in sections 4.4 and 4.5 can be used for an arbitrary velocity–depth function $v_j(z)$. We find values of r_j^* and τ_j^*, and insert these into (4.35) and (4.38). Then we can use the general expressions, given in section 4.1, which are very simple and do not cause any difficulties. The expressions $\tau_j{}^*$ and $r_j{}^*$ are known for many velocity–depth functions (see, e.g., Kaufman, 1953). We can also use numerical methods for finding these values.

However, it is often simpler to simulate arbitrary velocity variations inside the layers by a *step* approximation or a *linear* approximation than by an integrable analytic function $v_j(z)$ (see figure 4.6). We must remember that these approximations are used only for the calculation of kinematic parameters, $\tau_j{}^*$ and $r_j{}^*$. Hence, we need not calculate R-coefficients for the artificially introduced interfaces; nor do we need to consider these interfaces when calculating δ^*. For step and linear approximations, we can use the formulae of section 4.4 or 4.5 in each artificially introduced sub-layer (see also Gupta, 1966c, and Ravindra, 1967).

It is well known that a linear approximation causes many difficulties when calculating the displacement vector of the wave which has a smooth minimum inside layer (see figure 4.7). Second-order interfaces may cause false caustics, loops on the time-distance curves, or singularities on the amplitude–distance curves, for example. Therefore, for the calculation of this wave, we must use more complicated functions $v_j(z)$, with three or four parameters for each j, which guarantee continuity of velocity as well as of velocity gradient. However, this situation does not arise in the

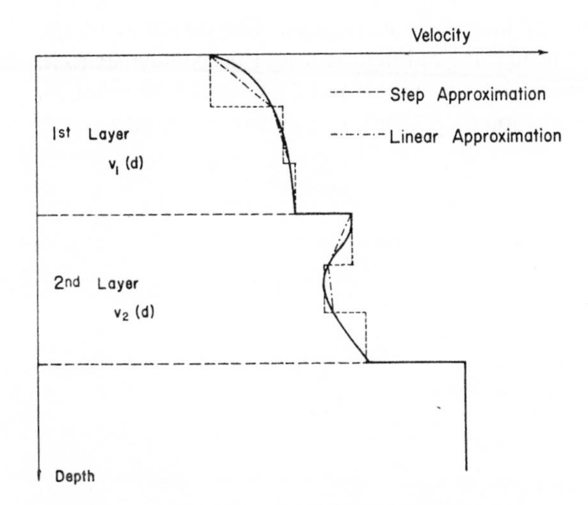

FIGURE 4.6 Simulation of an arbitrary velocity–depth function (solid line) by a step and a linear approximation.

calculation of *head waves* (generated at a sharp interface) where a linear approximation is very useful. If we want to obtain more exact results, we must increase the number of sublayers simulating a given velocity–depth graph.

4.7

KINEMATIC AND DYNAMIC ANALOGUES AND MULTIFOLD HEAD WAVES

In this section, we shall assume for simplicity that the multilayered medium consists only of homogeneous layers. An infinite number of head waves can propagate in this medium. The head waves which have the same time-distance curve are *kinematically analogous* to each other. All these waves, of course, must exist at the same epicentral distances and must have the same critical point. An example of fifteen kinematically analogous head waves, propagating in one layer, is given in figure 4.8, where the source is assumed to be radiating P-waves. All these waves arrive at the receiver at the same time and form one generalized head wave. If we want to find the displacement vector $\mathbf{W}^{*K}$ of this generalized wave, produced by the superposition of all the kinematic analogues, we must add all the individual waves. Thus, for a source radiating either P or SV waves, we obtain

$$\mathbf{W}^{*K} = \frac{\bar{v} \tan \vartheta_0^*}{i\omega r^{1/2}(r-r^*)^{3/2}} \exp[i\omega(t-\tau^*)] \sum_{(N_K)} \left(\Gamma_k \prod_{\substack{j=1 \\ j\neq k}}^{s-1} R_j^* \mathbf{n}^* \right), \tag{4.92}$$

where N_K is the number of kinematic analogues. The direction of the displacement vector $\mathbf{W}^{*K}$ need not be along the ray or perpendicular to it, as this vector may be composed of different P and SV waves at the point M. The summation in the expression (4.92) leads to a constant complex vector

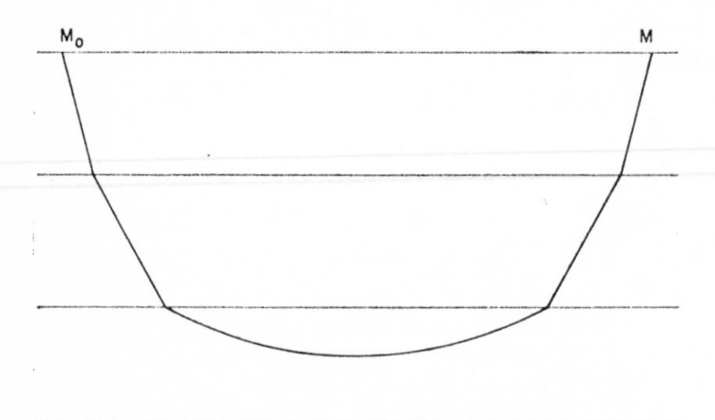

FIGURE 4.7 Ray of a wave with a smooth minimum inside a layer.

which does not change with the epicentral distance of the receiver. Therefore, the amplitude–distance curve of the generalized head wave and that of any of its constituent kinematic analogues differ only by a multiplication constant.

We can find smaller groups of waves, in a group of kinematic analogues, which are equivalent from the dynamic point of view also, i.e. waves which are *dynamically analogous* to each other. In figure 4.8, for example, the groups of waves 1–8, 9–12, and 13–14 are dynamically analogous; and the last wave (number 15) has no dynamic analogues. For the displacement vector $\mathbf{W}^{*D}$ of a generalized wave, produced by the superposition of all dynamically analogous waves, we get

$$\mathbf{W}^{*D} = N_D \mathbf{W}^*, \tag{4.93}$$

where N_D is the number of dynamically analogous waves in a given group and $\mathbf{W}^*$ is the displacement vector of any individual head wave in that group.

Note that head waves with *two or more* ray elements parallel to any interface cannot exist in a layered medium which consists of homogeneous layers. Examples of impossible head waves are given in figure 4.9. The reason for their impossibility is clear from wave front considerations: once

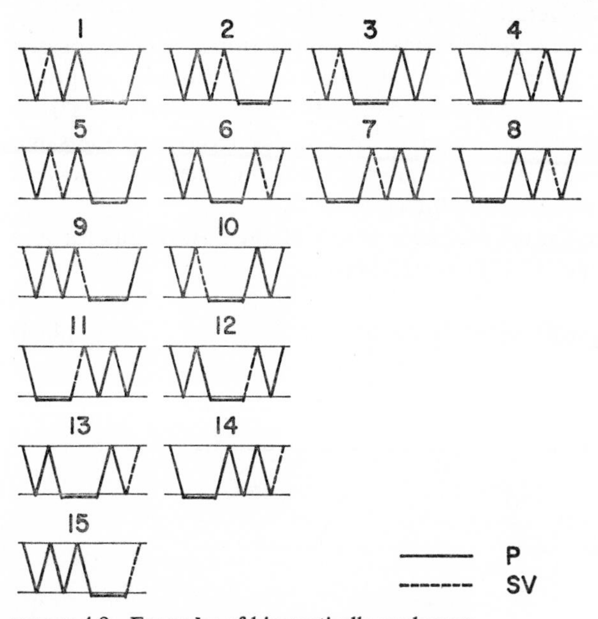

FIGURE 4.8 Examples of kinematically analogous head waves in a layer, for a P-wave source.

a *curved wave front* has produced a ray element parallel to any interface, all the subsequent wave fronts are straight lines in the *rz* plane and therefore cannot produce another head wave.

The R-waves propagating in layered media can also be divided into groups of dynamically and kinetically analogous waves. The theory and number of these waves are discussed in Vavilova and Petrashen (1966) and Vavilova and Pugach (1966). An example of kinematically analogous R-waves propagating in a single layer is given in figure 4.10 for a source of P-waves. The waves 1, 2, 3, and 4 are also dynamically analogous.

As we have seen, many different R and H waves exist in a layered medium With each head wave is associated one R-wave, which we shall call the associated R-wave or A-wave. The A-wave has the same succession of elements of the ray as the head wave under consideration, but the element of the ray parallel to the interface is excluded. The time–distance curve of the head wave is tangential to that of the A-wave at the critical point. An A-wave can be associated not only with one head wave but with a few groups of kinematically analogous head waves (which need not be dynamically analogous). The velocity of propagation along the element of the ray parallel to the interface, $\bar{v}$, is the same for all head waves in a given group. This element may, of course, exist at different parts of the ray of the head waves under consideration (see figure 4.11). For other groups, if they exist, $\bar{v}$ is different (for example, $\bar{v}$ may be compressional velocity for one group and shear velocity for another). We shall now study only one group of head waves, corresponding to a given velocity $\bar{v}$. All the head waves from this group associated with an A-wave arrive at the receiver at the same time and form one wave. We shall call the resultant interference wave a *multifold head wave* (i.e. twofold, threefold, etc.).

The expression for the displacement vector of the A-wave (as for any R-wave), $\mathbf{W}^{\mathrm{A}}(M)$, is (see (2.126) and (2.144))

$$\mathbf{W}^{\mathrm{A}}(M) = \frac{1}{L}\prod_{j=1}^{s-1} R_j \exp\{i\omega[t-\tau^{\mathrm{A}}(M)]\}\mathbf{n}^{\mathrm{A}}(M). \tag{4.94}$$

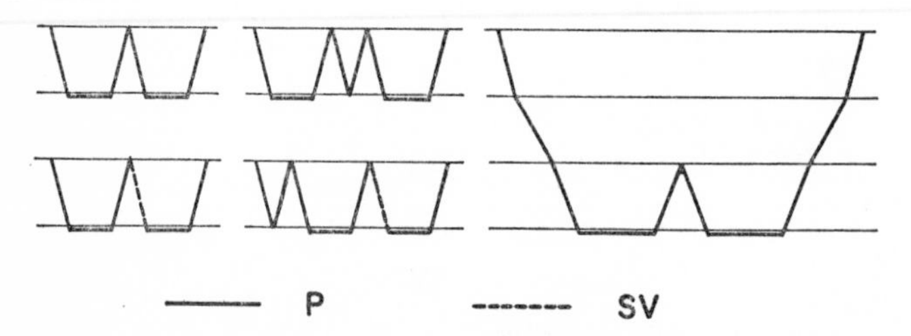

FIGURE 4.9 Examples of impossible head waves.

For the displacement vector of a member of the group of head waves associated with the A-wave, we find

$$\mathbf{W}^*(M) = \frac{\bar{v}\Gamma_{k_1} \tan \vartheta_0{}^*}{i\omega r^{1/2}(r-r^*)^{3/2}} \prod_{\substack{j=1 \\ j \neq k_1}}^{s-1} R_j{}^* \exp\{i\omega[t-\tau^*(M)]\}\mathbf{n}^*(M), \tag{4.95}$$

where k_1 is the number of ray elements between the source and the element of the ray parallel to the B-interface. The succession of R-coefficients $R_1{}^*$, $R_2{}^*$, …, R_{s-1}^* is the same as in (4.94), but the R-coefficient $R_{k_1}^*$ is excluded. If we use the relations (3.108) and (4.18), we find, from (4.95), that

$$\mathbf{W}^*(M) = \frac{\bar{v} \tan \vartheta_0{}^*}{i\omega r^{1/2}(r-r^*)^{3/2}} \left(-\frac{\mathrm{d}R_{k_1}}{\mathrm{d}[1-(\bar{v}\Theta)^2]^{1/2}} \prod_{\substack{j=1 \\ j \neq k_1}}^{s-1} R_j(\Theta) \right)_{\Theta=1/\bar{v}} \times \exp\{i\omega[t-\tau^*(M)]\}\mathbf{n}^*(M). \tag{4.96}$$

For other head waves in this group, only k_1 changes. If we add all the head waves in a given group, we find for the displacement vector ($\mathbf{W}^{*\mathrm{F}}$) of the multifold head wave

$$\mathbf{W}^{*\mathrm{F}}(M) = \frac{\bar{v} \tan \vartheta_0{}^*}{i\omega r^{1/2}(r-r^*)^{3/2}} \left(-\frac{\mathrm{d}\prod_{j=1}^{s-1} R_j(\Theta)}{\mathrm{d}[1-(\bar{v}\Theta)^2]^{1/2}} \right)_{\Theta=1/\bar{v}} \times \exp\{i\omega[t-\tau^*(M)]\}\mathbf{n}^*(M). \tag{4.97}$$

This is the general formula for the multifold head wave associated with a given A-wave. A similar expression was published by Pod"yapol'skiy (1966a).

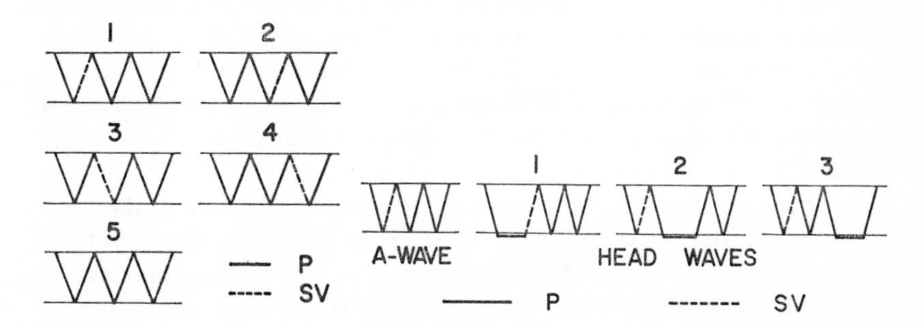

FIGURE 4.10 Examples of kinematically analogous R-waves in a layer, for a P-wave source.

FIGURE 4.11 Example of a group of kinematically analogous head waves with the associated R-wave (A-wave).

4.8 AMPLITUDES OF FUNDAMENTAL HEAD WAVES AND COMPARISON WITH AMPLITUDES OF R-WAVES

Fundamental head waves are the most important head waves in seismic prospecting and in seismic investigations of the earth's crust (see section 4.3.3). They are *compressional* along the whole ray and do not have any segment formed by reflection at an intermediate interface. In this section, we shall study the amplitudes of fundamental head waves and the ratio of the amplitudes of these waves with some other compressional waves. We shall investigate mainly the influence of inhomogeneity of the overburden on the amplitude properties of these waves. We shall not discuss here the mode of variation of the head wave amplitudes with all the parameters of the media, as there are a large number of parameters and they can vary considerably. We shall only seek some general conclusions. Head waves at various distances from the source pass through the layers of the overburden at the same angles. Therefore, horizontal stratification of the overburden will not affect their amplitude–distance curves very much. The character of the amplitude-distance curve remains the same as for a homogeneous overburden (see section 3.6). The variation of the head wave amplitudes with the epicentral distance is again given by the function $I(r) = r^{-1/2} (r-r^*)^{-3/2}$. Thus, the amplitude–distance curve is smooth and monotonically decreasing, without any maxima or minima. The slope of the amplitude–distance curve at a given epicentral distance r is higher for larger r^*. Therefore, more detailed conclusions about the slope and shape of the amplitude–distance curve can be easily derived from the formula for r^*.

Generally, we can expect that the absolute values of the amplitudes of head waves at a given epicentral distance r will depend only slightly on the horizontal stratification of the overburden if the critical distance r^* does not change much. On the other hand, amplitudes of R-waves are affected strongly by the structure of the overburden, especially at large epicentral distances. We shall present below amplitude–distance curves of some important waves propagating in different types of media. In all the following models, the depth of the B-interface is 30 km, the velocity of the compressional waves below the B-interface ($\bar{v}$) is 8 km/sec, the density is identical in all layers, the ratio of the velocity of shear and compressional waves is $1/\sqrt{3}$ in all layers, and the frequency of the incident wave is 10 cps. The parameters of the B-interface correspond, approximately, to those of the Mohorovičić discontinuity. In order that we may clearly evaluate the effect of stratification, we have – for methodological reasons – also included a few models which obviously do not correspond to the conditions prevailing in the earth's crust (e.g., the velocity in the high-velocity layer is taken too high, or the velocity in the low-velocity layer too low). By *amplitudes* here we understand the amplitudes of the *vertical* components of the displacement vector. The source of compressional wave is symmetric and lies near the earth's surface (at zero depth).

Note that, in this section, only the leading term in the ray series has been used for the calculation of wave amplitudes. Therefore, "properties of waves" refers here to properties of the leading term in the ray series for these waves. All the results are, of course, inaccurate in the critical region (see more details in chapter 7), and also in those cases where the angle between the ray and an interface is very small. (This applies especially to reflected waves at large epicentral distance.) As the ray formulae can be used only for non-interfering waves, the following results may be invalid in the regions of interference of waves under consideration with themselves or with other, more complicated, waves (e.g., multiply reflected and refracted waves) which have not been considered here (see Petrashen, 1957a).

We shall consider first an overburden composed of homogeneous layers and later an overburden with inhomogeneous layers.

4.8.1 *Overburden Composed of Homogeneous Layers*

The effect of stratification of the overburden on the amplitude properties of reflected and head waves was studied in detail by Berry and West (1966), Červený (1967a, 1968) and Červený and Yepinat'yeva (1967, 1968); the reader is referred to these works for a thorough discussion. Two examples presented below are taken from Červený (1967a) (where one numerical error has been corrected).

We shall assume here that the overburden is composed of three layers, each 10 km thick. The average velocity of the overburden is the same for all models (6 km/sec). The amplitudes of the compressional symmetric wave reflected from the interface, which lies at a depth of 30 km, are denoted by A_z^{R}, and the amplitudes of the fundamental symmetric head wave propagating along the same interface are denoted by $A_z{}^*$. Rays of both waves are given schematically in figure 4.12.

Example 1 Five models of a three-layer overburden are given in figure 4.13. Models 1 and 2 are characterized by an increase of velocity with depth, model 3 has a homogeneous overburden, and models 4 and 5 are characterized by a decrease of velocity with depth. The amplitude-distance curves for reflected and head waves for these models are given in figure 4.14. We can see from this figure that the amplitude–distance curves of all the head waves are practically the same, in spite of the large differences in the velocity–depth graphs. The reflected wave amplitudes differ from one another to a greater extent. The behaviour of the reflected waves is different both before and after the critical point. Before the critical point, the largest amplitudes are for models 4 and 5 and the smallest for model 1 (this behaviour is connected mainly with the properties of the B-interface, which is of higher velocity-contrast for models 4 and 5 than for models 1 and 2). Beyond the critical point, the largest amplitudes are those for a homogeneous overburden (model 3). At an epicentral distance of 400 km, the amplitudes for the homogeneous model 3 differ from the amplitudes of model 1 more than six times ($\frac{3}{4}$ order of magnitude). The ratio of the amplitudes of reflected and head waves at 250 km from the source is highest for the homogeneous model 3 (about 90) and smallest for model 1 (about 25). This ratio would be smaller if the layers were thinner. At large epicentral distances, the head wave amplitudes can even be of the same order of magnitude as the reflected wave amplitudes for a thinly layered overburden if velocity increases with depth.

Example 2 The five models of a three-layer overburden which are given in figure 4.15 are characterized by the existence of a high-velocity layer (models 1 and 2) or a low-velocity layer (models 4 and 5). The homogeneous model 3 is included for comparison. The models studied here are not representative of the real earth's crust. However, similar models can be of immediate interest for seismic prospecting, where layers with pronounced high or low velocities frequently exist in the overburden. The amplitude–distance curves of the head waves differ only slightly from one model to

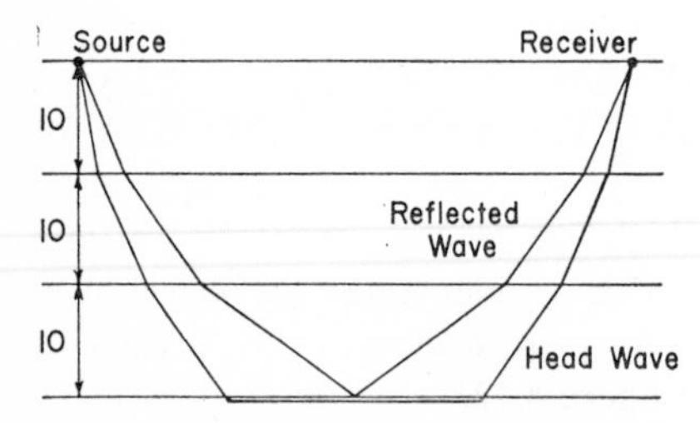

FIGURE 4.12 Schematic presentation of the rays of the fundamental symmetric reflected and head waves the amplitudes of which are given in figures 4.14 and 4.16 The overburden is composed of three homogeneous layers, each of 10 km thickness

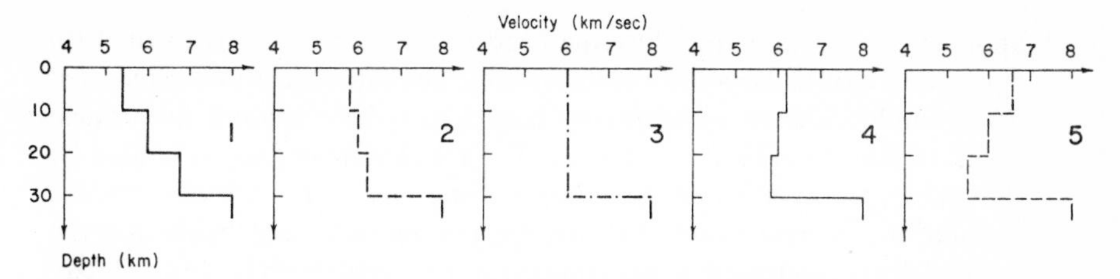

FIGURE 4.13 Velocity–depth graphs for five models of a three-layer overburden: velocity increasing with depth (1, 2), uniform velocity (3), and velocity decreasing with depth (4, 5).

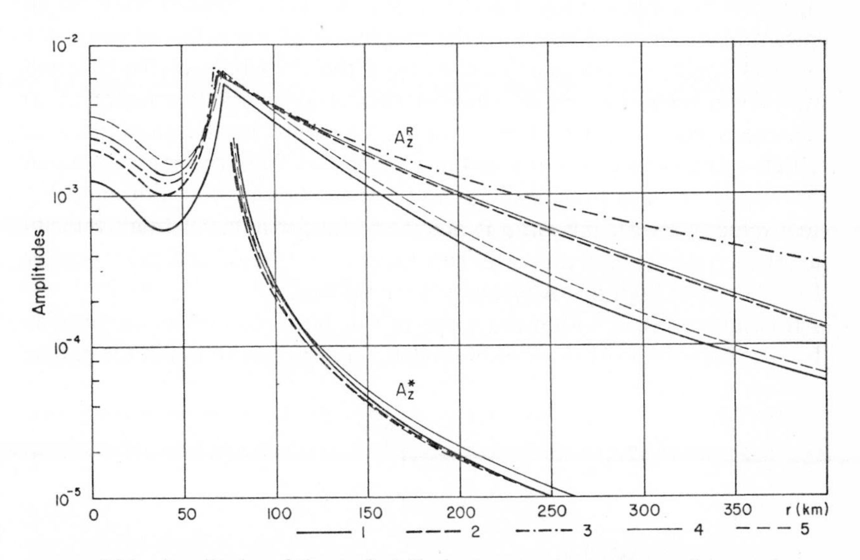

FIGURE 4.14 Amplitudes of the vertical displacement components vs. distance, for the two waves shown in figure 4.12 (corresponding to the five models in figure 4.13).

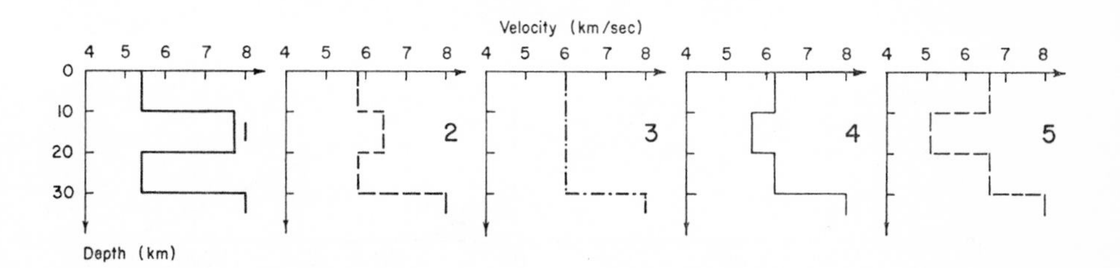

FIGURE 4.15 Velocity–depth graphs for five models of a three-layer overburden: high-velocity layer (1, 2), uniform velocity (3), low-velocity layer (4, 5).

another (see figure 4.16). Only the head wave for model 1 is distinct (in consequence of its larger critical distance). The differences in the reflected wave amplitudes are substantially larger, especially beyond the critical point. Before the critical point, the differences are small and are caused by the different properties of the reflecting interface in the various models. Beyond the critical point, the strongest amplitudes are again for the homogeneous model 3. The amplitudes for the models with the low-velocity layer in the overburden (models 4 and 5) differ only slightly from those for model 3. A high-velocity layer in the overburden (models 1 and 2) brings about far greater changes of amplitudes. The higher the velocity in the high-velocity layer, the lower is the amplitude of the reflected wave. At an epicentral distance of 400 km, the amplitude of the reflected wave for model 1 (high-velocity layer) differs from the amplitude of the reflected wave for a homogeneous overburden (model 3) by about a factor of 50 (i.e. more than 1 order of magnitude). The ratio of the amplitudes of reflected and head waves at a distance of 250 km from the source is again the largest for the homogeneous overburden (about 90). For model 5 (low-velocity layer), this ratio is about 35 and for model 1 (high-velocity layer) the ratio is only about 8. This ratio would be substantially smaller if the layers of high or low velocity were thinner.

It is not possible, within the scope of this book, to present amplitude–distance curves for other series of models. They can be found in the papers

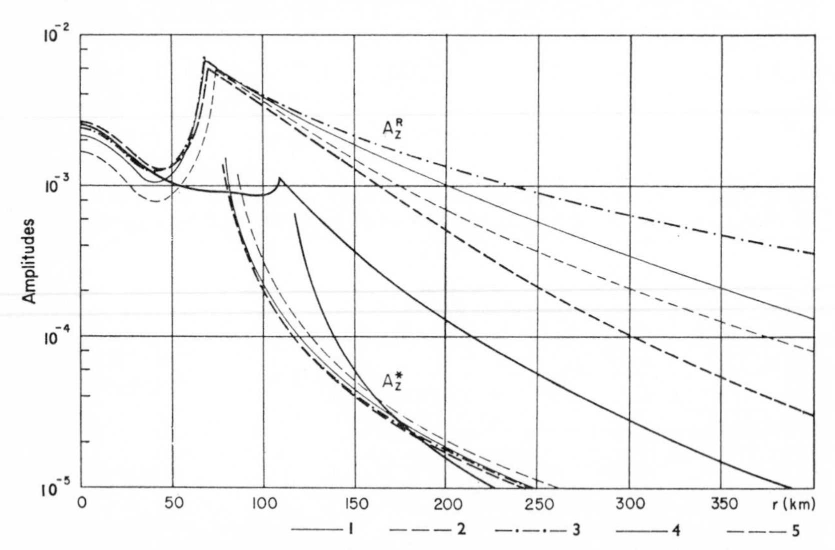

FIGURE 4.16 Amplitudes of the vertical displacement components vs. distance, for the two waves shown in figure 4.12 (corresponding to the five models in figure 4.15).

mentioned above. For example, the dependence of the reflected wave amplitudes on the thickness of the high-velocity layer in the overburden is very interesting. The amplitudes beyond the critical point are very sensitive to this thickness and are smaller for thinner layers. (Of course, we do not consider here *thin* layers of thickness comparable with, or less than, the wavelength.)

Generally, we can say that stratification of the overburden causes a decrease in the amplitudes of the reflected waves beyond the critical point (supercritical reflected waves) and an increase in their damping with the epicentral distance, in comparison with the case of a homogeneous overburden. Properties of supercritical reflected waves depend only slightly on the parameters of the reflecting interface and very greatly on stratification of the overburden (when the velocity contrast of the reflecting interface is not too large). They are affected very substantially by the presence of a thin high-velocity layer in the overburden. (The thinner the layer and greater the velocity in this layer, the smaller are the amplitudes of supercritical reflected waves.) The effect of a low-velocity layer is much smaller. The amplitudes of supercritical reflected waves are also influenced considerably by a monotonic stratification (i.e. only an increase or decrease of velocity with depth) of the overburden.

Note that the amplitudes of the subcritical reflected waves depend greatly on the parameters of the reflecting boundary and only slightly on the stratification of the overburden. Thus, the situation for subcritical reflected waves is just the reverse of that for supercritical waves.

The amplitudes of head waves usually depend only slightly on the inner structure of the overburden. Stratification of the overburden leads to a decrease in the ratio of the amplitudes of reflected and head waves in comparison with the case of a homogeneous overburden. This ratio is smaller for greater velocity differentiation in the layers of the overburden, smaller thickness of the high-velocity layer, and smaller difference between the maximum velocity in the overburden and $\bar{v}$. The main variations of this ratio are related to the behaviour of the reflected wave amplitudes, since they are more affected by a stratification of the overburden. With increasing epicentral distance, the ratio either (a) first increases, attains a maximum value, and then slowly decreases, or (b) increases and asymptotically approaches some constant value.

4.8.2 *Overburden with Inhomogeneous Layers*

We shall now assume that the velocity increases linearly with the depth inside individual layers. The layers are separated by the first-order or second-order discontinuities. The first-order discontinuities lie at depths of 15 and 30 km. We may consider these discontinuities as Conrad and Moho discontinuities and the material up to the depth of 30 km as the earth's crust. The velocity of compressional waves below the Moho discontinuity remains constant at 8 km/sec. In the above-described medium, in addition to reflected and head waves, we can also have waves which have a smooth minimum (turning point) inside the medium (see figure 4.7). In this section, we shall call these waves *refracted* waves. Other possible waves will not be investigated here.

Example 3 In this example we shall assume that the earth's crust is composed of three layers in each of which the velocity increases with depth according to a linear law (see figure 4.17). The interface between the first and second layer is a second-order interface (i.e. the velocity is continuous there, whereas the velocity gradient is discontinuous; see point A in the figure).

Only one fundamental head wave can propagate in the given medium, viz., the head wave propagating along the Moho interface (at a depth of 30 km). No pure head wave is generated at the Conrad discontinuity, as there is a positive velocity gradient below this interface and a ray penetrates into the third layer and becomes a refracted wave instead of a head wave. This refracted wave corresponds to the C_0-wave in chapter 6, where more details are given (see also section 3.9). The interference nature of this wave is not considered here.

The amplitude–distance curves of some important waves propagating in

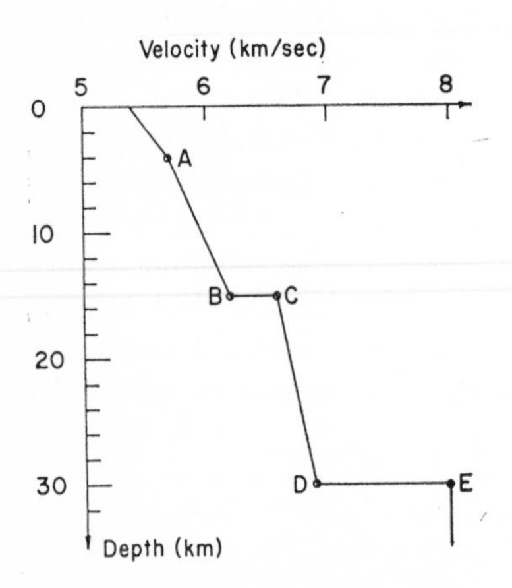

FIGURE 4.17 Velocity–depth graph for example 3.

this medium are given in figure 4.18. Figure 4.18A shows the corresponding travel time–distance curves. The amplitude–distance curve of the fundamental head wave (connected with the Moho discontinuity) is shown by the heavy line and is denoted by the number 3. This curve has the same character as in the previous cases: it begins at the critical point (which lies at the epicentral distance 79.5 km) and decreases monotonically with increasing epicentral distance.

The reflected waves from the Conrad and Moho discontinuities are denoted by the numbers 1 and 2, respectively. An important characteristic is that they do not exist at large epicentral distance, since the positive velocity gradients above the corresponding interfaces cause shadow zones for reflected waves. The reflected wave from the Conrad discontinuity exists only at epicentral distances less than 123.5 km, and the reflected wave from the Moho interface only at epicentral distances less than 249.9 km. Thus, at epicentral distances larger than 123.5 km, we should record no reflected waves from the Conrad discontinuity. (This conclusion follows from ray theory; in fact diffracted waves will exist.) We can see from figure 4.18 that the character of the amplitude–distance curve of the reflected wave is very similar to that of the curve for homogeneous layers in the overburden (see examples 1 and 2) only up to some epicentral distance, and then the amplitude suddenly begins to decrease very rapidly to vanish at the limiting distance mentioned earlier.

The amplitude–distance curve of the refracted wave is shown in figure 4.18 by the dashed line. This wave is registered as the first arrival at distances up to 132 km (at larger distances, the head wave from the Moho discontinuity arrives first). The refracted wave is very sensitive to interfaces of second order. If the ray of the refracted wave has a minimum just below this interface, the amplitude of this wave vanishes (see point *A* in figure 4.18, corresponding to the interface of second order at the depth of 4 km in the velocity–depth graph in figure 4.17). If the minimum of the ray is near the interface of first order between the two inhomogeneous media, the situation may be more complicated. In our case, two branches of the refracted wave exist between the distances 59.3 and 123.5 km (between points *C* and *B* in figure 4.18). Note that the part of the amplitude–distance curve from zero to the point *A* corresponds to the refracted wave with the minimum of the ray lying in the first layer (at depths less than 4 km), the part of the curve between *A* and *B* corresponds to the minimum lying in the second layer, and the part between *C* and *D* to the minimum lying in the third layer. The boundary rays, *B* and *D*, of the refracted wave are the same as the boundary rays for reflected waves (see also figure 4.18A).

We see from figure 4.18 that the head wave (curve 3) is again very weak

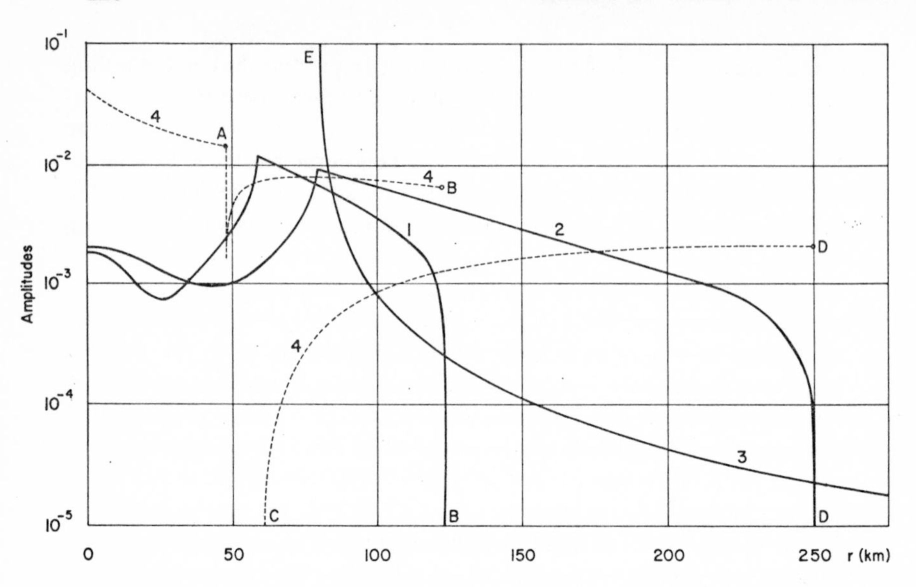

FIGURE 4.18 Amplitudes of the vertical displacement components vs. distance, for the fundamental head wave (3), reflected waves (1, 2), and refracted wave (4) in the medium shown in figure 4.17.

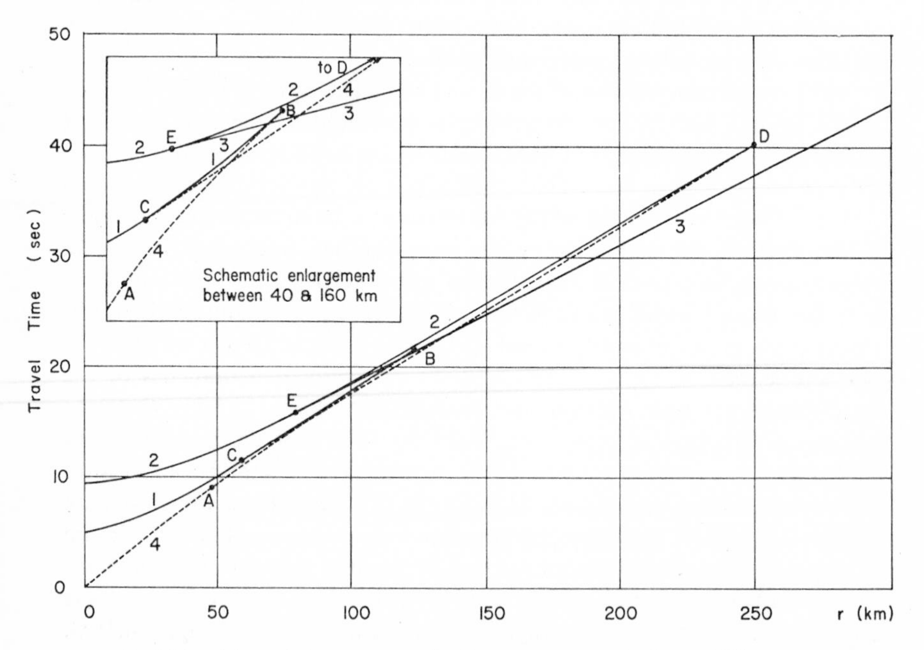

FIGURE 4.18A Time–distance curves for the waves shown in figure 4.18.

in comparison with the corresponding reflected wave (curve 2). The ratio of the amplitude of the reflected wave to that of the head wave is about 30 at epicentral distances from 120 to 230 km. However, between distances of about 230 and 250 km, this ratio drops rapidly, and beyond 250 km only the head wave exists. Thus, again as in the case of homogeneous layers, the inhomogeneity of the overburden leads to a decrease in the ratio of the amplitudes of reflected and head waves in comparison with the case of a homogeneous overburden. The inhomogeneity of the overburden has a strong influence on the reflected wave amplitudes, but only a slight influence on the head wave amplitudes.

The amplitudes of the refracted wave are, in our example, of the same order of magnitude as or even larger than those of the reflected waves. At distances beyond 250 km, the refracted wave does not exist (because of the shadow zone).

The refracted wave amplitudes, with the minimum of the ray lying in the third layer (between the depths of 15 and 30 km), are given by the dashed curve (no. 4) between the points C and D. The amplitudes increase with increasing epicentral distance. This fact is very interesting. If there were no increase of velocity in the third layer, this refracted wave would not exist and, instead of it, the head wave would propagate along the interface at the depth 15 km. The kinematic properties of this head wave would be very close to those of the refracted wave since the velocity gradient is small in the present case. The amplitude properties would, however, be just the reverse of the amplitude properties of the refracted wave. The amplitude–distance curve of the head wave would be very similar to curve 3 in figure 4.18, except that it would be shifted a little to the left. Whereas the amplitude of the head wave decreases rapidly with increasing distance, the amplitude of the refracted wave increases. The differences in the amplitudes of the two waves would be about 1–2 orders of magnitude at distances of 150–200 km. Thus, even a very small positive velocity gradient in the substratum transforms the head waves into refracted waves and increases their amplitudes considerably. More details will be given in chapter 6.

Example 4 In this example we shall assume the earth's crust to be similar to that in example 3 (see the velocity–depth graph in figure 4.19). The only difference lies above the Conrad discontinuity. The velocity near the earth's surface is much less than in the preceding example, and increases rapidly with increasing depth. The velocity variation above the Conrad discontinuity is simulated by a linear approximation (five layers, each 3 km thick). The points A, B, C, D represent second-order discontinuities. The amplitude–distance curves of the same waves as in example 3 are given in figure 4.20 (1 and 2 are the waves reflected from the Conrad and Moho discontinuities, 3 is the fundamental head wave propagating along the Moho discontinuity, and 4 is the refracted wave; curve 5 will be discussed later).

The amplitude–distance curve of the head wave (curve 3) is practically the same as in figure 4.18, and we can conclude that properties of the head wave from the Moho discontinuity are not substantially affected by a change in the velocity distribution above the Conrad discontinuity. The same is true of the wave reflected from the Moho (curve 2). The shadow zone for this wave now begins at the epicentral distance 241.6 km (in the previous case, it was 249.9 km). However, the wave reflected from the Conrad discontinuity has changed considerably: the shadow zone is much larger, the reflected wave exists only up to the distance of 90.6 km. Hence, at distances greater than 90 km, no waves can be registered which are reflected from the Conrad discontinuity for the velocity–depth graph in figure 4.19.

The refracted wave is now more complicated than in the previous case, as there are other second-order discontinuities in the overburden (see points *A*, *B*, *C*, *D* in figure 4.19). Each of these discontinuities causes a singularity in the amplitude–distance curve (see figure 4.20). If we want to get a smooth amplitude–distance curve for the refracted wave we must use more complicated simulations of the velocity–depth graph so that the second-order discontinuities are removed. It seems that the best representation of the amplitude–distance curve of the refracted wave in figure 4.20 is that given by a smooth curve connecting points *A*, *B*, *C*, *D*, *E*.

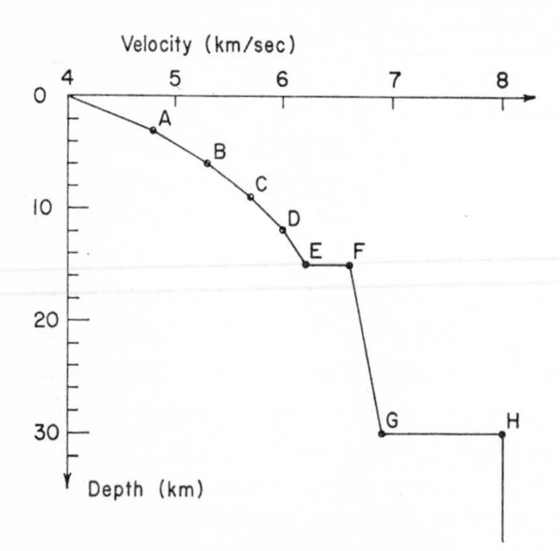

FIGURE 4.19 Velocity–depth graph for example 4.

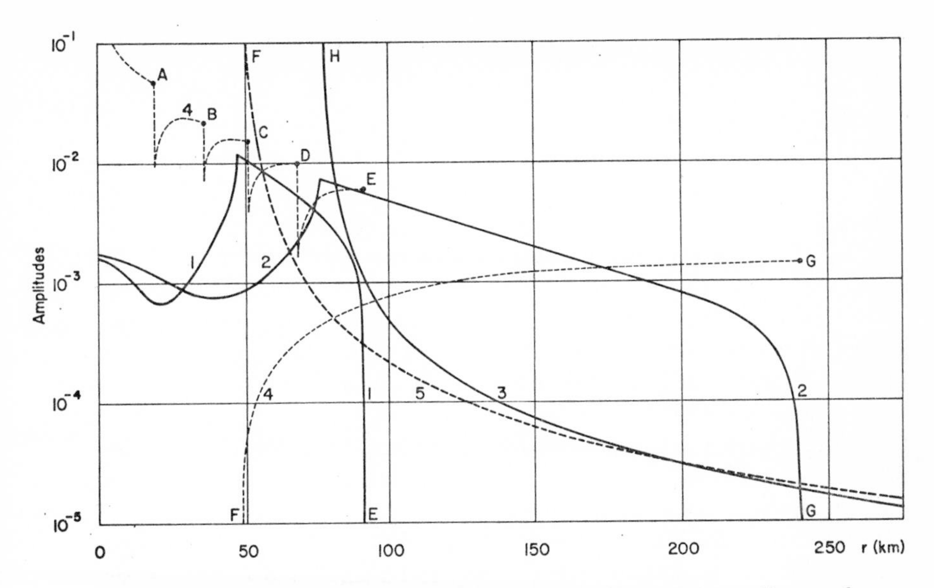

FIGURE 4.20 Amplitudes of the vertical displacement components vs. distance, for the fundamental head wave (3), reflected waves (1, 2), and refracted wave (4) in the medium shown in figure 4.19. Curve 5 corresponds to the head wave that would propagate along the Conrad discontinuity if there were a homogeneous layer between the Conrad and Moho.

The amplitude–distance curve 4 from F to G corresponds to the refracted wave with the minimum of the ray lying between the Conrad and Moho discontinuities. If there were a homogeneous layer between the Conrad and Moho, this wave would not exist, and, instead of it, a head wave would propagate along the Conrad. For comparison, the amplitude–distance curve of this head wave is also given in figure 4.20 (see the dashed curve 5). We can now see very clearly the substantial differences in the amplitude–distance curves of the two waves (4 and 5). Note that the theoretical travel time–distance curves of the two waves differ only very slightly from each other. Thus, we must emphasize again the great influence of the positive velocity gradient in the substratum on the amplitude properties of some waves. (For more details, see chapter 6.)

4.9

CONSTRUCTION OF THEORETICAL SEISMOGRAMS

The construction of theoretical seismograms for a multilayered medium is considerably more complicated than for the one-interface problem (see section 3.7). A great number of different R and H waves arrive at the receiver. As we know the expressions for individual waves (see sections 2.6.5 and 4.2), we can in theory construct theoretical seismograms by a superposition of these waves.

If the travel times of the various waves are different and the waves do not interfere, the possible inaccuracy of the ray formulae for individual waves does not necessarily prevent the construction of theoretical seismograms. However, the arrival times for the waves in a group can be very close together. Then the resulting interference wave can depend substantially on higher terms of the ray series, which have not been considered in our derivations. In other words, the (leading term) ray approximations for individual waves, considered here, can cancel each other. Therefore, the ray theory cannot in general be used for the construction of theoretical seismograms in the case of thin layers of arbitrary type, as the waves from these layers have a definite interference character. Moreover, it would be practically impossible to use the ray theory in this case, as the number of individual waves would be extremely high.

In the case of thick layers, the situation is different for small and large epicentral distances, small and large times, and a small and large number of interfaces. We can say, in general, that the ray theory cannot be used for large times, large epicentral distances, and a larger number of interfaces. There are again two reasons: first, the waves have a clear interference character so that higher terms in the series must be considered; second, the number of waves becomes extremely high. The best conditions for the application of ray theory for the construction of theoretical seismograms exist when there are only a few interfaces, when we are interested only in early-arriving phases (small t), and when the epicentral distance is not large (i.e. the waves characterized by a small angle between the ray and an interface do not exist). Also, the construction of a theoretical seismogram is inaccurate in the neighbourhood of the critical and other singular points, since the formulae for individual waves are inaccurate there.

The conditions under which theoretical seismograms can be constructed by ray theory cannot be found by the ray theory itself; this problem must be solved by comparison with the results of exact methods (see section 7.2). For example, the exact theoretical seismograms for a one-layered earth's crust calculated by Pekeris *et al.* (1965) were compared with theoretical seismograms constructed by ray methods by Smirnova (1966c). For the first-arriving phases, the agreement between the two theoretical seismograms is satisfactory.

5
CURVED AND DIPPING INTERFACES IN THE OVERBURDEN

5.1
GENERAL CURVED INTERFACES

In the previous chapter, we investigated head waves propagating in a medium in which all the interfaces are plane and parallel to each other. In reality, however, often the interfaces are dipping and curved. In this chapter we shall study the properties of non-interfering head waves propagating in media with curved and dipping interfaces. The interfaces are assumed to be of first order (i.e., the elastic parameters are discontinuous across them) and the layers are assumed to be *homogeneous* and thick in comparison with the wavelength. Interface effects connected with thin layers are not considered.

We shall make some simplifying assumptions because the general problem is either not tractable by the ray method or leads to very cumbersome expressions. The ray method is not applicable when the principal radii of curvature of the interfaces, in the close neighbourhood of the points where the ray strikes the interface, are discontinuous or are not considerably larger than the wavelength, or when the interfaces intersect in the vicinity of these points. Moreover, we must make an important assumption that the B-interface (at which the head wave is generated; see figure 5.1) is not curved, but *plane*. (When the B-interface is curved, head waves cannot be investigated by the ray method; see figure 3.41 for interference and damped head waves.)

When the above assumptions are fulfilled, the ray theory can be used, except at some singular points. The procedure for determining the displacement vector of a head wave, at any point M lying on its ray, is as follows:

(1) First we determine the displacement vector of the incident wave at the point O_k^* (see figure 5.1) by the methods described in sections 2.5 and 2.3.2.

(2) Then we use the formula (3.161) and determine the displacement vector of the head wave at the point $\bar{O}_k^*$.

(3) Finally, we use the formulae for continuation of the displacement vector along a ray, described in sections 2.4.4 and 2.3.2, and determine the displacement vector of the head wave, at the point M.

Note that the formulae of sections 2.5 and 2.4.4 are applicable only when the ray is a *plane curve*; but it is a purely geometrical problem to write similar formulae for a *three-dimensional ray path.* The only complication is that the planes of incidence at the points O_j^* $(j = 1, 2, \ldots, s-1)$ need not coincide, and, therefore, we must determine, for every point O_j^*, a local system of unit vectors $\mathbf{n}_{\mathrm{P}}$, $\mathbf{n}_{\mathrm{SV}}$, and $\mathbf{n}_{\mathrm{SH}}$. The SH-wave, in this case, will not be separated from P and SV waves. Apart from this complication, the procedure remains the same as described in the sections mentioned above. We shall not investigate this general situation of a three-dimensional ray path here, but should mention that the main difficulty lies in finding the kinematic parameters of the wave, i.e. its ray. As soon as the ray path is found, the determination of the amplitudes and phase shifts does not involve any major difficulties.

When the ray is a plane curve, the formulae given in sections 2.5 and 2.4.4 may be used and the procedure described above leads to simpler results. For a general orientation of the principal lines of the interfaces at the points O_j^*, we must use (2.59), taken from Gel'chinskiy (1961), for

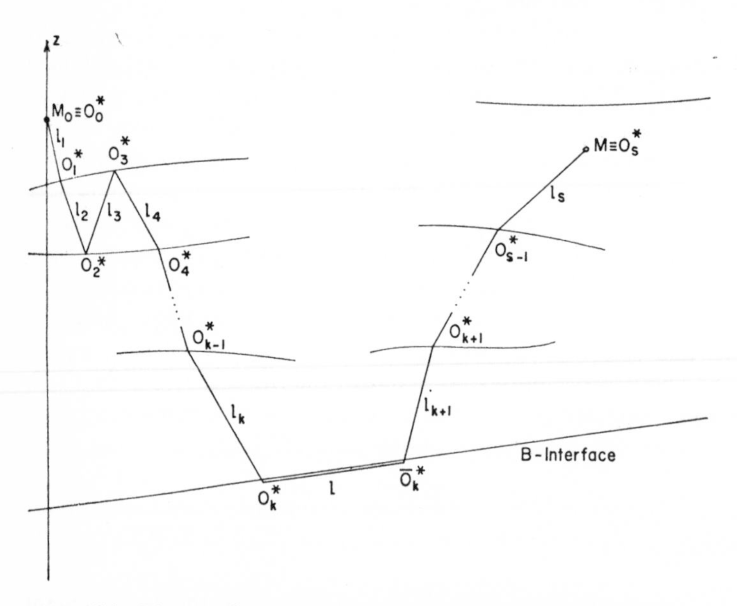

FIGURE 5.1 The ray of a head wave in a medium with curved and dipping interfaces. The B-interface is assumed to be plane.

successive calculation of the principal radii of curvature of the wave front of the wave under consideration.

We shall investigate here in detail one situation which is of considerable importance in seismic prospecting and in seismology, and which leads to relatively simpler final formulae for the displacement vector of head waves. This situation is characterized by the following assumptions:

(1) The ray is a plane curve (it lies in a vertical plane $E_{\parallel}$).

(2) The angles ϕ_j between the vertical plane $E_{\parallel}$ and a principal normal section of the interface at $O_j{}^*$ ($j = 1, 2, \ldots, k$) are zero (see more details about ϕ_j in section 2.3).

Note that R-waves propagating in a medium characterized by similar assumptions were investigated in section 2.6.4. The reader is advised to work through that section before proceeding with this chapter. It was also mentioned there that the second assumption (above) means that the trend of geological structures is either parallel or at right angles to the seismic profile.

The procedure to be followed below is very similar to that in the preceding chapter. We shall first obtain the solution for a harmonic source and then generalize the results for an impulse source. The source, which may be of P, SV, or SH type, lies at the point M_0 on the z-axis in any layer (see figure 5.1). The displacement vector of the spherical wave radiating from the source is given by (2.110). Our problem is to find expressions for the displacement vector $\mathbf{W}^*(M)$ of a head wave of a given type at a point M located in any layer (see figure 5.1). The ray of the head wave may be reflected or refracted any number of times along its path.

As in the preceding chapter, the ray of the H-wave is composed of $s+1$ elements: k elements from the source to the B-interface, $s-k$ elements from the B-interface to the receiver, and one element along the B-interface. The jth element of the ray ($j = 1, 2, \ldots, k, k+2, \ldots, s$) lies between the points O^*_{j-1} and $O_j{}^*$, the $(k+1)$th element between the points $\bar{O}_k{}^*$ and O^*_{k+1}, and the element along the B-interface lies between the points $O_k{}^*$ and $\bar{O}_k{}^*$ (see figure 5.1). As in sections 2.4.4 and 2.6.4, we shall use the following notation:

l_j = the length of the jth element of the ray ($j = 1, 2, \ldots, s$);

l = the length of the element of the ray along the B-interface;

v_j, ρ_j = the velocity of propagation and the density in the jth layer (between points O^*_{j-1} and $O_j{}^*$), which are constant along the jth element of the ray;

$\bar{v}$ = the velocity of propagation in the medium in which the ray of the head wave runs along the B-interface (between points $O_k{}^*$ and $\bar{O}_k{}^*$; see figure 5.1), where $\bar{v}$ is assumed constant;

$\sigma(O_j^*)$, $\sigma'(O_j^*)$ = the cross-sectional areas of an elementary ray tube at the point O_j^*, from the side of the incident wave and from the side of the reflected (refracted) wave, respectively;

$\vartheta(O_j^*)$, $\vartheta'(O_j^*)$ = the angle of incidence and the angle of reflection (refraction) at the point O_j^*, respectively;

$R_{\parallel}^{(j)}$, $R_{\perp}^{(j)}$ = the principal radii of curvature of the interface at the point O_j^*;

$r_{\parallel}(O_j^*)$, $r_{\perp}(O_j^*)$ = the principal radii of curvature of the wave front of the incident wave at O_j^*;

$r_{\parallel}'(O_j^*)$, $r_{\perp}'(O_j^*)$ = the principal radii of curvature of the wave front of the reflected (refracted) wave at O_j^*.

ε_j has the same meaning as in section 2.4.4. Note that the ray lies in the plane $E_{\parallel}$. One principal radius of curvature ($R_{\parallel}^{(j)}$) of the interface at O_j^* is associated with the intersection of the interface with $E_{\parallel}$, and the other ($R_{\perp}^{(j)}$) corresponds to the intersection of the interface with the plane perpendicular to $E_{\parallel}$ and the interface. The principal radii ($r_{\parallel}(O_j^*)$, $r_{\parallel}'(O_j^*)$, $r_{\perp}(O_j^*)$, and $r_{\perp}(O_j^*)$) of the wave fronts have similar interpretations ($r_{\parallel}(O_j^*)$ and $r_{\parallel}'(O_j^*)$ are associated with the intersection of the wave front of the corresponding wave with $E_{\parallel}$). Because the wave front of the wave generated by the source is spherical, the angles φ and φ', defined in section 2.3.2, vanish at all points O_j^*. (That means, in the notation of section 2.3.2, that $r_{\parallel} = r_1$, $r_{\perp} = r_2$, $R_{\parallel} = R_1$, $R_{\perp} = R_2$, $r_{\parallel}' = r_1'$, and $r_{\perp}' = r_2'$. Therefore, in this chapter we need only one set of symbols for the radii of curvature.) The basic coordinate system and the local coordinate systems at the points O_j^* are the same as in section 2.4.4.

To determine the displacement vector of the head wave at the point M we shall use the following procedure:

First step: Construction of the ray.

Second step: Determination of the principal radii of curvature of the wave front at the points $O_1^*, O_2^*, \ldots, O_k^*, \bar{O}_k^*, O_{k+1}^*, \ldots, O_s^*$.

Third step: Determination of the displacement vector of the incident wave at O_k^*.

Fourth step: Determination of the displacement vector of the head wave at $\bar{O}_k^*$.

Fifth step: Determination of the displacement vector of the head wave at M.

First Step

For a given medium with interfaces Σ_j, the construction of the ray of the head wave is only a geometric-kinematic problem. Therefore we shall not describe it here in detail. We can obtain positions of the points O_j^* ($j = 1, 2, \ldots, s-1$) and $\bar{O}_k^*$, values of l_j and l, and the angles $\vartheta(O_j^*)$ and $\vartheta'(O_j^*)$ (these values cannot be calculated for curved interfaces by the usual methods employed for plane-parallel interfaces; numerical methods must be used). The angles $\vartheta(O_k^*)$ and $\vartheta'(\bar{O}_k^*)$ must satisfy the relations

$$\sin \vartheta(O_k^*) = v_k/\bar{v}, \qquad \sin \vartheta'(\bar{O}_k^*) = v_{k+1}/\bar{v}. \tag{5.1}$$

Note that the value of $\sin \vartheta(O_j^*)/v_j$ is not constant along the whole ray (as it was in the preceding chapter) for the local coordinate systems are not identical (i.e., the angles between the basic z-axis and the local z-axis are different at the points O_j^*). Therefore, for the determination of the ray, we must also know the slopes of the Σ_j interfaces. It should again be noted that construction of the ray is the most complicated part of the whole procedure.

Second Step

To determine the radii of curvature of the wave front of head waves at the points O_j^*, we must first find the principal radii, $R_{\parallel}^{(j)}$ and $R_{\perp}^{(j)}$, of the Σ_j interfaces at the points O_j^*. This is not difficult when analytic expressions for the interfaces are known (formulae for $R_{\parallel}^{(j)}$ and $R_{\perp}^{(j)}$ can be found in any standard textbook of differential geometry). If we know all these values, we can determine the principal radii of curvature of the wave front at the points O_j^*, starting from the following values of $r_{\parallel}(O_1^*)$ and $r_{\perp}(O_1^*)$:

$$r_{\parallel}(O_1^*) = l_1, \qquad r_{\perp}(O_1^*) = l_1. \tag{5.2}$$

Values of $r_{\parallel}(O_j^*)$, $r_{\perp}(O_j^*)$, $r_{\parallel}'(O_j^*)$, and $r_{\perp}(O_j^*)$ for $j = 1, 2, \ldots, k$ are then given by the following relations (see (2.163) to (2.165)):

$$\begin{aligned} r_{\parallel}(O_j^*) &= r_{\parallel}'(O_{j-1}^*)+l_j, & r_{\perp}(O_j^*) &= r_{\perp}'(O_{j-1}^*)+l_j, \\ r_{\parallel}'(O_j^*) &= r_{\parallel}(O_j^*)/\Delta_{\parallel}^{(j)*}, & r_{\perp}(O_j^*) &= r_{\perp}(O_j^*)/\Delta_{\perp}^{(j)*}, \end{aligned} \tag{5.3}$$

where

$$\begin{aligned} \Delta_{\parallel}^{(j)*} &= \frac{v_{j+1} \cos^2 \vartheta(O_j^*)}{v_j \cos^2 \vartheta'(O_j^*)} \\ &\quad + \frac{r_{\parallel}(O_j^*)}{R_{\parallel}^{(j)} \cos^2 \vartheta'(O_j^*)} \left(\frac{v_{j+1}}{v_j} \cos \vartheta(O_j^*) \pm \cos \vartheta'(O_j^*) \right), \\ \Delta_{\perp}^{(j)*} &= \frac{v_{j+1}}{v_j} + \frac{r_{\perp}(O_j^*)}{R_{\perp}^{(j)}} \left(\frac{v_{j+1}}{v_j} \cos \vartheta(O_j^*) \pm \cos \vartheta'(O_j^*) \right), \end{aligned} \tag{5.4}$$

the positive sign to be used for reflected waves and the negative for refracted waves.

For the calculation of $r_{\|}(O_j^*)$, $r_{\|}'(O_j^*)$, $r_{\perp}(O_j^*)$, and $r_{\perp}'(O_j^*)$ for $j = k+1, \ldots, s-1$, we can again use the standard procedure given by equations (5.3) and (5.4), except that we must consider the point $\bar{O}_k^*$. We shall denote the values of $\Delta_{\|}^*$ and $\Delta_{\perp}^*$ at the point $\bar{O}_k^*$ by $\bar{\Delta}_{\|}^{(k)*}$ and $\bar{\Delta}_{\perp}^{(k)*}$, respectively. Then we can write, from (5.3) and (5.4),

$$\begin{aligned} r_{\|}'(O_k^*) &= r_{\|}(O_k^*)/\Delta_{\|}^{(k)*}, & r_{\perp}'(O_k^*) &= r_{\perp}(O_k^*)/\Delta_{\perp}^{(k)*}, \\ r_{\|}(\bar{O}_k^*) &= r_{\|}'(O_k^*)+l, & r_{\perp}'(\bar{O}_k^*) &= r_{\perp}(O_k^*)+l, \\ r_{\|}'(\bar{O}_k^*) &= r_{\|}(\bar{O}_k^*)/\bar{\Delta}_{\|}^{(k)*}, & r_{\perp}'(\bar{O}_k^*) &= r_{\perp}(\bar{O}_k^*)/\bar{\Delta}_{\perp}^{(k)*}, \end{aligned} \tag{5.5}$$

where

$$\begin{aligned} \Delta_{\|}^{(k)*} &= \infty, & \Delta_{\perp}^{(k)*} &= \bar{v}/v_k, \\ \bar{\Delta}_{\|}^{(k)*} &= 0, & \bar{\Delta}_{\perp}^{(k)*} &= v_{k+1}/\bar{v}. \end{aligned} \tag{5.6}$$

The values of $r_{\|}'(\bar{O}_k^*)$ and $r_{\perp}'(\bar{O}_k^*)$ determined by (5.5) serve as starting values for the determination of all other $r_{\|}(O_j^*)$, $r_{\|}'(O_j^*)$, $r_{\perp}(O_j^*)$, and $r_{\perp}'(O_j^*)$, $j = k+1, \ldots, s-1$.

From (5.5) and (5.6), we see that $r_{\|}'(\bar{O}_k^*) = \infty$. This follows from simple geometrical considerations also, as the wave front of the head wave is conical in the $(k+1)$th layer. If the $(k+1)$th interface, Σ_{k+1}, is plane (may be dipping), $r_{\|}(O_{k+1}^*)$ and $r_{\|}'(O_{k+1}^*)$ remain infinite. Suppose that the interfaces $\Sigma_{k+1}, \ldots, \Sigma_{k+p-1}$ are plane and that the first non-plane interface is Σ_{k+p} (i.e., $R_{\|}^{(k+p)} \neq \infty$). Then $r_{\|}(O_j^*) = r_{\|}'(O_j^*) = \infty$ for $j = k+1, \ldots, k+p-1$ and $r_{\|}(O_{k+p}^*) = \infty$, too. The principal radius $r_{\|}'(O_{k+p}^*)$, however, is finite. We find from (5.3) and (5.4) that

$$r_{\|}'(O_{k+p}^*) = \frac{R_{\|}^{(k+p)} \cos^2 \vartheta'(O_{k+p}^*)}{(v_{k+p+1}/v_{k+p}) \cos \vartheta(O_{k+p}^*) \pm \cos \vartheta'(O_{k+p}^*)}, \tag{5.7}$$

where the positive sign is to be used for the reflected wave and the negative one for the refracted wave at O_{k+p}^*. The other values of $r_{\|}(O_j^*)$ and $r_{\|}'(O_j^*)$ for $j = k+p+1, \ldots, s-1$ can be calculated from (5.3) and (5.4).

For $r_{\perp}'(\bar{O}_k^*)$, we obtain simply, from (5.5) and (5.6), the expression

$$r_{\perp}'(\bar{O}_k^*) = \frac{\bar{v}}{v_{k+1}} \left(l + \frac{v_k}{\bar{v}} r_{\perp}(O_k^*) \right). \tag{5.8}$$

As the value of $r_{\perp}(O_k^*)$ was determined earlier, the value of $r_{\perp}'(\bar{O}_k^*)$ can be calculated from (5.8) without difficulty. Successive calculation of $r_{\perp}(O_j^*)$ and $r_{\perp}'(O_j^*)$ $(j = k+1, k+2, \ldots, s-1)$, from (5.3) and (5.4), is also easy.

Finally, in recapitulation, values of $r_\perp(O_j^*)$ and $r_\perp'(O_j^*)$ may be calculated successively from (5.2) to (5.4) for all points O_j^* ($j = 1, 2, \ldots, s-1$), except that we must include the point $\bar{O}_k^*$. In numerical determinations of $r_\parallel(O_j^*)$ and $r_\parallel'(O_j^*)$, we must be careful because these values may be infinite for some points O_j^*. We must first calculate them for O_j^* ($j = 1, 2, \ldots, k$) and then for O_j^* ($j = k+p+1, \ldots, s-1$), using $r_\parallel'(O_{k+p}^*)$ given by (5.7) as a starting value.

Third Step

We shall now denote the displacement vector of the wave incident on the B-interface by $\mathbf{W}^B$. We can write for it the following expression:

$$\mathbf{W}^B = U^B \exp[i\omega(t-\tau^B)]\mathbf{n}^B. \tag{5.9}$$

The functions U^B, τ^B, and $\mathbf{n}^B$ can be easily determined by using the methods described in section 2.6.4. From (2.169) to (2.171) and (2.96), we find that at the point O_k^*

$$U^B(O_k^*) = \frac{(-1)^{\epsilon'}}{\left(r_\parallel(O_k^*)r_\perp(O_k^*)\prod_{j=1}^{k-1}\Delta_\parallel^{(j)*}\Delta_\perp^{(j)*}\right)^{1/2}}\prod_{j=1}^{k-1}R_j^*, \tag{5.10}$$

$$\tau^B(O_k^*) = \sum_{j=1}^{k}(l_j/v_j), \tag{5.11}$$

where

$$\epsilon' = \sum_{j=1}^{k}\epsilon_j$$

(see section 2.4.4) and R_j^* are the R-coefficients for Σ_j interfaces, calculated for the angle of incidence $\vartheta(O_j^*)$. (If we calculate R_j^* using formulae (2.74), we must substitute there $\Theta = \sin\vartheta(O_j^*)/v_j$. As mentioned earlier, the parameter Θ is not constant along the whole ray as it was in the preceding chapter, for the local coordinate systems are not identical. Therefore we must calculate the value of Θ separately for each j.)

Fourth Step

If we know the incident wave $\mathbf{W}^B$ at the point O_k^*, we can determine the displacement vector of the head wave $\mathbf{W}^*(\bar{O}_k^*)$ by using the general formulae (3.151) and (3.161). We find that

$$\mathbf{W}^*(\bar{O}_k^*) = U^*(\bar{O}_k^*)\exp\{i\omega[t-\tau^*(\bar{O}_k^*)]\}\mathbf{n}^*(\bar{O}_k^*), \tag{5.12}$$

where

$$U^*(\bar{O}_k^*) = \frac{1}{i\omega}U^{B}(O_k^*)\frac{\bar{v}\Gamma_k \tan\vartheta(O_k^*)[r_{\|}(O_k^*)r_{\perp}(O_k^*)]^{1/2}}{l^{3/2}[l+r_{\perp}(O_k^*)v_k/\bar{v}]^{1/2}}, \tag{5.13}$$

$$\tau^*(\bar{O}_k^*) = \tau^{B}(O_k^*)+l/\bar{v}, \tag{5.14}$$

Γ_k being the head wave coefficient for the wave under consideration on the B-interface and l the distance between O_k^* and $\bar{O}_k^*$.

Inserting (5.10) into (5.13) yields

$$U^*(\bar{O}_k^*) = \frac{1}{i\omega}\frac{\bar{v}\Gamma_k \tan\vartheta(O_k^*)(-1)^{\epsilon'}}{l^{3/2}\left((l+r_{\perp}(O_k^*)v_k/\bar{v})\prod_{j=1}^{k-1}\Delta_{\|}^{(j)*}\Delta_{\perp}^{(j)*}\right)^{1/2}}\prod_{j=1}^{k-1}R_j^*. \tag{5.15}$$

This is the final expression for the head wave under consideration at the point $\bar{O}_k^*$ lying on the B-interface.

Fifth Step

The displacement vector of the head wave at the point M can be found by using the formulae for continuation of the displacement vector along a ray, viz. (2.99) and (2.166). We obtain

$$\mathbf{W}^*(M) = U^*(M)\exp\{i\omega[t-\tau^*(M)]\}\mathbf{n}^*(M), \tag{5.16}$$

where $\tau^*(M)$ is given by the relation

$$\tau^*(M) = \frac{l}{\bar{v}}+\sum_{j=1}^{s}\frac{l_j}{v_j}, \tag{5.17}$$

and $U^*(M)$ can be expressed in terms of $U^*(\bar{O}_k^*)$, as follows:

$$U^*(M) = U^*(\bar{O}_k^*)\frac{(-1)^{\epsilon''}}{L(M,\bar{O}_k^*)}\prod_{j=k+1}^{s-1}R_j^*, \tag{5.18}$$

where

$$\epsilon'' = \sum_{j=k+1}^{s}\epsilon_j, \tag{5.19}$$

$$L(M,\bar{O}_k^*) = \left(\frac{r_{\|}'(O_{s-1}^*)+l_s}{r_{\|}'(\bar{O}_k^*)}\prod_{j=k+1}^{s-1}\Delta_{\|}^{(j)*}\right)^{1/2} \times\left(\frac{r_{\perp}'(O_{s-1}^*)+l_s}{r_{\perp}'(\bar{O}_k^*)}\prod_{j=k+1}^{s-1}\Delta_{\perp}^{(j)*}\right)^{1/2}. \tag{5.20}$$

The unit vector $\mathbf{n}^*(M)$ has the usual meaning. It lies along the ray when the wave arrives at M as a P-wave, is perpendicular to $E_{\|}$ for an SH-wave, and is perpendicular to the ray and lies in $E_{\|}$ for an SV-wave. The first term on the right-hand side in the above expression for L can be written as

$$\left(\frac{r_{\|}'(O_{s-1}^*)+l_s}{r_{\|}(\bar{O}_k^*)}\prod_{j=k+1}^{s-1}\Delta_{\|}^{(j)*}\right)^{1/2} = \left[\prod_{j=k+2}^{s}\left(1+\frac{l_j}{r_{\|}'(O_{j-1}^*)}\right)\right]^{1/2}. \tag{5.21}$$

If the interfaces $\Sigma_{k+1}, \ldots, \Sigma_{k+p-1}$ are plane and the first non-plane interface is Σ_{k+p}, we can write

$$\left(\frac{r_{\|}'(O_{s-1}^*)+l_s}{r_{\|}(\bar{O}_k^*)}\prod_{j=k+1}^{s-1}\Delta_{\|}^{(j)*}\right)^{1/2} = \left[\prod_{j=k+p+1}^{s}\left(1+\frac{l_j}{r_{\|}'(O_{j-1}^*)}\right)\right]^{1/2}, \tag{5.21'}$$

as $l_j/r_{\|}'(O_{j-1}^*) = 0$ for $j = k+1, k+2, \ldots, k+p$. The starting value for the calculation of the principal radii of curvature, $r_{\|}'(O_{k+p}^*)$, is given by (5.7).

Inserting (5.15) into (5.18) yields

$$U^*(M) = \frac{\bar{v}\Gamma_k \tan \vartheta(O_k^*)(-1)^{\epsilon}}{i\omega l^{3/2}(v_k/\bar{v})^{1/2}L_{\|}^*L_{\perp}^*}\prod_{\substack{j=1\\ j\neq k}}^{s-1} R_j^*, \tag{5.22}$$

where

$$\epsilon = \sum_{j=1}^{s} \epsilon_j, \tag{5.23}$$

$$L_{\|}^* = \left\{\left(\prod_{j=1}^{k-1}\Delta_{\|}^{(j)*}\right)\left[\prod_{j=k+p+1}^{s}\left(1+\frac{l_j}{r_{\|}'(O_{j-1}^*)}\right)\right]\right\}^{1/2}, \tag{5.24}$$

$$L_{\perp}^* = \left\{\frac{v_{k+1}}{v_k}[r_{\perp}'(O_{s-1}^*)+l_s]\left(\prod_{\substack{j=1\\ j\neq k}}^{s-1}\Delta_{\perp}^{(j)*}\right)\right\}^{1/2}. \tag{5.25}$$

Equations (5.16) and (5.22) to (5.25) give us the final expressions for the displacement vector of the head wave at the point M.

The expression for $L_{\perp}^*$ can be rewritten in a simpler form also:

$$L_{\perp}^* = \left\{[r_{\perp}'(O_{s-1}^*)+l_s]\prod_{j=1}^{s-1}{}' \Delta_{\perp}^{(j)*}\right\}^{1/2}, \tag{5.26}$$

where the symbol Π' means that we must take a product of $\Delta_{\perp}^{(j)*}$ determined successively for all points O_j^*, including the point $\bar{O}_k^*$. In other words,

$$\prod_{j=1}^{s-1}{}' \Delta_{\perp}^{(j)*} = \bar{\Delta}_{\perp}^{(k)*}\prod_{j=1}^{s-1}\Delta_{\perp}^{(j)*}.$$

The value of $\bar{\Delta}_{\perp}^{(k)*}$ is given by (5.6).

If an impulse wave is generated by the source, the displacement vector $\mathscr{W}^*(M)$ of the resulting head wave is given by

$$\mathscr{W}^*(M) = |\omega U^*(M)|F(t-\tau^*(M);\chi^*(M))\mathbf{n}^*(M), \tag{5.27}$$

where $\chi^*(M) = \arg U^*(M)+\frac{1}{2}\pi$ and F is given by (3.119).

We shall not write down the expressions for the horizontal and vertical components of the displacement vector of the head wave as they can be found easily by using the usual methods, viz., replacing the unit vector $\mathbf{n}^*(M)$ by its horizontal or vertical component. If the receiver lies on the earth's surface, the components of the unit vector have to be replaced by the conversion coefficients, as described in section 2.4.2.

5.2 PLANE DIPPING INTERFACES

The expressions (5.22) to (5.25) and (5.16) are applicable for plane dipping interfaces also. The formulae for $L_\parallel^*$ and $L_\perp^*$, however, become simpler in this case. We find, from (5.4),

$$\Delta_\parallel^{(j)*} = \frac{v_{j+1}}{v_j}\frac{\cos^2 \vartheta(O_j^*)}{\cos^2 \vartheta'(O_j^*)}, \qquad \Delta_\perp^{(j)*} = \frac{v_{j+1}}{v_j}. \tag{5.28}$$

Inserting this into (5.24) and (5.26) yields

$$L_\parallel^* = \left(\frac{v_k}{v_1}\right)^{1/2} \prod_{j=1}^{k-1} \frac{\cos \vartheta(O_j^*)}{\cos \vartheta'(O_j^*)}, \qquad L_\perp^* = \left(l\frac{\bar{v}}{v_1} + \sum_{j=1}^{s} l_j \frac{v_j}{v_1}\right)^{1/2}. \tag{5.29}$$

Note also that if the interfaces Σ_j are continuous and do not intersect each other in the region under consideration, the value of ϵ in (5.23) is always zero.

For parallel interfaces, the results are, of course, the same as in chapter 4; (5.29) reduces to

$$L_\parallel^* = \left(\frac{v_k}{v_1}\right)^{1/2} \frac{\cos \vartheta_1^*}{\cos \vartheta_k^*}, \qquad L_\perp^* = \left(\frac{r}{\sin \vartheta_1^*}\right)^{1/2}. \tag{5.30}$$

Inserting this into (5.22) yields an expression which is the same as (4.43); note that $\epsilon = 0$ in this case.

6
INTERFERENCE HEAD WAVES

In previous chapters we have investigated only non-interfering, pure head waves. However, head waves which have an interference character are very important in many applications in seismic prospecting and in refraction studies of the earth's crust. Theoretical studies of properties of these waves are still in their initial stage, and only recently have a few papers dealing with some types of waves appeared (see section 1.1 for references).

Here we shall investigate only one type of head wave which has an interference character. This head wave exists when the velocity below the plane interface is not constant, but increases slightly with depth. It has already been described briefly in section 3.9. We shall assume, for simplicity, that there is only one plane interface and that the source (which radiates the compressional wave (2.110)) and receiver lie in the first medium, at distances h and H from the interface, respectively. The compressional velocity in the first medium, α_1, is assumed to be constant, whereas the compressional velocity in the second medium increases linearly with depth according to the relation

$$\alpha(d) = \alpha_2(1+bd), \qquad \alpha_2 > \alpha_1, \tag{6.1}$$

where d is the depth below the interface ($d = |z|$). We shall assume that $b \geqslant 0$ and refer to b as the *velocity gradient* (although, strictly speaking, the velocity gradient is $b\alpha_2$). The dimension of b is (distance)$^{-1}$. For $b = 0$, the velocity in the second medium is constant, viz., equal to α_2 (and the situation will be the same as in chapter 3). We shall assume in this chapter that b is very small; large gradients will not be considered here.

The head waves for $b = 0$ were investigated in chapter 3. Some of the results from that chapter will be needed here, for purposes of comparison. We shall call the compressional head wave (of the type $mnp = 131$) which would exist for $b = 0$ a *pure head wave* (or simply H-wave). We must remember that this wave does not exist in the medium under consideration (for $b \neq 0$), but is used only for comparison. We found in chapter 3 that the arrival time of the pure head wave, τ^*, is given by

$$\tau^* = \frac{r-r^*}{\alpha_2} + \frac{h+H}{\alpha_1[1-(\alpha_1/\alpha_2)^2]^{1/2}}, \tag{6.2}$$

where r is the epicentral distance of the receiver and r^* is the critical distance, given by

$$r^* = (h+H)\alpha_1/(\alpha_2{}^2-\alpha_1{}^2)^{1/2}. \tag{6.2'}$$

The pure head wave exists only at epicentral distances larger than r^*, and at these distances it is always the first wave to reach the receiver (when we do not consider the direct wave). Its ray is given schematically in figure 6.1a. The angle of incidence corresponds to the critical angle θ^*, which is given by the relation $\sin\theta^* = \alpha_1/\alpha_2$, and the second element is always parallel to the interface.

When $b \neq 0$, the situation is quite different. The first wave to reach the receiver is the compressional wave, which has a turning point below the

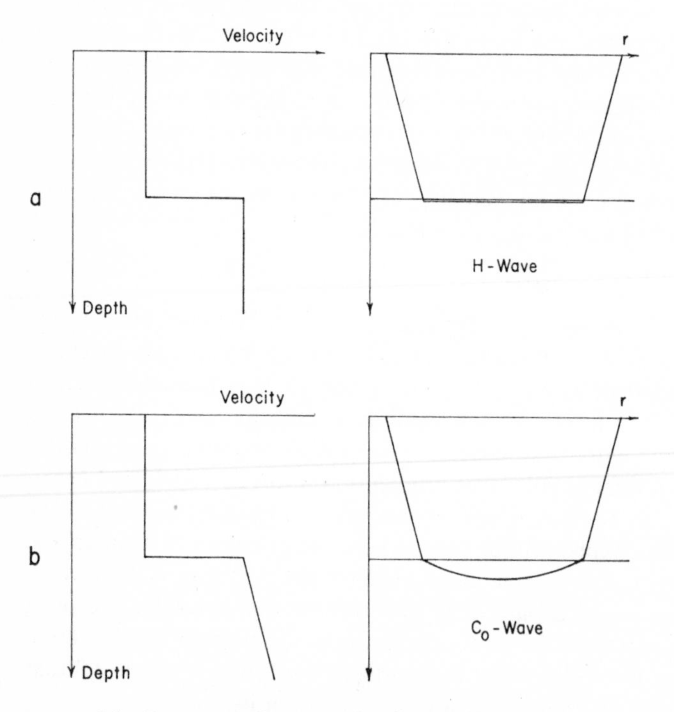

FIGURE 6.1 Two types of velocity distribution below the B-interface, and the corresponding waves.

interface in the second medium (see figure 6.1b). We shall call this the C_0-wave. (In seismic prospecting and in explosion studies of the earth's crust, the C_0-wave is sometimes called the *diving wave* or *refracted wave*.) The angle of incidence for the C_0-wave, θ_1, is less than the critical angle θ^*, and the second element of the ray is not parallel to the interface. The other waves which will successively arrive at the receiver are the compressional waves which are reflected s-times ($s = 1, 2, \ldots$) from the interface from below (see figure 6.2). We shall call these C_s-waves. (Transformed P–S waves are not considered here, as they will reach the receiver considerably later.) The angles of incidence of C_s-waves ($s = 1, 2, \ldots$) will lie between θ_1 and θ^*. For small gradients b, and not too large epicentral distances, these angles will differ from one another only slightly, and, consequently, their times of arrival τ_{C_s} will be very close together. It will be shown in section 6.1 (see (6.13)) that we can write, approximately,

$$\tau_{C_s} = \tau^* - \frac{b^2(r-r^*)^3}{24\alpha_2(1+s)^2}, \tag{6.3}$$

where τ^* is the arrival time of the pure head wave, given by (6.2).

Table 6.1 gives values of $\tau^* - \tau_{C_s}$, calculated from (6.3), for $s = 0, 1, 2, 3$ for a model of the earth's crust in which $h = H = 30$ km, $\alpha_1 = 6.4$ km/sec, $\alpha_2 = 8$ km/sec. The critical point for this model lies at the epicentral distance $r^* = 80$ km. The values of $\tau^* - \tau_{C_s}$ are determined at epicentral

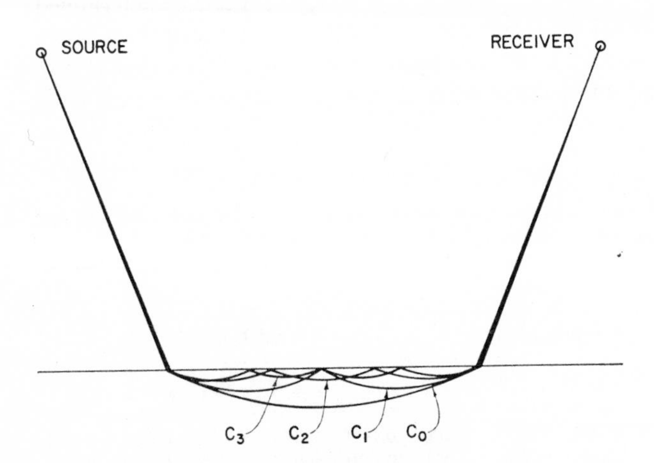

FIGURE 6.2 Ensemble of C_s-waves ($s = 0, 1, 2, \ldots$) for a positive velocity gradient below the interface.

distances $r = 120$, 160, and 200 km for gradients $b = 0.002\ \text{km}^{-1}$ and $0.005\ \text{km}^{-1}$ (for these gradients, the velocity increase within a depth of 1 km is 16 m/sec and 40 m/sec, respectively).

From (6.3) and table 6.1, we can see that the differences $\tau^* - \tau_{C_s}$ decrease very rapidly with increasing s, with decreasing epicentral distance, and with decreasing gradient b. When the epicentral distance is not large and/or the gradient is small, all C_s-waves ($s = 0, 1, \ldots$) will interfere with one another. For example, for gradients $b = 0.002\ \text{km}^{-1}$ and $0.005\ \text{km}^{-1}$ (and for epicentral distances $r \leqslant 200$ km), all waves will interfere when the duration δt of the pulse of the incident wave is longer than 0.04 sec and 0.23 sec, respectively (see table 6.1). With increasing epicentral distance, $\tau^* - \tau_{C_s}$ increases and the individual waves C_0, C_1, ... will successively separate from the interference wave and exist independently. The wave C_0 will separate first, at the epicentral distance r_0 at which $\tau_{C_1} - \tau_{C_0} = \delta t$; the wave C_1 will separate next, at the epicentral distance r_1 where $\tau_{C_2} - \tau_{C_1} = \delta t$; and so on. Generally, the wave C_s will separate from the interference wave at epicentral distance r_s where $\tau_{C_{s+1}} - \tau_{C_s} = \delta t$. From (6.3), we easily obtain the following approximate expression for r_s:

$$r_s \sim r^* + \left(\frac{32\alpha_2 \delta t}{b^2}\right)^{1/3} \left(\frac{(1+s)^2(1+\frac{1}{2}s)^2}{1+\frac{2}{3}s}\right)^{1/3}. \tag{6.4}$$

Thus, at epicentral distances r for which $r_i < r < r_{i+1}$, $i+2$ independent waves will exist: one interference wave, which is formally given by $C_{i+1} + C_{i+2} \ldots$, and $i+1$ separated, non-interfering waves $C_0, C_1, \ldots, C_i$. Properties of *non-interfering* C_s-waves ($s = 0, 1, \ldots, i$) can be determined by standard zero-order ray formulae (see section 6.1), but properties of the interference wave cannot be studied by the ray theory. (Our expression in section 6.1 for the interference wave as a superposition of waves C_{i+1}

TABLE 6.1

Differences in travel times of pure head waves and C_s-waves for two values of the velocity gradient b ($H = h = 30$ km, $\alpha_1 = 6.4$ km/sec, $\alpha_2 = 8$ km/sec)

		$\tau^* - \tau_{C_s}$ (sec)							
Epicentral distance	τ^*	$b = 0.002\ \text{km}^{-1}$				$b = 0.005\ \text{km}^{-1}$			
r (km)	(sec)	$s = 0$	$s = 1$	$s = 2$	$s = 3$	$s = 0$	$s = 1$	$s = 2$	$s = 3$
120	20.62	0.001	0.000	0.000	0.000	0.008	0.002	0.001	0.000
160	25.62	0.011	0.003	0.001	0.000	0.066	0.017	0.007	0.004
200	30.62	0.036	0.009	0.004	0.002	0.225	0.056	0.025	0.014

$+C_{i+2}+\ldots$ is only formal; in fact it is not possible to investigate it in this way.) This interference wave, which is kinematically very similar to the pure head wave, will be called the *interference head wave*, or C^+-wave.

For $r < r_0$, the wave C_0 is not yet separated and we receive only the C^+-wave. In this chapter we shall investigate properties of the C^+-wave only for $r < r_0$. It should be noted that r_0 can be rather large, especially when b is small (see (6.4) or table 6.1). From (6.4), we obtain

$$r_0 = r^* + (32\alpha_2 \delta t/b^2)^{1/3}. \tag{6.4'}$$

For example, for the model of the earth's crust considered in table 6.1 and for gradients $b = 0.002\ \text{km}^{-1}$ and $0.005\ \text{km}^{-1}$, we obtain (when $\delta t = 0.2$ sec) $r_0 = 314$ km and 207 km, respectively.

Theoretical investigations into the nature of the interference head waves were carried out by Chekin (1965, 1966), who derived the expressions for the C^+-wave in the region $r^* < r < r_0$. Chekin (1964) also derived expressions for head waves for the situation in which the gradient below the interface is *negative* (in this case, we obtain so-called *damped head waves*; see figure 3.41). We shall not consider this situation here. More detailed studies of interference head waves have been made by Buldyrev and Lanin (1965, 1966) and Lanin (1966, 1968), who investigated these waves also in regions where some C_s-waves became separated ($r > r_0$). Moreover, they considered not only an inhomogeneous substratum below a plane interface, but also non-plane interfaces (see figure 3.41). We have noted in section 3.9 that interference head waves (of the same mechanism as above) can also arise from curved interfaces, when the interface is convex for a receiver situated on the side of the incident wave. The interested reader is referred to Lanin (1968), who has given all the formulae and tables necessary for numerical calculation of the properties of interference head waves from curved (spherical and cylindrical) interfaces.

We shall derive zero-order ray formulae for C_s-waves ($s = 0, 1, 2, \ldots$) in section 6.1. These formulae can be used only when the C_s-wave under consideration is separated from the interference wave. However, we shall also formally write down the zero-order ray formulae for a C^+-wave (adding all C_s-waves), for they can provide asymptotic values even in the regions of their inapplicability. To distinguish clearly between the asymptotic and correct values for the C^+-wave, we shall call the fictitious interference wave which is described by the superposition of all C_s-waves (in their zero-order ray approximation) the C-wave. In section 6.2, we shall present the results of Chekin's investigations for the C^+-wave in the region $r^* < r < r_0$. Numerical examples will be given in section 6.3. Only the most important case, viz., the one-interface problem for compressional interference head waves, will be discussed here. For other types of head waves, and more complicated media, a similar procedure can be used.

6.1
THE C-WAVE: ZERO-ORDER EXPRESSIONS

We shall denote the displacement vector of the C_s-wave at the point M (where the receiver lies) by $\mathbf{W}^{C_s}$. Using (2.126) to (2.128), (2.153) to (2.162), and (2.138), we find that

$$\mathbf{W}^{C_s} = U^{C_s} \exp[i\omega(t-\tau_{C_s})]\mathbf{n}_{C_s}, \tag{6.5}$$

where

$$U^{C_s} = R_{13}R_{31}(R_{33})^s \, e^{is\pi/2}/L_s, \tag{6.6}$$

$$L_s = \left| r \cot \theta_1 \left(\frac{h+H}{\cos^2 \theta_1} - \frac{2(s+1)\alpha_2 \cos \theta_1}{b\alpha_1 \cos \theta_3 \sin^2 \theta_3} \right) \right|^{1/2}, \tag{6.7}$$

$$r = (h+H) \tan \theta_1 + \frac{2}{b}(s+1) \cot \theta_3, \tag{6.8}$$

$$\tau_{C_s} = \frac{h+H}{\alpha_1 \cos \theta_1} + \frac{2(s+1)}{\alpha_2 b} \ln \frac{1+\cos \theta_3}{\sin \theta_3}. \tag{6.9}$$

The angle of incidence, θ_1, is different for different values of s. If the epicentral distance r is given, we can find θ_1 by numerical solution of equation (6.8). The angle of refraction, θ_3, is connected with θ_1 by Snell's law, $\sin \theta_3 = \alpha_2 \sin \theta_1/\alpha_1$. The unit vector $\mathbf{n}_{C_s}$ lies along the ray (i.e., $n_r^{C_s} = \sin \theta_1$, $n_z^{C_s} = \cos \theta_1$). The R-coefficients R_{13}, R_{31}, and R_{33} have the usual meaning (see chapter 2). Note that L_s, R_{13}, R_{31}, and R_{33} are always real.

These are the final expressions for harmonic C_s-waves. If an impulse wave is radiated by the source, we determine the displacement vector $\mathcal{W}^{C_s}$ of the C_s-wave from the expression

$$\mathcal{W}^{C_s} = |U^{C_s}| f(t-\tau_{C_s}; \arg U^{C_s})\mathbf{n}_{C_s}. \tag{6.10}$$

The expressions for C_s-waves, (6.5) to (6.10), can be simplified considerably when b is small and the epicentral distance is not large. We shall try to express $\mathbf{W}^{C_s}$ in terms of the displacement vector of the pure head wave, $\mathbf{W}^*$ (which would arise if $b = 0$). $\mathbf{W}^*$ is given by the formula (see chapter 3)

$$\mathbf{W}^* = U^* \exp[i\omega(t-\tau^*)]\mathbf{n}^*, \tag{6.11}$$

where

$$U^* = \frac{\alpha_2 \Gamma \tan \theta^*}{i\omega r^{1/2} l^{3/2}}, \qquad l = r - r^*. \tag{6.11'}$$

In these expressions, Γ is the H-coefficient of the type $mnp = 131$, r^* is the critical distance given by (6.2′), l is the distance from the critical point (i.e. the length of the element of the ray of the pure head wave along the interface in the second medium), τ^* is the arrival time given by (6.2), and $\mathbf{n}^*$ is a unit vector along the ray ($n_r^* = \sin \theta^*$, $n_z^* = \cos \theta^*$).

From (6.8), we find easily that an approximate expression for $\cos \theta_3$ (when b is small and r is not too large) is

$$\cos \theta_3 \sim \frac{bl}{2(s+1)}\left[1 - \frac{lb^2}{8(s+1)^2}\left(l - \frac{\alpha_1(h+H)}{\alpha_2[1-(\alpha_1/\alpha_2)^2]^{3/2}}\right)\right]. \tag{6.12}$$

Inserting this into (6.9) and (6.7) yields approximately

$$\tau_{C_s} \sim \tau^* - \frac{\pi}{12\omega(s+1)^2}\zeta^3, \tag{6.13}$$

$$L_s \sim \frac{2(s+1)\cot \theta^*}{b}\left(\frac{r}{l}\right)^{1/2}, \tag{6.14}$$

where

$$\zeta = lb^{2/3}/\Lambda_2^{1/3} \tag{6.15}$$

and Λ_2 is the wavelength of the wave under consideration just below the interface, i.e.

$$\Lambda_2 = 2\pi\alpha_2/\omega. \tag{6.16}$$

(Note that (6.13) can also be written in the form (6.3).) The quantity ζ, given by (6.15), plays a very important role in the following investigation of the interference head wave. It increases for increasing epicentral distance, increasing gradient b, and for decreasing wavelength Λ_2.

Using (3.107) and (6.12), we find that

$$R_{13}R_{31} = \frac{1}{2}\frac{R_{13}R_{31}}{\cos \theta_3} 2\cos \theta_3 \sim \frac{bl\Gamma}{s+1}. \tag{6.17}$$

Taking into account that $R_{33} \to -1$ for small b (for $\theta_3 \to \frac{1}{2}\pi$), and inserting (6.17) and (6.14) into (6.5), we get

$$\mathbf{W}^{C_s} \sim \mathbf{W}^* \frac{\pi\zeta^3}{(s+1)^2} \exp\left(-i(s-1)\frac{\pi}{2} + \frac{i\pi\zeta^3}{12(s+1)^2}\right). \tag{6.18}$$

We can see from this expression that the displacement vector of the C_s-

wave can be expressed in terms of the displacement vector of the pure head wave when b is small and the epicentral distance is not large (i.e. ζ is small). The multiplication factor depends only on ζ and s.

It was shown above (see (6.3) or (6.13)) that the arrival time of the C_s-wave differs only slightly from the arrival time τ^* of the pure head wave when b is small and the epicentral distance is not large. However, its dynamic characteristics, such as the amplitude, spectrum, and wave-form, differ considerably, even when b is very small. The amplitude of the pure head wave is inversely proportional to the frequency ω, whereas the amplitude of the C_s-wave is independent of frequency (see (6.6)). In other words, the H-wave is a first-order wave and the C_s-wave is of zeroth order (in terms of ray theory). In the high-frequency limit, the pure head wave vanishes, but the amplitude of the C_s-wave remains constant, different from zero, even for very small b. Consequently, the spectrum and the wave-form of the C_s-wave differ from those of the pure head wave. The difference between these waves is particularly marked in their amplitude–distance curves. The amplitudes of pure head waves decrease very rapidly with increasing epicentral distance (proportional to $1/r^{1/2}l^{3/2}$; see (6.11′)); the amplitudes of C_s-waves, on the other hand, increase with increasing epicentral distance when b is small and r is not too large (as $l^{3/2}/r^{1/2}$; see (6.18) and (6.15)). Examples of amplitude curves of pure head waves and C_0-waves are given in figures 6.6 and 4.20. The shape of the amplitude curve of the C_0-wave does not change when b changes; it merely shifts down (for b decreasing) or up (for b increasing) on the logarithmic scale. It remains the same also for $b \to 0$. (Note that we are considering zero-order ray expressions for the C_0-wave.)

The differences in the behaviour of C_s-waves and the pure head wave can be explained as follows. The pure head wave is a first-order wave (in terms of the ray theory). The zeroth coefficient in the ray series for the refracted wave (which is responsible for the generation of the pure head wave; see figure 3.2) vanishes along the ray which is parallel to the interface. Therefore, the zeroth coefficient in the ray series for the pure head wave also vanishes, and the leading term contains a factor $(i\omega)^{-1}$. When $b \neq 0$, the zeroth coefficient of the ray series for the waves under consideration does not vanish. Therefore, C_s-waves are zero-order waves. However, the zero-order formulae for C_s-waves become less exact when b decreases (for a given epicentral distance), as the ray of the incident wave approaches the critical ray, which has singular properties. Therefore, for $b \to 0$, we cannot obtain the expression for pure head waves from the expression for any of the C_s-waves or their combination, and it is necessary to carry out a special investigation (see section 6.2).

We shall now write the formulae for the displacement vector of the C-wave. According to our definition, this wave is given by the superposition of all C_s-waves (in their zero-order ray approximation), i.e.,

$$\mathscr{W}^{C} = \sum_{s=0}^{\infty} \mathscr{W}^{C_s}, \tag{6.19}$$

where $\mathscr{W}^{C_s}$ is given by (6.10). In the frequency domain, we can write

$$\mathbf{W}^{C} = \sum_{s=0}^{\infty} \mathbf{W}^{C_s}. \tag{6.19'}$$

When b is small and r is not too large, we can use the approximate formula (6.18) for $\mathbf{W}^{C_s}$, and obtain

$$\mathbf{W}^{C} \sim \mathbf{W}^{*} Q_1(\zeta), \tag{6.20}$$

where

$$Q_1(\zeta) = \pi\zeta^3 \sum_{s=0}^{\infty} (s+1)^{-2} \exp\left(-i(s-1)\frac{\pi}{2} + \frac{i\pi\zeta^3}{12(s+1)^2}\right). \tag{6.21}$$

In (6.20), $\mathbf{W}^*$ is the displacement vector of the pure head wave, given by (6.11). The values of $|Q_1(\zeta)|$ and $\arg Q_1(\zeta)$ are given in figure 6.3.

Thus, the approximate zero-order ray expressions for the displacement vector of the C-wave can be found (in a manner similar to the C_s-wave) as a multiple of the displacement vector of the H-wave. The multiplication factor is $Q_1(\zeta)$ for small ζ. We can see from (6.18) and (6.21) that the

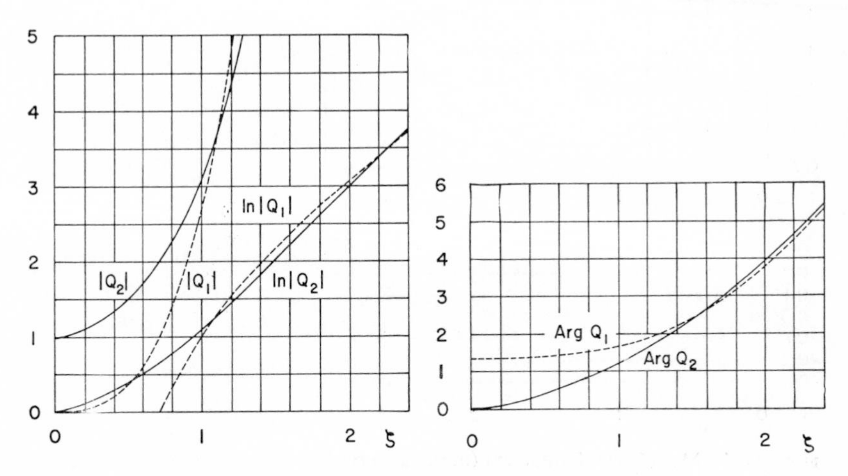

FIGURE 6.3 Modulus and argument (in radians) of Q_1 and Q_2 versus ζ.

amplitude of the C_s-wave is $(s+1)^2$ times less than that of the C_0-wave. Thus, the largest part of the energy of the C-wave is contributed by the C_0-wave, and the C_s-waves for $s > 2$ make a very small contribution. In discussing the C-wave, it is enough to consider only the waves C_0, C_1, and C_2 for practical purposes (see also figure 6.6 for the differences between the amplitudes of the C_0 and C waves). Since the wave C_0 contributes most of the energy of the C-wave, and the contribution of other C_s-waves $(s \geqslant 1)$ is substantially smaller, we can find a better approximation for the C-wave if we use (6.5) for the C_0-wave in (6.19′), and express other C_s-waves $(s \geqslant 1)$ only approximately in terms of $\mathbf{W}^{C_0}$. From (6.18) it follows that

$$\mathbf{W}^{C_s} \sim (s+1)^{-2} \exp\left\{-is\frac{\pi}{2}+\frac{i\pi\zeta^3}{12}\left(\frac{1}{(s+1)^2}-1\right)\right\} \mathbf{W}^{C_0}.$$

Inserting the above into (6.19′), we obtain

$$\mathbf{W}^{C} = \mathbf{W}^{C_0}\bar{Q}_1(\zeta), \tag{6.22}$$

where

$$\bar{Q}_1(\zeta) = \frac{1}{\pi\zeta^3} \exp\left(-i\frac{\pi}{2}-i\frac{\pi\zeta^3}{12}\right) Q_1(\zeta), \tag{6.23}$$

or

$$\bar{Q}_1(\zeta) = \sum_{s=0}^{\infty} (s+1)^{-2} \exp\left\{-i\pi\left[\frac{s}{2}-\frac{\zeta^3}{12}\left(\frac{1}{(s+1)^2}-1\right)\right]\right\}. \tag{6.23′}$$

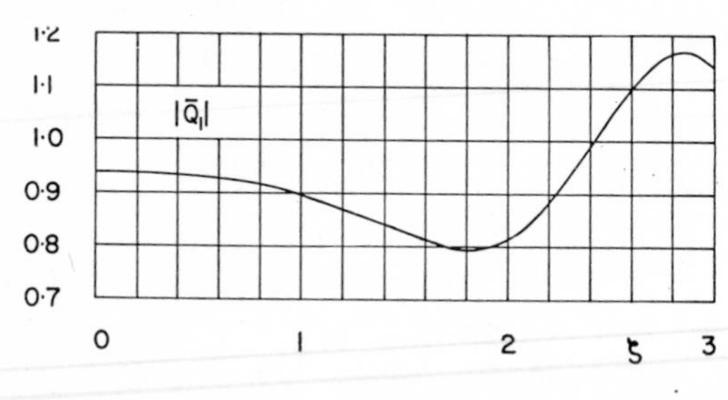

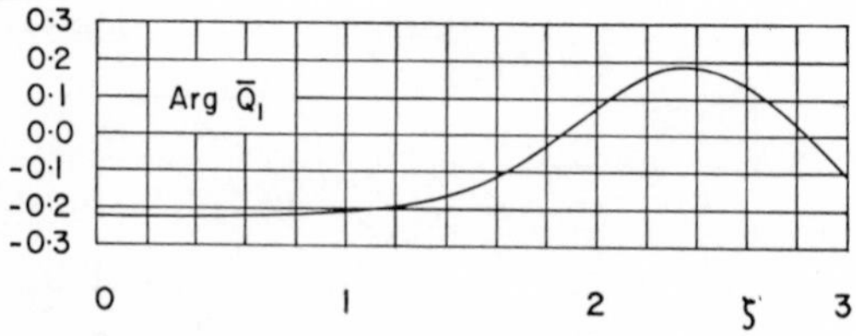

FIGURE 6.4 Modulus and argument (in radians) of $\bar{Q}_1$ versus ζ.

Values of $|\bar{Q}_1|$ and arg $\bar{Q}_1$ are given in figure 6.4. We can see that $|\bar{Q}_1(\zeta)|$ is very close to one and arg $\bar{Q}_1(\zeta)$ is very close to zero. It follows that the interference C-wave is determined mainly by the C_0-wave (as shown earlier also), and that the properties of the C-wave remain very different from those of pure head waves.

We have shown earlier that the zero-order ray formulae for C_s-waves cannot be used when θ_1 approaches θ^*, as the critical ray $\theta_1 = \theta^*$ is a singular ray. Therefore, the formulae (6.20) and (6.22) are inapplicable when ζ is very small. They will serve only as asymptotic values for increasing ζ, but will be again inapplicable when ζ is large, because then the C_s-waves with small values of s will separate out from the interference wave. (The maximum distance up to which all C_s-waves interfere is r_0). For r near r_0, the zero-order formulae for the C_0-wave can be used approximately. It follows that, for r close to r_0, the formulae for the C-wave can also be used approximately, as the other C_s-waves ($s \geqslant 1$) make only a small contribution to the interference wave.

6.2 THE INTERFERENCE HEAD WAVE C^+

It is to be expected from the above discussion that properties of the interference head wave C^+ will be similar to those of the pure head wave for very small ζ, and to those of the C-wave for larger ζ (for epicentral distances close to r_0). The behaviour of the C^+ wave for small and intermediate ζ (at epicentral distances $r^* < r < r_0$; see (6.4′)) was investigated by Chekin (1965, 1966), who found expressions for the potentials of the C^+-wave using the integral method (see section 7.3). We shall not repeat his derivations here, but simply write down the results. For the displacement vector of the C^+-wave, we can write

$$\mathbf{W}^{C^+} = \mathbf{W}^* Q_2(\zeta), \tag{6.24}$$

where ζ is given by (6.15); the modulus and argument of the function $Q_2(\zeta)$ are given in figure 6.3. (We shall not give the analytic expressions for $Q_2(\zeta)$ as they are very cumbersome.) We can see from figure 6.3 that the values of $|Q_2(\zeta)|$ are close to unity for small ζ. From this it follows that the properties of the C^+-wave are nearly the same as those of the pure head wave for small ζ. For large ζ, the function $Q_2(\zeta)$ approaches the function $Q_1(\zeta)$ asymptotically. The formula of Chekin (6.24), is applicable in the immediate vicinity of the critical point also, where the expressions for the pure head wave $\mathbf{W}^*$ found in chapter 3 (see (6.11) and (6.11′)) become invalid. In the neighbourhood of the critical point, we must use, for $\mathbf{W}^*$ in

(6.24), the expression given in chapter 7 (see (7.87)). It should be emphasized again that (6.24) can be used only for epicentral distances $r^* < r < r_0$ (where all the C_s-waves interfere). This interval is large for small b, but small for larger b. The formula is not exact; it is asymptotic and is valid for high frequencies.

Using the same procedure as in the derivation of (6.22), we can express $\mathbf{W}^{C^+}$ also in terms of $\mathbf{W}^{C_0}$ as follows:

$$\mathbf{W}^{C^+} = \mathbf{W}^{C_0}\bar{Q}_2(\zeta), \tag{6.25}$$

where

$$\bar{Q}_2(\zeta) = \frac{1}{\pi\zeta^3}\exp\left(-i\frac{\pi}{2}-i\frac{\pi\zeta^3}{12}\right)Q_2(\zeta). \tag{6.26}$$

For large ζ, the function $\bar{Q}_2$ approaches $\bar{Q}_1$ asymptotically. Values of $\bar{Q}_2(\zeta)$ can be easily obtained from the values of $Q_2(\zeta)$ given in figure 6.3.

The two expressions (6.24) and (6.25) give us a continuous transition from the first-order ray formula for pure head waves (for $\zeta \sim 0$) to the zero-order formula for C-waves (for larger ζ).

If the source radiates an impulse wave, the expression for the transient interference head wave must be found from (6.24) or (6.25) by numerical Fourier analysis. With increasing distance, the waves C_0, C_1, C_2, ... will successively separate from the interference head wave at epicentral distances r_0, r_1, r_2, ... As soon as the kth wave, C_k, separates from the interference wave, we can use for it the zero-order ray formula (6.10). Properties of C^+-waves for epicentral distances $r > r_0$, however, have not been studied here; the interested reader is referred to Lanin (1968). It seems that, even for very large epicentral distances, the interference head wave can be rather strong, and that in theoretical seismograms two dominant phases will appear – one corresponding to the C_0-wave and the other corresponding to the interference head wave (with times of arrival

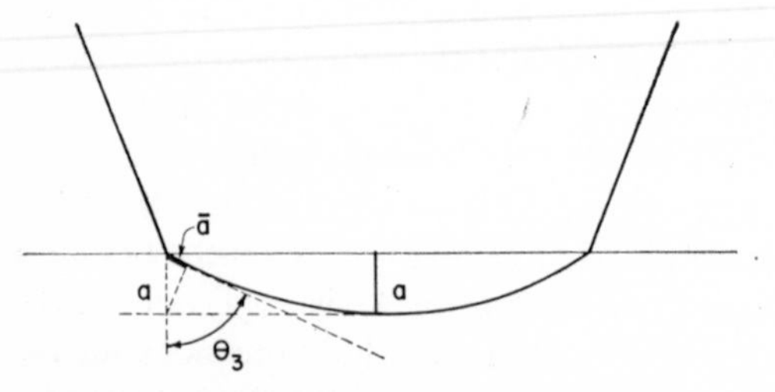

FIGURE 6.5 Determination of $\bar{a}$ from the ray diagram of the C_0-wave.

close to those for pure head waves); the phases corresponding to other separated, non-interference C_s-waves will be weaker.

The value of ζ may be calculated by three different procedures:

(1) If we know the wavelength Λ_2, the velocity gradient b, and the distance from the critical point l, we may use (6.15).

(2) If we know the frequency f, the time–distance curve of the C_0-wave (τ_{C_0}), and the theoretical time–distance curve of the H-wave (τ^*), we find from (6.13) that

$$\zeta \sim [24f(\tau^* - \tau_{C_0})]^{1/3}. \tag{6.27}$$

(3) We can calculate ζ from a known ray of the C_0-wave and known wavelength Λ_2. The depth a of a turning point of the ray of the C_0-wave below the interface is given by the expression

$$a \sim \tfrac{1}{8}l^2 b \tag{6.28}$$

which follows simply from geometrical considerations. The projection ($\bar{a}$) of a onto the direction of the ray, at the point of intersection of the ray with the interface, is given by the relation

$$\bar{a} = a \cos\theta_3 \sim b^2 l^3/16. \tag{6.29}$$

From (6.29) and (6.15), we obtain

$$\zeta \sim (16\bar{a}/\Lambda_2)^{1/3}. \tag{6.30}$$

The value of $\bar{a}$ can be found directly from the ray diagram of the C_0-wave (see figure 6.5). Similarly, inserting (6.28) into (6.15) yields

$$\zeta \sim (8a/\Lambda_2)^{1/2}(b\Lambda_2)^{1/6}. \tag{6.31}$$

This formula can be used when we know the depth (a) of the turning point of the ray and the parameters of the medium and the source, b and Λ_2. For $0.001 < b\Lambda_2 < 0.01$, we have

$$0.9(a/\Lambda_2)^{1/2} < \zeta < 1.3(a/\Lambda_2)^{1/2}. \tag{6.32}$$

For $b\Lambda_2 = 1/512 \sim 0.002$, we have simply $\zeta \sim (a/\Lambda_2)^{1/2}$.

6.3
NUMERICAL EXAMPLES

It can be seen from figure 6.3 that the function $|Q_2(\zeta)|$ increases very rapidly as ζ increases. Therefore, the amplitudes of the interference head wave, in comparison with the amplitudes of the pure head waves, are larger for larger ζ (i.e. for larger l and b and smaller Λ_2). When $\zeta = 1.65$, the amplitudes of C^+-waves are approximately one order of magnitude greater than the amplitudes of pure head waves (as $|Q_2(\zeta)| = 10$ for $\zeta = 1.65$). We can easily appreciate, from (6.32), how deep the ray of the C_0-wave must penetrate below the interface to cause an increase in the head wave amplitudes by one order of magnitude. Inserting $\zeta = 1.65$ into (6.32), we find that even a penetration of 1.5–3.5 wavelengths below the interface causes this increase (when $0.001 < b\Lambda_2 < 0.01$). Thus, even a very small positive velocity gradient below the interface has a very strong influence on the head wave amplitudes. For example, in seismic investigations of the earth's crust, the ray has only to penetrate below the Moho discontinuity by about 1.5–3.5 km for the amplitudes of compressional head waves to increase by one order of magnitude (when $f \sim 8$ cps and $0.001\ \text{km}^{-1} < b < 0.01\ \text{km}^{-1}$).

The amplitude–distance curve of the C^+-wave for one model of the one-layer earth's crust is given in figure 6.6. It is assumed that $b = 0.00283$ km^{-1}, $h = H = 30$ km, $\alpha_1 = 6.4$ km/sec, $\alpha_2 = 8$ km/sec, $\beta_1/\alpha_1 = \beta_2/\alpha_2 = 1/\sqrt{3}$, $\rho_1/\rho_2 = 1, f = 6.4$ cps. Note that the gradient $b = 0.00283\ \text{km}^{-1}$

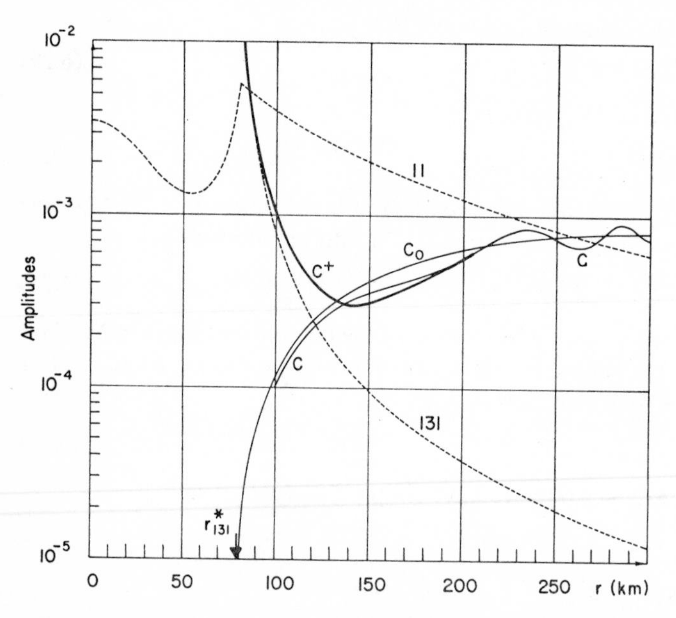

FIGURE 6.6 The amplitude of the vertical displacement component vs. distance, for the C^+-wave ($H = h = 30$ km, $\alpha_1 = 6.4$ km/sec, $\alpha_2 = 8$ km/sec, $b = 0.00283\ \text{km}^{-1}$, $\gamma_1 = \gamma_2 = 0.577$, $\rho_1/\rho_2 = 1, f = 6.4$ cps). The amplitude-distance curves for the pure head wave 131 ($b = 0$), the reflected wave 11, and the C and C_0 waves are given for comparison.

corresponds to an increase in velocity of 22.6 m/sec per km below the Moho. The critical point for this model lies at the epicentral distance $r = 80$ km. Figure 6.6 also gives the amplitude–distance curves of the pure head wave 131 (propagating in the same medium with $b = 0$), the reflected wave 11, and waves C and C_0. The C and C^+ waves were calculated using (6.22) and (6.25). All the amplitude curves are inexact in the immediate neighbourhood of the critical point ($r = 80$ km). For example, in the figure the amplitudes of the pure head waves and interference head wave C^+ are infinite just at the critical point. The true values for these waves in the critical region can be obtained simply by multiplying our curves, given in figure 6.6, by a special function $|\mu_2(y^*)|$ defined by (7.86) (see chapter 7).

The amplitude curve of the C^+-wave is given only between the epicentral distances of 80 and 210 km. For distances larger than 210 km, it oscillates around the amplitude–distance curve of the C_0-wave (its amplitude being close to that of the C-wave). We consider here amplitudes of *harmonic* waves; we must remember that for impulse waves the wave C_0 will separate from the interference wave at the distance r_0. (When the length of the impulse is, for example, $\delta t = T = 1/6.4$ sec, the epicentral distance r_0 is approximately 251 km.)

The amplitude–distance curve of the C^+-wave has its minimum value at

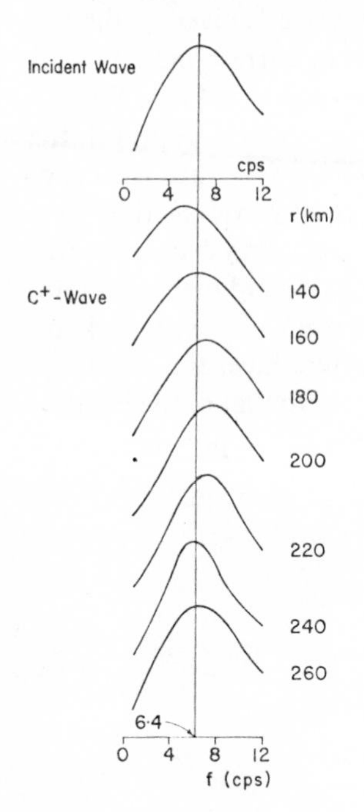

FIGURE 6.7 Amplitude spectra of the C+-wave at various epicentral distances, for the same model as in figure 6.6.

K

about 140 km. For smaller values of b, the minimum lies at larger distances. At large distances, the amplitudes of C^+-waves are more than one order of magnitude greater than those of pure head waves. Thus, we can see that the head wave amplitudes are, in general, very sensitive to velocity gradients in the substratum. Even a very small positive gradient can increase the amplitudes by one order of magnitude or more. More detailed discussion of the amplitude–distance curves of interference head waves is given in Červený and Janský (1966). The interference of the C^+-wave with the reflected wave in the critical region has not been considered in this chapter. The methods of section 3.8 can be used for this purpose.

Formulae (6.24) or (6.25) enable us to calculate the spectrum of the C^+-wave. This spectrum is similar to that of the pure head wave for small epicentral distances and to that of the C_0-wave for large distances. Suppose that the amplitude spectrum of the incident wave is $|S(\omega)|$. Then the amplitude spectrum of the pure head wave is $|S(\omega)|/\omega$ and the amplitude spectrum of the C_0-wave is the same as that of the incident wave, $|S(\omega)|$. However, the spectrum of the C^+-wave does not change from $|S(\omega)|/\omega$ to $|S(\omega)|$ monotonically with increasing distance. The amplitude spectra of the C^+-wave at epicentral distances $r = 140, 160, \ldots, 260$ km are given in figure 6.7 (for the same model of the earth's crust considered in figure 6.6). The assumed amplitude spectrum of the incident wave is given at the top of the figure. The dominant frequency (the frequency corresponding to the maximum of the amplitude spectrum) of this spectrum is 6.4 cps. We can see from the figure that for small epicentral distances ($r = 140$ km), the dominant frequency of the amplitude spectrum of the C^+-wave is smaller than 6.4 cps. (Note that the same holds for pure head waves, as they have the spectrum $|S(\omega)|/\omega$.) However, the dominant frequency increases with distance, reaching about 8 cps at a distance of 200 km, and then decreases. For large epicentral distances, the dominant frequency is close to 6.4 cps. Thus, for a positive velocity gradient below the interface, the dominant frequency of the interference head wave may be close to that of the incident wave, or even higher (while the dominant frequency of the pure head wave is always smaller).

7
HEAD WAVES BY WAVE METHODS

7.1
INTRODUCTION

In the previous chapters (except chapter 6), properties of head waves were studied by means of the ray theory. This theory is very useful because of its simplicity and conceptual clarity and enables us to solve a large variety of problems connected with the propagation of head waves in different types of media consisting of thick layers. However, the ray method is only an approximate method and has a number of serious limitations. There are circumstances – particularly in the neighbourhood of points where the phase function is singular (such as the critical point) – when more exact methods must be used.

A formal solution to the problem of a point source in an elastic medium has been obtained so far only for a few types of media, the most important being horizontally or spherically stratified media in which all the layers are homogeneous. For dipping and curved interfaces, or for generally inhomogeneous media, formal solutions have not yet been found (except in some very simple cases), and head waves propagating in these media can be investigated at present only by the ray method. In the following we shall consider only a horizontally layered medium with homogeneous layers.

The formal solution, which is usually in the form of a single integral for steady state response and a double integral for impulse response, can be found by different methods. All these methods will not be described here; a short outline with many references was given in section 1.1. The integrand of the above-mentioned integrals, which can be obtained by satisfying interface conditions, is relatively simple for the one-interface problem, but becomes very cumbersome for a large number of layers.

For the one-interface problem, numerical methods for the evaluation of the integrals are well known. The integrals can be evaluated approximately (for high frequencies) by suitable transformation of integration contours. If we use the method of steepest descent, the results correspond to those found by the ray theory. Various modifications of this method enable us to obtain solutions for the situations where the steepest descent method itself fails. (More details are given in section 7.2). Methods for the exact numerical evaluation of the integrals are also very well developed (see, e.g., Cagniard, 1962; Dix, 1963; Bortfeld, 1962a, 1962b, 1964) for the one-interface problem.

The situation is considerably more complicated if the medium is n-layered. Then the methods used in the one-interface problem usually fail; the integrand has many singularities and does not vary smoothly along the integration contours. Therefore, direct numerical methods for evaluation of the integrals can hardly be used, and even the approximate methods are very complicated. This problem will be dealt with briefly in section 7.3.

7.2
THE ONE-INTERFACE PROBLEM

In this section, we shall use a wave (integral) method to derive the expressions for head waves which arise when a spherical wave is incident on a plane interface between two homogeneous media (which are in welded contact). This problem has been solved many times by different authors; references can be found in section 1.1. In chapter 3 this problem was solved by the ray method. Here we wish to appreciate the accuracy of the ray method for head waves and to give new formulae for the critical region where this method is not valid. It would be very cumbersome, and indeed unnecessary, to study all possible types of head waves again. We shall, therefore, assume that the harmonic source radiates P-waves and lies in the first medium. The main attention will be devoted to compressional head waves of the type 131. We introduce a cylindrical coordinate system r, z, φ by placing the source at the point $r = 0$, $z = h > 0$ and by assuming that the interface coincides with the plane $z = 0$. The velocities and densities α_1, β_1, ρ_1 and α_2, β_2, ρ_2 have the same meaning as in chapter 3. Similarly, as in that chapter, $V_1 = \alpha_1$, $V_2 = \beta_1$, $V_3 = \alpha_2$, $V_4 = \beta_2$. The displacement components along the r and z axes are denoted by $(W_r)_1$ and $(W_z)_1$ in the first medium, and by $(W_r)_2$ and $(W_z)_2$ in the second medium. We introduce the wave numbers k_i by the relation

$$k_i = \omega / V_i \qquad (i = 1, 2, 3, 4). \tag{7.1}$$

We can express (see, e.g., Ewing *et al.*, 1957, p. 107) the displacement components $(W_r)_j$ and $(W_z)_j$ $(j = 1, 2)$ in terms of the potentials of compressional and shear waves, Φ_j and Ψ_j, as follows:

$$(W_r)_j = \frac{\partial \Phi_j}{\partial r} + \frac{\partial^2 \Psi_j}{\partial r \partial z}, \qquad (W_z)_j = \frac{\partial \Phi_j}{\partial z} + \frac{\partial^2 \Psi_j}{\partial z^2} + \frac{\omega^2}{\beta_j^2} \Psi_j \tag{7.2}$$

($j = 1, 2$), where the potentials Φ_j and Ψ_j are solutions of the wave equations

$$\nabla^2\Phi_j = \frac{1}{\alpha_j^2}\frac{\partial^2\Phi_j}{\partial t^2}, \qquad \nabla^2\Psi_j = \frac{1}{\beta_j^2}\frac{\partial^2\Psi_j}{\partial t^2} \qquad (j = 1, 2). \tag{7.3}$$

We shall now assume that the potential Φ^0 of the spherical harmonic wave radiated by the source is given by

$$\Phi^0 = \frac{1}{-ik_1R_0}\exp[i\omega(t-R_0/\alpha_1)], \tag{7.4}$$

where R_0 is the distance from the source, i.e. $R_0 = [r^2+(h-z)^2]^{1/2}$. For the displacement components of this wave, we find, from (7.2),

$$\begin{bmatrix} W_r^0 \\ W_z^0 \end{bmatrix} = \frac{1}{R_0}\left(1+\frac{1}{ik_1R_0}\right)\exp[i\omega(t-R_0/\alpha_1)]\begin{bmatrix} r/R_0 \\ -(h-z)/R_0 \end{bmatrix}. \tag{7.5}$$

If $k_1R_0 \gg 1$, we can write

$$\begin{bmatrix} W_r^0 \\ W_z^0 \end{bmatrix} = \frac{1}{R_0}\exp[i\omega(t-R_0/\alpha_1)]\begin{bmatrix} \sin\theta_0 \\ -\cos\theta_0 \end{bmatrix}, \tag{7.6}$$

where $\sin\theta_0 = r/R_0$, $\cos\theta_0 = (h-z)/R_0$. These equations are exactly the same as (3.1) when an appropriate medium and wave type are specified.

7.2.1 *Formal Solution*

The expression (7.4) for the potential Φ^0 can be rewritten as follows:

$$\Phi^0 = \frac{e^{i\omega t}}{-ik_1}\frac{e^{-ik_1R_0}}{R_0}. \tag{7.7}$$

The second term in the above equation can be expressed by means of the Sommerfeld integral (see Sommerfeld, 1909, and Ewing *et al.*, 1957, p. 13) as

$$\frac{e^{-ik_1R_0}}{R_0} = \int_0^\infty J_0(kr)\exp(-\nu_1|z-h|)\frac{k\,dk}{\nu_1}, \tag{7.8}$$

where J_0 is the zero-order Bessel function and $\nu_1^2 = k^2-k_1^2$. The formal solution of the equations (7.3) has been discussed in several books (see Ewing *et al.*, 1957, pp. 107–10, for example); therefore, we shall not describe the procedure in detail. We can write for Φ_1, Ψ_1, Φ_2, and Ψ_2 the following expressions:

$$
\begin{aligned}
\Phi_1 &= \frac{e^{i\omega t}}{-ik_1}\left\{\int_0^\infty \frac{k}{\nu_1} J_0(kr)\exp(-\nu_1|z-h|)dk \right.\\
&\qquad\qquad \left. + \int_0^\infty Q_1(k)J_0(kr)\exp[-\nu_1(z-h)]dk\right\},\\
\Psi_1 &= \frac{e^{i\omega t}}{-ik_1}\int_0^\infty Q_2(k)J_0(kr)\exp[-\nu_2(z-h)]dk,\\
\Phi_2 &= \frac{e^{i\omega t}}{-ik_1}\int_0^\infty Q_3(k)J_0(kr)\exp[\nu_3(z-h)]dk,\\
\Psi_2 &= \frac{e^{i\omega t}}{-ik_1}\int_0^\infty Q_4(k)J_0(kr)\exp[\nu_4(z-h)]dk.
\end{aligned}
\tag{7.9}
$$

The first two equations hold for $z > 0$ and the last two for $z < 0$. Q_i, which are functions of k, can be determined from the interface conditions, and ν_i are given by the relation

$$\nu_i = (k^2 - k_i^2)^{1/2} \qquad (i = 1, 2, 3, 4). \tag{7.10}$$

For $k < k_i$, $\nu_i = i(k_i^2 - k^2)^{1/2}$ along the integration contour.

Inserting (7.9) into the four interface conditions expressing continuity of displacement and stress at the interface, we find the expressions for Q_i. Finally, we can write

$$
\begin{aligned}
\Phi_1 &= \frac{e^{i\omega t}}{-ik_1}\left\{\int_0^\infty J_0(kr)\exp(-\nu_1|z-h|)\frac{kdk}{\nu_1} \right.\\
&\qquad\qquad \left. + \int_0^\infty \tilde{Q}_1(k)J_0(kr)\exp[-\nu_1(z+h)]\frac{kdk}{\nu_1}\right\},\\
\Psi_1 &= \frac{e^{i\omega t}}{-ik_1}\int_0^\infty \tilde{Q}_2(k)J_0(kr)\exp(-\nu_2 z-\nu_1 h)\frac{kdk}{\nu_1},\\
\Phi_2 &= \frac{e^{i\omega t}}{-ik_1}\int_0^\infty \tilde{Q}_3(k)J_0(kr)\exp(\nu_3 z-\nu_1 h)\frac{kdk}{\nu_1},\\
\Psi_2 &= \frac{e^{i\omega t}}{-ik_1}\int_0^\infty \tilde{Q}_4(k)J_0(kr)\exp(\nu_4 z-\nu_1 h)\frac{kdk}{\nu_1},
\end{aligned}
\tag{7.11}
$$

where the functions $\tilde{Q}_n(k)$ are the solutions of four linear equations

$$\sum_{n=1}^{4} \tilde{a}_{in}\tilde{Q}_n = \tilde{b}_i \qquad (i = 1, 2, 3, 4), \tag{7.12}$$

in which the coefficients $\tilde{a}_{in}$ are given by the matrix $\tilde{A}$ where

$$\tilde{A} = \begin{bmatrix} 1 & -\nu_2 & -1 & -\nu_4 \\ \nu_1 & -k^2 & \nu_3 & k^2 \\ -(2\mu_1 k^2 - \rho_1\omega^2) & 2\mu_1 k^2 \nu_2 & 2\mu_2 k^2 - \rho_2\omega^2 & 2\mu_2 k^2 \nu_4 \\ 2\mu_1\nu_1 & -(2\mu_1 k^2 - \rho_1\omega^2) & 2\mu_2\nu_3 & 2\mu_2 k^2 - \rho_2\omega^2 \end{bmatrix}, \tag{7.13}$$

and the $\tilde{b}_i$ are

$$\tilde{b}_1 = -1, \quad \tilde{b}_2 = \nu_1, \quad \tilde{b}_3 = 2\mu_1 k^2 - \rho_1\omega^2, \quad \tilde{b}_4 = 2\mu_1\nu_1. \tag{7.14}$$

The first term in the first equation (for Φ_1) in (7.11) represents the direct compressional wave (7.4), and the second term represents the reflected compressional wave in a broad sense, which includes *all* the *compressional* waves, propagating in the *first* medium, generated by the direct wave when it impinges on the interface (viz., reflected wave, head wave, and Stoneley wave). Similarly, the expressions for Ψ_1, Φ_2, and Ψ_2 in (7.11) represent the reflected shear wave and the refracted compressional and shear waves (all in a broad sense), respectively.

Many singularities of the integrands in (7.11) lie on the real k-axis. These singularities include the branch points $k = k_i$ $(i = 1, 2, 3, 4)$ and a possible pole at the point at which the determinant $||\tilde{A}||$ vanishes. The contour of integration passes over all the singularities from the side of positive real parts of ν_i (see figure 7.1).

Several different transformations of the original contours of integration have been employed. A review of more than twenty papers dealing with this problem up to 1954 was given by Honda and Nakamura (1954): other

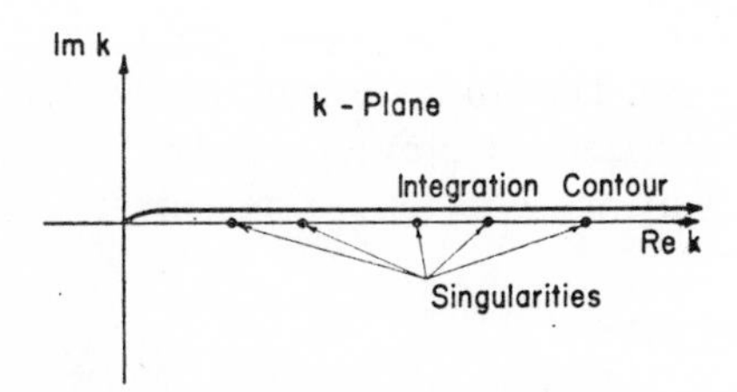

FIGURE 7.1 The contour of integration in the k-plane.

references can be found in Ewing *et al.* (1957) and in other books. The original contour of integration can be replaced by other contours, some of which are branch line contours. Approximate methods are usually used for the evaluation of the branch line integrals. Frequently, branch cuts are chosen parallel to the imaginary axis or along curves along which the real part of the corresponding radical ν_i is constant. It is also very simple to rewrite the integrals in a form suitable for solution by the method of *steepest descent* (in which the cuts are transformed to coincide with the paths of steepest descent). The reader is referred to Morse and Feshbach (1953, pp. 437–43), Jeffreys and Jeffreys (1956, pp. 503–6), and Brekhovskikh (1960, p. 245) for details of this method. In the paper of Honda and Nakamura referred to above, where a comparison of these approximate methods is made, it is shown that the most exact results are given by the method of steepest descent despite its simplicity.

In a number of recent papers, several modifications of the method of steepest descent have been used for the evaluation of the above integrals, with results that are more exact than those given by the original method itself. Most of these modifications have been used for investigation of the critical region where the method of steepest descent does not give correct results. See, for example, Brekhovskikh (1960), Červený (1959, 1961, 1966a), Hron (1962), and Smirnova and Yermilova (1959). Smirnova (1962, 1964a, 1966a, 1966b) used a modification of the steepest descent method for a number of other, hitherto uninvestigated, cases. Some other modifications of the method (see Červený, 1967b) make it possible to calculate the above integrals exactly.

Note that the method of steepest descent gives results which are exactly the same as those obtained by the ray method (see chapters 2 and 3). Therefore, if we want to appreciate the accuracy of the results given by the ray theory, we must use more exact modifications of this method. We shall follow here the technique used earlier by one of the authors (see Červený, 1959, 1966a, 1967b). The four integrals in (7.11) will not all be evaluated since the procedure is identical in all the cases. It will be enough to investigate only the integral for Φ_1. As noted before, the first integral in the first equation of (7.11) is the incident wave. Therefore, we shall study only the second term, which represents the reflected compressional wave (in its broad sense). We shall designate it only by one letter Φ for simplicity, i.e.,

$$\Phi = \frac{e^{i\omega t}}{-ik_1} \int_0^\infty \tilde{Q}_1(k) J_0(kr) \exp[-\nu_1(z+h)] \frac{k\mathrm{d}k}{\nu_1}. \tag{7.15}$$

Solving (7.12), we find the expression for $\tilde{Q}_1(k)$

$$\tilde{Q}_1(k) = \Delta_2/\Delta_1, \tag{7.16}$$

where

$$\begin{aligned}\Delta_{1,2} = & \pm k^2[k^2q-\omega^2(\rho_2-\rho_1)]^2+\nu_1\nu_2\nu_3\nu_4k^2q^2\mp\nu_2\nu_3\rho_1\rho_2\omega^4 \\ & -\nu_1\nu_4\rho_1\rho_2\omega^4\mp\nu_3\nu_4(k^2q+\rho_1\omega^2)^2-\nu_1\nu_2(k^2q-\rho_2\omega^2)^2,\end{aligned} \tag{7.17}$$

where q is given by (2.74″) and the upper sign applies for Δ_1 and the lower for Δ_2.

In several of the papers mentioned above, the case of an incident wave which is the complex conjugate of (7.4) was investigated, i.e. in these papers the incident wave is given as

$$\Phi^0 = \frac{1}{ik_1R_0}\exp[-i\omega(t-R_0/\alpha_1)]. \tag{7.4'}$$

For such an incident wave the value of Φ is the complex conjugate of (7.15), i.e.

$$\Phi = \frac{e^{-i\omega t}}{ik_1}\int_0^\infty \tilde{Q}_1(k)J_0(kr)\exp[-\nu_1(z+h)]\frac{kdk}{\nu_1}, \tag{7.15'}$$

where the expressions for $\tilde{Q}_1(k)$ and ν_i are the same as given above, but the values of ν_i are to be taken as the complex conjugates of those defined above (i.e. $\nu_i = -i(k_i^2-k^2)^{1/2}$ for $k_i > k$).

It is quite arbitrary which incident wave, (7.4) or (7.4′), is considered in the following. If we find the solution for one of them, the solution for the other can simply be written as the complex conjugate of the first. We shall assume in the following that the potential of the incident wave is given by (7.4′) as this form was used by Červený (1966a), whom we shall follow here.

7.2.2 *The Contour of Integration*

If the potential of the incident wave is given by (7.4′), the potential of the compressional wave, generated at the interface by it, is given by (7.15′). It is convenient to transform the expression (7.15′) by replacing the Bessel function by Hankel functions. For this purpose, we note that

$$J_0(kr) = \tfrac{1}{2}[H_0^{(1)}(kr)+H_0^{(2)}(kr)], \tag{7.18}$$

where $H_0^{(1)}$ and $H_0^{(2)}$ are Hankel functions of the first and second kind. Substituting (7.18) into (7.15′), we obtain a sum of two integrals. In the integral containing $H_0^{(2)}(kr)$, we replace k by $-k$ and make use of the fact that $H_0^{(2)}(e^{-i\pi}kr) = -H_0^{(1)}(kr)$, and also that $\tilde{Q}_1(k) = \tilde{Q}_1(-k)$. We then obtain two integrals with the same integrands, the limits of one integral being 0 to ∞, and the limits of the other being $-\infty$ to 0. Combining both integrals, we can write

$$\Phi = \frac{e^{-i\omega t}}{2ik_1}\int_{-\infty}^{+\infty} H_0^{(1)}(kr)\tilde{Q}_1(k)\exp[-\nu_1(z+h)]\frac{kdk}{\nu_1}. \tag{7.19}$$

Substituting $k = k_1 u$, we can write

$$\Phi = \tfrac{1}{2}e^{-i\omega t}\int_{-\infty}^{+\infty} H_0^{(1)}(k_1 ru)\exp[ik_1(z+h)(1-u^2)^{1/2}]\,Q(u)(1-u^2)^{-1/2}\,udu, \tag{7.20}$$

where

$$Q(u) = \frac{Q^{(2)}+\bar{Q}^{(2)}w_3}{Q^{(1)}+\bar{Q}^{(1)}w_3}, \tag{7.21}$$

$$Q^{(1,2)} = \pm\alpha_1\alpha_2\beta_1\beta_2(u/\alpha_1)^2[\rho_2-\rho_1-u^2q/\alpha_1^2]^2+w_1w_4\rho_1\rho_2\beta_1\alpha_2 +w_1w_2\beta_2\alpha_2(\rho_2-qu^2/\alpha_1^2)^2, \tag{7.22}$$

$$\bar{Q}^{(1,2)} = w_1w_2w_4q^2u^2/\alpha_1^2\pm\rho_1\rho_2\alpha_1\beta_2w_2\pm w_4\beta_1\alpha_1(\rho_1+qu^2/\alpha_1^2)^2,$$

and

$$w_i = [1-(V_iu/\alpha_1)^2]^{1/2} \qquad (i = 1, 2, 3, 4), \tag{7.23}$$

with q being given by (2.74″). The upper sign in (7.22) applies to $Q^{(1)}$ and $\bar{Q}^{(1)}$, and the lower to $Q^{(2)}$ and $\bar{Q}^{(2)}$. Comparing (7.21) with (2.74), we can see that $Q(u)$ is the reflection coefficient R_{11} for $u = \sin\theta_1$ (i.e., for $\Theta = u/\alpha_1$).

The sheet of the Riemann surface on which the integration contour lies is given by the relation

$$\arg w_i = \tfrac{1}{2}\pi \qquad \text{for } u > \alpha_1/V_i \qquad (i = 1, 2, 3, 4). \tag{7.23'}$$

For the displacement components, W_r and W_z, we obtain, from (7.20) and (7.2), for $z = H$,

$$W_r = -\tfrac{1}{2}k_1 e^{-i\omega t}\int_{-\infty}^{\infty} Q(u)\,H_1^{(1)}(k_1ru)\exp[ik_1(h+H)(1-u^2)^{1/2}] \times u^2(1-u^2)^{-1/2}\,du,$$

$$W_z = \tfrac{1}{2}ik_1 e^{-i\omega t}\int_{-\infty}^{+\infty} Q(u)\,H_0^{(1)}(k_1ru)\exp[ik_1(h+H)(1-u^2)^{1/2}]\,udu. \tag{7.24}$$

Note that the integration contour goes through the point $u = 0$, where the values of $H_0^{(1)}$ and $H_1^{(1)}$ become infinite. We shall, however, transform the integration contour from $-\infty$ to ∞ into a new contour Ω, which will be described later in detail. The new integration contour will not go through the point $u = 0$. We denote the point lying on the integration contour Ω with the minimum absolute value of $|u|$ by u_Ω. If

$$k_1 r |u_\Omega| \gg 1, \tag{7.25}$$

we can expand the Hankel functions asymptotically over the whole integration contour Ω as follows (see Menzel, 1955):

$$\begin{aligned} H_0^{(1)}(k_1 ru) &\sim \left(\frac{2}{\pi k_1 ru}\right)^{1/2} \exp(ik_1 ru - i\pi/4), \\ H_1^{(1)}(k_1 ru) &\sim \left(\frac{2}{\pi k_1 ru}\right)^{1/2} \exp(ik_1 ru - 3i\pi/4). \end{aligned} \tag{7.26}$$

Using this in (7.24), we obtain

$$\begin{aligned} W_r &= \exp(-i\omega t + i\pi/4)\left(\frac{k_1}{2\pi r}\right)^{1/2} \int_\Omega Q(u) e^{ik_1 a(u)} u^{3/2} (1-u^2)^{-1/2} du, \\ W_z &= \exp(-i\omega t + i\pi/4)\left(\frac{k_1}{2\pi r}\right)^{1/2} \int_\Omega Q(u)\, e^{ik_1 a(u)} u^{1/2} du, \end{aligned} \tag{7.27}$$

where

$$a(u) = ru + (h+H)(1-u^2)^{1/2}. \tag{7.28}$$

Details of Ω and u_Ω will be specified later. If $|u_\Omega|$ is not large, we cannot use (7.26) and we must either consider several terms in the expansion of the Hankel function (see Brekhovskikh, 1960, p. 250) or use another method. This situation arises if we are interested in properties of reflected waves at small epicentral distances (in comparison with the wavelength). The main purpose of our investigation, however, is to give expressions for head waves and for reflected waves in the critical region for which the assumption (7.25) is quite acceptable, as will be shown later.

The standard procedure for an approximate evaluation of the integrals similar to (7.27) is that of the method of steepest descent. The saddle point u_0 is given by the formula

$$\left(\frac{da(u)}{du}\right)_{u=u_0} = 0, \tag{7.29}$$

i.e.

$$u_0 = \frac{r}{[r^2 + (h+H)^2]^{1/2}} = \sin\theta_1, \tag{7.30}$$

where θ_1 is the angle of reflection (the acute angle between the ray of the reflected wave and the z-axis). Thus, we can see that u_0 is always real, $0 \leqslant u_0 < 1$.

The contour of steepest descent which passes through the saddle point $u = u_0$ is shown in figure 7.2a. This contour forms an angle of $\pi/4$ with the negative direction of the real axis at the saddle point. Along this contour, the modulus of the integrand decreases rapidly with decreasing distance from the saddle point and its oscillations are suppressed. It is then possible to limit the integrations to the small effective part of the contour in the vicinity of the saddle point. The length of the effective path is dependent on frequency, the path being longer for smaller frequencies. Thus, the method of steepest descent is useful only at high frequencies. When the function $Q(u)$ changes only slowly in the vicinity of the saddle point, it can be replaced by a constant $Q(u_0)$ ($Q(u_0)$ equals the coefficient of reflection R_{11}, as was mentioned above). However, if the function $Q(u)$ changes very rapidly in the neighbourhood of the saddle point, this method is not acceptable. This situation arises when u_0 is close to any of the branch points where the radicals w_i are nearly zero, i.e. u_0 is close to α_1/α_2, α_1/β_2, or 1. (The value of α_1/β_1 is always greater than 1 and u_0 cannot be close to it.) In these cases it is necessary to replace $Q(u)$ by a suitable superposition of simpler functions.

We can use several different modifications of the integration contour

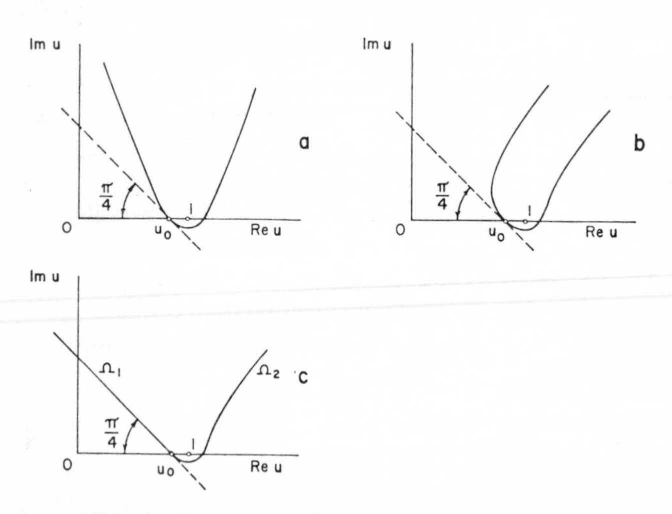

FIGURE 7.2 (a) The contour of integration for the method of steepest descent, passing through the saddle-point u_0. (b) and (c) are modifications of this contour.

described above. The significant assumption for these new contours is that they must coincide with the effective part of the path of steepest descent in the vicinity of the saddle point. We can then be sure that the integrand will decrease rapidly with increasing distance from the saddle point and will not oscillate to any appreciable extent. We shall present here two such contours (see figure 7.2b and 7.2c).

The contour in figure 7.2b is given by the parametric equation

$$(1-u^2)^{1/2} = (1-u_0{}^2)^{1/2} + y\,e^{-i\pi/4} \tag{7.31}$$

with the real parameter y varying from $-\infty$ to ∞. At the saddle point, $y = 0$. Evaluation of the integrals along this contour is convenient when u_0 is not too small ($1 > u_0 > 0.7$). Away from unity, on the left-hand branch of Ω, the integrand decreases but slowly with increasing y and even becomes divergent for $u_0 < 1/\sqrt{2}$. But the choice of the new contour eliminates all difficulties which normally arise when $u_0 \to 1$. (This is because it is no longer necessary to determine the derivatives of the radical $(1-u^2)^{1/2}$ in the neighbourhood of the point $u = u_0$; this derivative increases infinitely as $u_0 \to 1$). This contour makes an angle $\pi/4$ with the negative direction of the real axis at the point $u = u_0$. Use of this contour makes it possible to investigate head waves when the refractive index $n(= \alpha_1/\alpha_2)$ approaches unity (Červený, 1959).

The contour in figure 7.2c, on the other hand, consists of two parts. One part, Ω_2, corresponds to the contour in figure 7.2b and the other part, Ω_1, is given by the parametric equation

$$u = u_0 - y\,e^{-i\pi/4}, \tag{7.32}$$

with the real parameter y varying from 0 to ∞. This contour can be used for any value of u_0, even for $u_0 < 1/\sqrt{2}$. Both the contours mentioned above can simply be used for numerical integration by standard quadrature methods.

If we want to find the full solution, we must take into account the singular points which lie on the real axis. We have four branch points, $u_i{}^*$ ($i = 1, 2, 3, 4$), corresponding to the zero values of radicals w_i, and possibly one pole (which corresponds to the point at which the denominator of the function $Q(u)$ vanishes). The branch points $u_i{}^*$ are given by the expressions $u_i{}^* = \alpha_1/V_i$ ($i = 1, 2, 3, 4$), i.e.,

$$u_1{}^* = 1, \quad u_2{}^* = \alpha_1/\beta_1, \quad u_3{}^* = \alpha_1/\alpha_2, \quad u_4{}^* = \alpha_1/\beta_2. \tag{7.33}$$

The original integration contour, from $-\infty$ to ∞, is then transformed into one of steepest descent passing through the saddle point $u = u_0$, some branch line contours, and the contour around the pole. The physical meaning of the individual integrals from the asymptotic point of view is as follows:

(1) The integral along the path of steepest descent passing through the saddle point (or along any of the modified contours) represents the reflected compressional wave of the type 11.

(2) The branch line integrals represent head waves. If the branch point u_i^* lies right of the point $u = 1$, the head wave is inhomogeneous (see Brekhovskikh, 1960; Červený, 1957a), and if it lies between 0 and 1, we get regular head waves. The branch line integral corresponding to the radical w_1 (branch point $u_1^* = 1$) need not be calculated, as the contour of steepest descent passing through the saddle point always bypasses it (more details follow presently).

(3) The contribution of the pole represents the Stoneley wave.

We are not interested here in Stoneley waves or in inhomogeneous head waves; therefore, we shall study only the branch points $u = u_i^*$ lying between 0 and 1. Their number depends on the distribution of velocities, but for an incident P-wave, which is considered here, it cannot exceed 2. If $\alpha_2 < \alpha_1$, no branch point exists in this region. If $\alpha_2 > \alpha_1$ and $\beta_2 < \alpha_1$, one

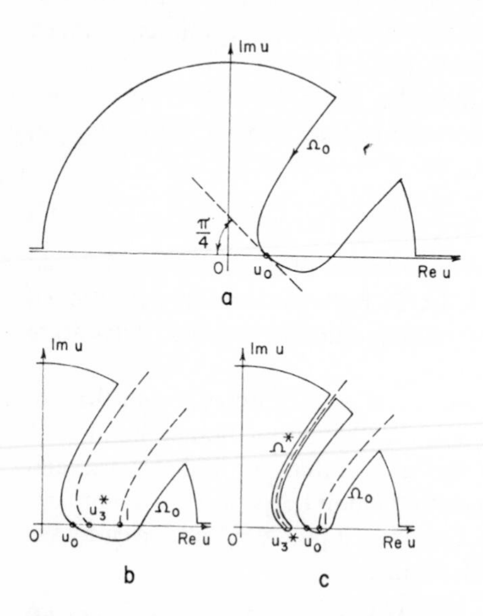

FIGURE 7.3 (a) The contour of integration Ω_0. (b) Situation for $u_0 < u_3^*$. (c) Situation for $u_0 > u_3^*$, when an additional (branch line) integral along Ω^* must be evaluated.

branch point exists there, viz., $u = u_3^*$. The corresponding branch line integral along the cut of the radical w_3 represents the head wave of the type 131. If $\beta_2 > \alpha_1$, two branch points exist in the region of interest, viz., u_3^* and u_4^*. The corresponding branch line integrals represent the head waves of the type 131 and 141, respectively.

We shall use the contour in figure 7.2b in the following for evaluation of the integral under consideration. The cuts of the radicals w_i may be defined in this case by the following parametric equation:

$$(1-u^2)^{1/2} = (1-u_i^*)^{1/2} + y\,\mathrm{e}^{-i\pi/4}, \tag{7.34}$$

where y is a real parameter such that $0 \leqslant y < \infty$, and $i = 3$ or 4 (the index i should be distinguished from the $i = \sqrt{(-1)}$ in the exponent). This equation can be used for $i = 1$ also (i.e. for the cut of the radical w_1). In this case, $u_1^* = 1$ and the first term on the right-hand side vanishes. For $u_i^* > 1$, we shall not construct the cuts, as we do not want to calculate inhomogeneous head waves.

We shall assume now that $\alpha_2 > \alpha_1$ and $\beta_2 < \alpha_1$. Then only the branch point $u = u_3^*$ lies between 0 and 1. The corresponding situation in the complex plane of $u = \mathrm{Re}\,u + i\,\mathrm{Im}\,u$ is given in figure 7.3. Ω_0 corresponds to the contour of integration (7.31) passing through the saddle point u_0. With the above choice of cuts, the integration contour Ω_0 never intersects either of the cuts of the radicals w_1 and w_3. The contour may coincide with the cut of the radical w_3 for $y > 0$, but it never intersects it. However, the cut of the radical w_3 can intersect the semicircle in the upper half-plane (see figure 7.3c). In this case we must add a branch line integral along this cut, which we denote by Ω^*. If we do not add this, a change in the integration contours would not be permissible since the new integration contour would end on a different sheet of the Riemann surface than the original one. Note that this branch line integral must be added if $u_0 > u_3^*$ $(= \alpha_1/\alpha_2)$, i.e. if

$$\sin\theta_1 > n, \tag{7.35}$$

where n is the refractive index given by the relation

$$n = \alpha_1/\alpha_2. \tag{7.35$'$}$$

The integral along the semicircle in the upper half-plane vanishes if its radius increases to infinity. From figure 7.3 we can see that the cut of the radical w_1 never intersects the new integration contour. Thus, for $\sin\theta_1 < n$, we can replace the original integration contour by Ω_0; and for $\sin\theta_1 > n$, by Ω_0 and Ω^*. Therefore, $u_0 = u_3^*$ ($\sin\theta_1 = n$) is a very important limiting case for the following investigation. It corresponds to the critical ray (see chapter 3). The points lying on the critical ray are called *critical points.* We are interested here mainly in the calculation of the above integrals in the critical region (where u_0 is close to $u_3{}^*$) where the ray theory is not applicable. For a given H, the critical point divides the profile into two sections: (1) $u_0 < u_3{}^*$ ($\sin\theta_1 < n$), before the critical point; and (2) $u_0 > u_3{}^*$ ($\sin\theta_1 > n$), beyond the critical point. Before the critical point, only the reflected wave exists; beyond it, both reflected and head waves are present. The reflected wave before the critical point merges into the *interference reflected-head wave* beyond it, the properties of both waves changing continuously across the critical point.

Note that when $u_4{}^* = \alpha_1/\beta_2 < 1$, we shall have to add one more branch line integral if $\sin\theta_1 > \alpha_1/\beta_2$. This integral represents the head wave of the type 141, which will not be studied here since the analysis for it is similar to that for the head wave 131.

We could construct the cuts for the contour in figure 7.2c in a manner similar to the above. The parametric equations for the cuts would be the same as (7.32), except that u_0 would be replaced by $u_i{}^*$.

For the calculation of W_r and W_z when u_0 is close to $u_3{}^*$, we could use the contour of integration given in figure 7.4 instead of the contours Ω_0 and Ω^* described above. This contour has been used for exact numerical integrations; some results are given in Červený (1967b). If we want to find an approximate analytic expression for the waves under consideration, this contour of integration is not so convenient as the contours mentioned before.

The entire contour Ω_0, described above, lies on the right side of the tangent to this contour at the saddle point $u = u_0$. Similarly, both integration contours Ω_0 and Ω^* lie on the right side of the tangent to the cut at the branch point $u = u_3{}^*$ when $u_3{}^* < u_0$. As this tangent forms an angle

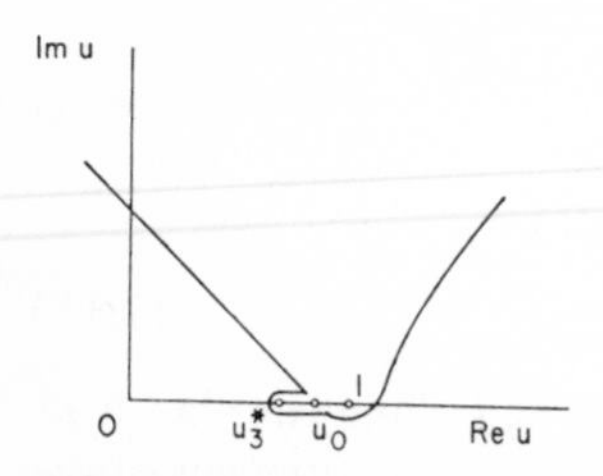

FIGURE 7.4 The modified contour of integration when the saddle-point u_0 is close to the branch point $u_3{}^*$ (for $u_0 > u_3{}^*$).

$\pi/4$ with the opposite direction of the real u-axis, it holds for all points u lying on contour Ω and Ω^* that

$$|u| > (1/\sqrt{2})\ \mathrm{Min}\ (\alpha_1/\alpha_2, \sin\theta_1).$$

The assumption (7.25) can now be written as follows:

$$(1/\sqrt{2})k_1 r\ \mathrm{Min}\ (\alpha_1/\alpha_2, \sin\theta_1) \gg 1. \tag{7.36}$$

In the following we shall find separately the expressions for the reflected wave 11 (corresponding to the integration contour Ω_0) and the head wave 131 (corresponding to the branch line integral along Ω^*). Before proceeding, however, we introduce another expression for the function $Q(u)$, given by (7.21), which will often be used in what follows. We can write

$$Q(u) = B_1(u) - B_2(u)w_3, \tag{7.37}$$

where

$$\begin{aligned} B_1(u) &= \frac{Q^{(2)}Q^{(1)} - \bar{Q}^{(2)}\bar{Q}^{(1)}w_3^{\,2}}{(Q^{(1)})^2 - (\bar{Q}^{(1)})^2 w_3^{\,2}}, \\ B_2(u) &= \frac{Q^{(2)}\bar{Q}^{(1)} - \bar{Q}^{(2)}Q^{(1)}}{(Q^{(1)})^2 - (\bar{Q}^{(1)})^2 w_3^{\,2}}, \end{aligned} \tag{7.38}$$

the functions $Q^{(1)}$, $Q^{(2)}$, $\bar{Q}^{(1)}$, $\bar{Q}^{(2)}$, and w_3 being given by (7.22) and (7.23). The functions $B_1(u)$ and $B_2(u)$ do not contain the radical w_3 and are smooth across the point $u = n$, where we can write

$$B_2(n) = -\left(\frac{\mathrm{d}Q}{\mathrm{d}w_3}\right)_{w_3=0}. \tag{7.39}$$

Comparing (7.39) with (3.108), or (7.38) with (3.110), we see that

$$B_2(n) = \Gamma_{131}, \tag{7.40}$$

where Γ_{131} is the head wave coefficient, the analytic expression for which is given by (3.99).

7.2.3 *The Reflected Wave*

In this section we shall study properties of the compressional R-wave of the type 11. Expressions for the displacement components are given by (7.27) except that, for the R-wave, the integration contour is only Ω_0 (see eq. (7.31)). At the point $u = u_0$ ($=\sin\theta_1$), arguments of w_1, w_2, and w_4 equal zero; $\arg w_3 = 0$ for $u_0 < n$, and $\arg w_3 = \pi/2$ for $u_0 > n$. If we introduce the variable y, defined by (7.31), we can write

$$ik_1a(u) = ik_1a(u_0)+ik_1ye^{-i\pi/4}\,[(h+H)-r(1-u_0^{\,2})^{1/2}/u_0] \\ -k_1ry^2/2u_0^{\,3} + \dots ,$$

in which the second term on the right vanishes because of (7.30). Therefore,

$$ik_1a(u) \sim ik_1a(u_0) \;-k_1ry^2/2u_0^{\,3}, \tag{7.41}$$

where we have ignored higher terms because of the assumption (7.36). Finally, with all the appropriate changes mentioned above, we can write the following expressions for the displacement components of the reflected wave:

$$\begin{bmatrix} W_r^{\,11} \\ W_z^{\,11} \end{bmatrix} = \exp[-i\omega(t-a(u_0)/\alpha_1)]\left(\frac{k_1}{2\pi r u_0}\right)^{1/2} \\ \times \int_{-\infty}^{+\infty} \tilde{Q}(y)\exp(-k_1ry^2/2u_0^{\,3})\mathrm{d}y \begin{bmatrix} u_0 \\ (1-u_0^{\,2})^{1/2} \end{bmatrix}, \tag{7.42}$$

where $\tilde{Q}(y)$ denotes the function obtained by substituting the new variable y for u in the expression (7.21) for $Q(u)$. Naturally, for $y = 0$ (i.e. for $u = u_0$), $\tilde{Q}(y)$ is the same as the appropriate R-coefficients, i.e.

$$\tilde{Q}(0) = Q(u_0) = R_{11}, \tag{7.43}$$

where R_{11} is given by (2.74).

The integrand in (7.42) decreases very rapidly with increasing $|y|$ owing to the presence of the function $\exp(-k_1ry^2/2u_0^{\,3})$ in it. The integrand has its largest values for $y = 0$, i.e. for u close to u_0. For those u_0 for which $\tilde{Q}(y)$ changes only very little with a change in y around the point $y = 0$, we can put $\tilde{Q}(y) \sim R_{11}$, and obtain from (7.42), for large $k_1r/2u_0^{\,3}$,

$$\begin{bmatrix} W_r^{\,11} \\ W_z^{\,11} \end{bmatrix} = \frac{R_{11}}{L}\exp[-i\omega(t-\tau_{11})]\begin{bmatrix} \sin\theta_1 \\ \cos\theta_1 \end{bmatrix}, \tag{7.44}$$

where

$$L = [r^2+(h+H)^2]^{1/2}, \tag{7.45}$$

$$\tau_{11} = \frac{a(u_0)}{\alpha_1} = \frac{1}{\alpha_1}[r\sin\theta_1+(h+H)\cos\theta_1] = \frac{h+H}{\alpha_1\cos\theta_1}. \tag{7.46}$$

The expression (7.44) for $W_r^{\,11}$ and $W_z^{\,11}$ is just the same as was found in section 2.6.2, using the ray theory. (It is in fact the complex conjugate, as mentioned above.)

If, however, u_0 is near u_3^* $(=n)$, the function $\tilde{Q}(y)$ changes very rapidly in the neighbourhood of the point $y = 0$, as it contains the radical w_3. In this case it is convenient to divide it into two parts: one part which changes rapidly in the neighbourhood of this point and the other which is a smooth function. The latter part can be replaced by a constant; and only the former part remains to be integrated. For this purpose it is possible to use the formula (7.37) for $Q(u)$.

It is clear from (7.42) that the expressions for the components W_r^{11} and W_z^{11} differ only in factors u_0 and $(1-u_0^2)^{1/2}$ from each other. We shall therefore discuss only the component W_z^{11}, merely writing down the final formula for W_r^{11}. From (7.42) and (7.37), we have

$$W_z^{11} = (1-u_0^2)^{1/2}\left(\frac{k_1}{2\pi r u_0}\right)^{1/2} \exp[-i\omega(t-\tau_{11})]$$
$$\times\left\{\int_{-\infty}^{+\infty} \tilde{B}_1(y)\exp(-k_1 r y^2/2u_0^3)\mathrm{d}y - \int_{-\infty}^{+\infty} \tilde{B}_2(y)\tilde{w}_3\exp(-k_1 r y^2/2u_0^3)\mathrm{d}y\right\}, \tag{7.47}$$

where the functions $\tilde{B}_1(y)$, $\tilde{B}_2(y)$, and $\tilde{w}_3$ are the same as $B_1(u)$, $B_2(u)$, and w_3 given by (7.38) and (7.23), with the new variable y substituted for u, using (7.31). The functions $\tilde{B}_1(y)$ and $\tilde{B}_2(y)$ do not contain the radical $\tilde{w}_3$, and therefore they do not change as rapidly, in the neighbourhood of the point $y = 0$, as the function $\tilde{Q}(y)$ when u_0 is close to n. (The derivatives of $\tilde{B}_1$ and $\tilde{B}_2$ remain finite for $u_0 = n$.) As a first approximation, we can write

$$\begin{aligned} \tilde{B}_1(y) &\sim \tilde{B}_1(0)\ (=B_1(u_0)), \\ \tilde{B}_2(y) &\sim \tilde{B}_2(0)\ (=B_2(u_0)). \end{aligned} \tag{7.48}$$

The accuracy of these approximations was discussed by Červený (1966a). It was shown that they give good results for high frequencies when the refractive index n is not too small and when γ_1 and γ_2 are not very different from each other. In other cases, it is better to use methods of exact numerical integration (see section 7.2.6).

For the radical $\tilde{w}_3$, however, such an approximation as (7.48) cannot be used when u_0 is close to n because at this point the derivative of $\tilde{w}_3$ has infinite values. Thus

$$W_z^{11} = \frac{\cos\theta_1}{L}\exp[-i\omega(t-\tau_{11})]$$
$$\times\left\{B_1(u_0) - B_2(u_0)L\left(\frac{k_1}{2\pi r u_0}\right)^{1/2}\int_{-\infty}^{+\infty}\tilde{w}_3\exp(-k_1 r y^2/2u_0^3)\mathrm{d}y\right\}. \tag{7.49}$$

The function $\tilde{w}_3$ can be rewritten in the form

$$\tilde{w}_3 = [1-(\alpha_2 u(y)/\alpha_1)^2]^{1/2} = n^{-1}[(1-u^2(y))^{1/2}-(1-n^2)^{1/2}]^{1/2} \times [(1-u^2(y))^{1/2}+(1-n^2)^{1/2}]^{1/2}$$
$$= in^{-1}[(1-n^2)^{1/2}+(1-u_0{}^2)^{1/2}+ye^{-i\pi/4}]^{1/2} \times [(1-n^2)^{1/2}-(1-u_0{}^2)^{1/2}-ye^{-i\pi/4}]^{1/2}. \qquad (7.50)$$

If it is assumed that

$$\left(\frac{k_1 r}{2u_0{}^3}\right)^{1/2}[(1-n^2)^{1/2}+(1-u_0{}^2)^{1/2}] \gg 1, \qquad (7.51)$$

then (7.50) yields, approximately,

$$\tilde{w}_3 = in^{-1}[(1-u_0{}^2)^{1/2}+(1-n^2)^{1/2}]^{1/2}[(1-n^2)^{1/2}-(1-u_0{}^2)^{1/2}-ye^{-i\pi/4}]^{1/2}. \qquad (7.52)$$

The argument of the second radical on the right of the above expression is given (for $y = 0$) by

$$\arg[(1-n^2)^{1/2}-(1-u_0{}^2)^{1/2}-ye^{-i\pi/4}]^{1/2} \begin{array}{ll} = -\frac{1}{2}\pi & \text{for } u_0 < n, \\ = 0 & \text{for } u_0 > n. \end{array} \qquad (7.53)$$

Inserting (7.52) into (7.49), we can write, after some simple rearrangements,

$$\begin{bmatrix} W_r{}^{11} \\ W_z{}^{11} \end{bmatrix} = \frac{1}{L}\exp[-i\omega(t-\tau_{11})]\left\{B_1(u_0) - \frac{i}{n}B_2(u_0)\left(\frac{2u_0{}^3}{k_1 r}\right)^{1/4} \times [(1-n^2)^{1/2}+(1-u_0{}^2)^{1/2}]^{1/2}g_1(y^0)\right\}\begin{bmatrix} \sin\theta_1 \\ \cos\theta_1 \end{bmatrix}, \qquad (7.54)$$

where

$$g_1(y^0) = \pi^{-1/2}\int_{-\infty}^{+\infty}[y^0-y\exp(-i\pi/4)]^{1/2}\,e^{-y^2}dy, \qquad (7.55)$$

$$y^0 = \left(\frac{k_1 r}{2u_0{}^3}\right)^{1/2}[(1-n^2)^{1/2}-(1-u_0{}^2)^{1/2}]. \qquad (7.56)$$

The argument of the radical in (7.55) is given (for $y = 0$) by the relation

$$\arg[y^0-y\exp(-i\pi/4)]^{1/2} \begin{array}{ll} = -\frac{1}{2}\pi & \text{for } y^0 < 0 \text{ (i.e. for } u_0 < n), \\ = 0 & \text{for } y^0 > 0 \text{ (i.e. for } u_0 > n). \end{array} \qquad (7.57)$$

The value of $|y^0|$ is a measure of the distance from the critical point. $|y^0|$ equals zero just at the critical point ($u_0 = n$) and increases with increasing distance from it. Beyond the critical point, $y^0 > 0$; before it, $y^0 < 0$.

The expression (7.54) can be used at an arbitrary distance from the critical point, including the critical point itself. This expression can be simplified if the function $g_1(y^0)$ is multiplied and divided by $(y^0)^{1/2}$. Then

$$\begin{bmatrix} W_r^{11} \\ W_z^{11} \end{bmatrix} = \frac{1}{L}\exp[-i\omega(t-\tau_{11})] \times \{B_1(u_0) - B_2(u_0)[1-(\alpha_2 u_0/\alpha_1)^2]^{1/2}\mu_1(y^0)\}\begin{bmatrix} \sin\theta_1 \\ \cos\theta_1 \end{bmatrix}, \quad (7.58)$$

where

$$\mu_1(y^0) = \pi^{-1/2}\int_{-\infty}^{+\infty} [1-y\exp(-i\pi/4)/y^0]^{1/2}e^{-y^2}dy \quad (7.59)$$

and

$$\arg[1-y\exp(-i\pi/4)/y^0]^{1/2} = 0 \qquad \text{for } y = 0. \quad (7.60)$$

The expression (7.58) cannot be used for $y^0 = 0$, but for non-zero y^0 it is simpler than (7.54). Values of the functions $g_1(y^0)$ and $\mu_1(y^0)$ are given in figures 7.5a and 7.5b. $\mu_1(y^0)$ approaches unity asymptotically for large $|y^0|$.

The formula (7.58) can be simplified further. Since $R_{11} = Q(u_0) = B_1(u_0)-B_2(u_0)[1-(\alpha_2 u_0/\alpha_1)^2]^{1/2}$, we obtain from (7.58)

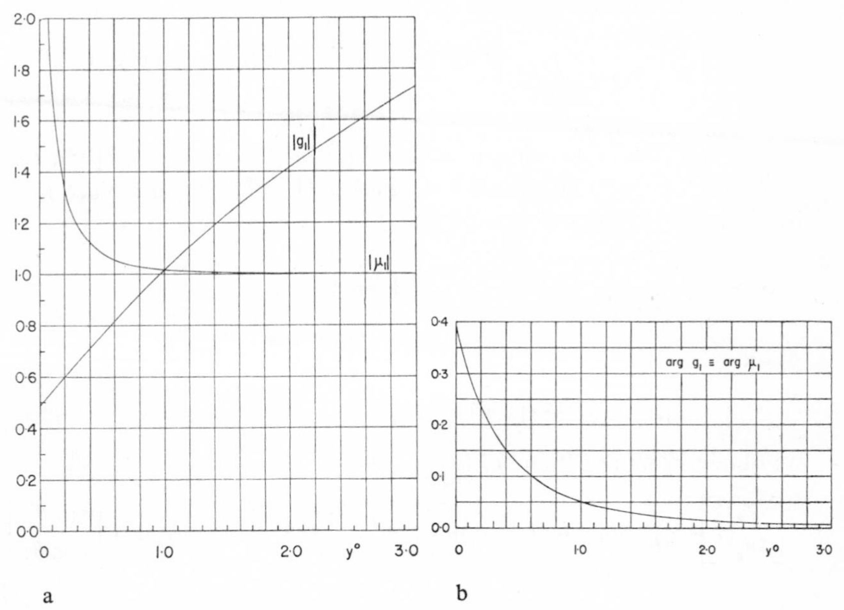

FIGURE 7.5 Moduli and arguments (in radians) of g_1 and μ_1 versus y^0.

$$\begin{bmatrix} W_r^{11} \\ W_z^{11} \end{bmatrix} = \frac{1}{L}\exp[-i\omega(t-\tau_{11})]\left\{R_{11} - B_2(u_0)\left[1-\left(\frac{\alpha_2 u_0}{\alpha_1}\right)^2\right]^{1/2} \times [\mu_1(y^0)-1]\right\}\begin{bmatrix} \sin\theta_1 \\ \cos\theta_1 \end{bmatrix}. \quad (7.61)$$

Comparing this with (7.44), we can see that the first term corresponds to the ray theory, the second one being a correction to it. Note that for large distances from the critical point (i.e. for large $|y^0|$), $\mu_1(y^0) \sim 1$ and (7.61) reduces to (7.44).

The second term in the curly brackets in (7.61) attains its maximum value in the neighbourhood of the critical point, and its significance decreases with increasing distance from this point. We shall, therefore, replace some slightly varying functions in it directly by their values at $u_0 = n$. We can then write

$$\begin{bmatrix} W_r^{11} \\ W_z^{11} \end{bmatrix} = \frac{1}{L}\exp[-i\omega(t-\tau_{11})]\{R_{11}+N\}\begin{bmatrix} \sin\theta_1 \\ \cos\theta_1 \end{bmatrix}, \quad (7.62)$$

where

$$N = -B_2(u_0)[1-(\alpha_2 u_0/\alpha_1)^2]^{1/2}(\mu_1(y^0)-1). \quad (7.63)$$

By rearranging, we obtain

$$N = -\frac{i}{n}\left(\frac{2u_0{}^3}{k_1 r}\right)^{1/4}[(1-n^2)^{1/2}+(1-u_0{}^2)^{1/2}]^{1/2}B_2(u_0)[g_1(y^0)-(y^0)^{1/2}]. \quad (7.64)$$

If we use the approximations $u_0{}^{3/4}[(1-n^2)^{1/2}+(1-u_0{}^2)^{1/2}]^{1/2}B_2(u_0) \sim 2^{1/2}n^{3/4}(1-n^2)^{1/4}B_2(n)$ and $r \sim (h+H)n/(1-n^2)^{1/2}$, and replace $B_2(n)$ by Γ_{131} (see (7.40)), we obtain from (7.64)

$$N \sim -\left(\frac{h+H}{2\Lambda}\right)^{-1/4}(4\pi)^{-1/4}n^{-1/2}(1-n^2)^{3/8}\Gamma_{131} \times [2^{3/4}ig_1(y^0) - 2^{3/4}i(y^0)^{1/2}]. \quad (7.65)$$

Putting

$$g^*(n) = (4\pi)^{-1/4}n^{-1/2}(1-n^2)^{3/8}\Gamma_{131}, \quad (7.66)$$

we obtain, from (7.62) and (7.65),

$$\begin{bmatrix} W_r^{11} \\ W_z^{11} \end{bmatrix} = \frac{1}{L}\exp[-i\omega(t-\tau_{11})]\left\{R_{11}-\left(\frac{h+H}{2\Lambda}\right)^{-1/4}g^*(n)G_1(y^0)\right\}\begin{bmatrix} \sin\theta_1 \\ \cos\theta_1 \end{bmatrix}, \quad (7.67)$$

where

$$G_1(y^0) = 2^{3/4}i(g_1(y^0)-(y^0)^{1/2}) \tag{7.68}$$

and $\Lambda = \alpha_1/f$.

The function $G_1(y^0)$ can be written in another form which will be useful later. By substituting $y-y^0 \exp(i\pi/4) = y_1$ in the integral $g_1(y^0)$, we obtain

$$G_1(y^0) = -2^{3/4} \exp(7i\pi/8-i(y^0)^2)\pi^{-1/2}$$
$$\times \int_{-\infty}^{\infty} y_1^{1/2} \exp[-y_1^2-2y^0y_1\exp(i\pi/4)]\mathrm{d}y_1-2^{3/4}i(y^0)^{1/2}, \tag{7.69}$$

where, for $y_1 > 0$,

$$\begin{aligned} \arg y_1^{1/2} &= 0 \qquad \text{for } u_0 \leqslant n, \\ &= \pi \qquad \text{for } u_0 > n. \end{aligned} \tag{7.70}$$

Before the critical point, (7.67) gives the final expression for the displacement component of the reflected wave. Beyond the critical point, the expressions for the head wave must be added to it, because the two waves interfere there. Therefore, the expressions for W_r^{11} and W_z^{11} given above are discontinuous through the critical point.

7.2.4 *The Head Wave*

In this section we shall study the properties of the compressional head wave of the type 131. Expressions for its displacement components W_r^{131} and W_z^{131} are as follows (see (7.27):)

$$W_r^{131} = \left(\frac{k_1}{2\pi r}\right)^{1/2} \exp(-i\omega t+i\pi/4) \int_{\Omega^*} Q(u) \exp[ik_1 a(u)]u^{3/2} \times(1-u^2)^{-1/2}\,\mathrm{d}u, \tag{7.71}$$
$$W_z^{131} = \left(\frac{k_1}{2\pi r}\right)^{1/2} \exp(-i\omega t+i\pi/4) \int_{\Omega^*} Q(u) \exp[ik_1 a(u)]u^{1/2}\mathrm{d}u,$$

where $a(u)$ and $Q(u)$ are given by (7.28) and (7.37). The integration contour Ω^* bypasses the cut of the radical w_3, which is given (see eq. (7.34)) by the parametric equation

$$(1-u^2)^{1/2} = (1-n^2)^{1/2}+y \exp(-i\pi/4) \qquad (0 \leqslant y < \infty). \tag{7.72}$$

The radical w_3 has different signs on either side of the cut given above. We now introduce a new integration variable y, given by the relation (7.72), and write, from (7.23),

$$w_3 = n^{-1}y^{1/2}[y+2(1-n^2)^{1/2}\exp(i\pi/4)]^{1/2}\exp(-i\pi/4), \tag{7.73}$$

where

$$\begin{aligned} \arg y^{1/2} &= 0 \quad \text{on the left side of the cut,} \\ &= \pi \quad \text{on the right side of the cut,} \end{aligned} \tag{7.74}$$

and

$$\arg[y+2(1-n^2)^{1/2}\exp(i\pi/4)]^{1/2} = \tfrac{1}{8}\pi, \quad \text{for } y = 0. \tag{7.75}$$

Thus, the radical $y^{1/2}$ has different signs on the two sides of the cut. All other quantities have the same values on either side.

The function $ik_1a(u)$ in the exponent in the integrals (7.71) can be expanded in powers of y:

$$ik_1a(u) = ik_1a(n) - k_1n^{-1}y(1-n^2)^{1/2}l\exp(i\pi/4) - k_1ry^2/2n^3 + \ldots, \tag{7.76}$$

where

$$l = r - r^*, \qquad r^* = (h+H)n/(1-n^2)^{1/2}. \tag{7.77}$$

In the above expressions, l is the horizontal distance from the critical point and r^* is the critical distance. l also gives the length of the segment of the ray of the head wave in the second medium, parallel to the interface.

Inserting (7.73) and (7.76) into (7.71), and taking into account that the values of $B_1(u)$ and $B_2(u)$ are the same on both sides of the cut, we obtain

$$\begin{aligned} W_z^{131} = &-2\left(\frac{k_1(1-n^2)}{2\pi rn^3}\right)^{1/2} \exp[-i\omega(t-a(n)/\alpha_1)-i\pi/4] \\ &\times \int_0^\infty \tilde{B}_2(y)y^{1/2}[y+2(1-n^2)^{1/2}\exp(i\pi/4)]^{1/2} \\ &\times \exp\{-k_1y(1-n^2)^{1/2}l\exp(i\pi/4)/n - k_1ry^2/2n^3\}\mathrm{d}y. \end{aligned} \tag{7.78}$$

The function $\tilde{B}_2(y)$ is the same as $B_2(u)$ given by (7.38), with the variable y substituted for u, using (7.72).

We see from (7.78) that the integrand will attain its maximum value for y close to zero. The function $\tilde{B}_2(y)$ can be expanded in powers of y. For the first approximation, as in the case of the reflected wave, we take only the first term in the expansion, i.e., we put

$$\tilde{B}_2(y) \sim \tilde{B}_2(0) = B_2(n) = \Gamma_{131} \tag{7.79}$$

(see (7.40)). This approximation gives good results in the same situations as the approximation (7.48); see above.

Assuming that

$$[k_1 r(1-n^2)/2n^3]^{1/2} \gg 1, \tag{7.80}$$

we obtain

$$W_z^{131} = \left[2^7 \frac{n^3(1-n^2)^3}{\pi^2 k_1 r^5}\right]^{1/4} \Gamma_{131} \exp[-i\omega(t-\tau_{131})+7i\pi/8] \times \int_0^\infty y^{1/2} \exp[-y^2-2^{1/2}y(1+i)y^*]dy, \tag{7.81}$$

where

$$y^* = l[k_1 n(1-n^2)/2r]^{1/2}, \tag{7.82}$$

$$\tau_{131} = \frac{a(n)}{\alpha_1} = \frac{1}{\alpha_1}[rn+(h+H)(1-n^2)^{1/2}] = \frac{r}{\alpha_2}+\frac{(h+H)}{\alpha_1}\left[1-\left(\frac{\alpha_1}{\alpha_2}\right)^2\right]^{1/2}. \tag{7.83}$$

y^* (as y_0) is a measure of the distance from the critical point. Just at the critical point, $y^* = 0$, and with increasing distance from the critical point y^* increases. y^* is always real and positive. For large y^* we find easily that

$$\int_0^\infty y^{1/2} \exp[-y^2-2^{1/2}(1+i)yy^*]dy \sim \int_0^\infty y^{1/2} \exp[-2^{1/2}(1+i)yy^*]dy = 2^{-5/2}\pi^{1/2}y^{*-3/2} \exp(-3i\pi/8). \tag{7.84}$$

Inserting this into (7.81), and into a similar expression for W_r^{131}, we get

$$\begin{bmatrix} W_r^{131} \\ W_z^{131} \end{bmatrix} = \frac{\Gamma_{131}}{k_1 r^{1/2} l^{3/2}(1-n^2)^{1/2}} \exp[-i\omega(t-\tau_{131})+i\pi/2] \begin{bmatrix} n \\ (1-n^2)^{1/2} \end{bmatrix}. \tag{7.85}$$

We can easily see that this corresponds to the expressions derived in chapter 3 using the ray theory. (As noted earlier, the expression (7.85) is in fact the complex conjugate of the one in chapter 3.) For the purpose of comparison, we use the relation $1/[k_1(1-n^2)^{1/2}] = \alpha_2 \tan\theta_1^*/\omega$, where $\sin\theta_1^* = \alpha_1/\alpha_2$.

In the critical region, however, y^* is small and (7.85) is not correct. We introduce the function $\mu_2(y^*)$ such that

$$\mu_2(y^*) = 2^{5/2}y^{*3/2}\pi^{-1/2} \exp(7i\pi/8) \int_0^\infty y^{1/2} \exp\{-y^2-yy^*2^{1/2}(1+i)\}dy. \tag{7.86}$$

From this and (7.81) we then obtain

$$\begin{bmatrix} W_r^{131} \\ W_z^{131} \end{bmatrix} = \frac{\Gamma_{131}\mu_2(y^*)}{k_1 r^{1/2} l^{3/2}(1-n^2)^{1/2}} \exp[-i\omega(t-\tau_{131})] \begin{bmatrix} n \\ (1-n^2)^{1/2} \end{bmatrix}. \quad (7.87)$$

For large y^*, $\mu_2(y^*) \sim \exp(i\pi/2)$, and therefore (7.87) gives the same results as (7.85). For small y^* (in the critical region), however, the two expressions differ. Numerical values of $|\mu_2(y^*)|$ and $\arg \mu_2(y^*)$ are given in figure 7.6, a and b.

The expression (7.87) can be used at an arbitrary distance from the critical point; however, at the critical point, it is inconvenient (as $l \to 0$ and $\mu_2(y^*) \to 0$). We can simply write the new expression

$$\begin{bmatrix} W_r^{131} \\ W_z^{131} \end{bmatrix} = \frac{n^{3/4}(1-n^2)^{1/4}\Gamma_{131}}{2^{3/4}k_1^{1/4}r^{5/4}} g_2(y^*) \exp[-i\omega(t-\tau_{131})] \begin{bmatrix} n \\ (1-n^2)^{1/2} \end{bmatrix}, \quad (7.88)$$

where

$$g_2(y^*) = 2^{5/2}\pi^{-1/2} \exp(7i\pi/8) \int_0^\infty y^{1/2} \exp\{-y^2 - 2^{1/2}y(1+i)y^*\} dy. \quad (7.89)$$

Numerical values of $|g_2(y^*)|$ and $\arg g_2(y^*)$ are given in figure 7.6 a and b.

For the purpose of studying the interference of reflected and head waves

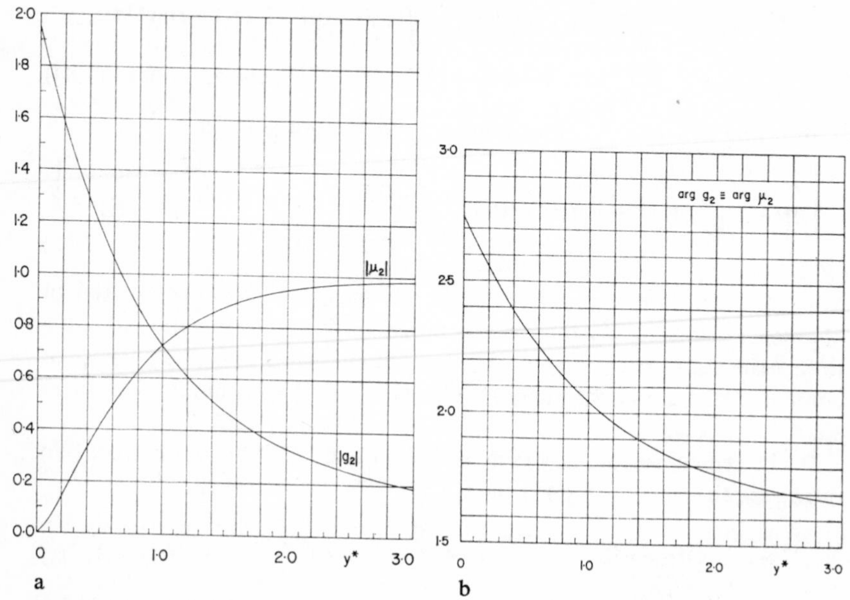

FIGURE 7.6 Moduli and arguments (in radians) of g_2 and μ_2 versus y^*.

in the next section, it will be useful to write one more expression for W_r^{131} and W_z^{131}, which follows from (7.88):

$$\begin{bmatrix} W_r^{131} \\ W_z^{131} \end{bmatrix} = \frac{1}{2r}\left(\frac{h+H}{2\Lambda}\right)^{-1/4} (1-n^2)^{1/4} n^{3/4} (1-u_0^2)^{1/8} (2\pi)^{-1/4} u_0^{-1/4} \times \Gamma_{131} g_2(y^*) \exp[-i\omega(t-\tau_{131})] \begin{bmatrix} n \\ (1-n^2)^{1/2} \end{bmatrix}. \tag{7.90}$$

In the neighbourhood of the critical point, $u_0 \sim n$ and we can write

$$\begin{bmatrix} W_r^{131} \\ W_z^{131} \end{bmatrix} = \frac{1}{r}\left(\frac{h+H}{2\Lambda}\right)^{-1/4} n g^*(n) g_2(y^*) 2^{-3/4} \exp[-i\omega(t-\tau_{131})] \times \begin{bmatrix} n \\ (1-n^2)^{1/2} \end{bmatrix}, \tag{7.91}$$

where $g^*(n)$ is given by (7.66).

Note that the expression (7.90) is completely equivalent to (7.88) and (7.87). The expression (7.91), however, can be used only in the neighbourhood of the critical point because its derivation depends on the assumption that $u_0 \sim n$.

The Accuracy of the Ray Formulae for Head Waves

We can see from (3.61) and (7.87) that the only difference between the formulae for head waves found by the integral and the ray methods is the function $\mu_2(y^*)$ in (7.87). When $\mu_2(y^*) = \exp(i\pi/2)$, the formulae are the same. For small y^* (close to the critical point), $\mu_2(y^*)$ differs substantially from $\exp(i\pi/2)$; with increasing y^*, however, they asymptotically approach each other (see figure 7.6, a and b). Far enough from the critical point, the two formulae give practically the same results. We shall first consider the function $|\mu_2(y^*)|$, as we are interested mainly in the head wave amplitudes. This function differs from unity by less than 10 per cent for $y^* > 1.6$, and by less than 5 per cent for $y^* > 2.4$. Thus, when we study the head wave amplitudes, we can use the ray formulae with a maximum error of 10 per cent for $y^* > 1.6$. The difference between arg $\mu_2(y^*)$ and $\frac{1}{2}\pi$ (=1.57) is a little larger, it is about 18 per cent for $y^* = 1.6$ and less than 10 per cent for $y^* \geqslant 2.5$. Generally, we can say that for $y^* \geqslant 2.5$ we can use the ray formulae for head waves with a maximum error of 10 per cent.

In section 3.8, the length of the interference zone of the reflected and head waves beyond the critical point was evaluated. It was shown that the two waves interfere when $\tau_{11} - \tau_{131} < \delta t$, where δt is the duration of the pulse. For a sinusoidal pulse of the duration of one period, T_0, we have

$\tau_{11}-\tau_{131} < T_0$. We can easily see that $y^* \sim \omega^{1/2}(\tau_{11}-\tau_{131})$. From this it follows that the head wave separates from the reflected wave for about $y^* \sim (2\pi)^{1/2} = 2.5$ (for a sinusoidal pulse of the length of one period). In seismic measurements, the duration of seismic pulses is usually even larger than one period, and, therefore, the head wave separates from the reflected wave for $y^* > 2.5$. Since we know that we can use the ray formulae for head waves for $y^* > 2.5$ (with a maximum error of 10 per cent), we can derive an important conclusion: *Beyond the interference zone of reflected and head waves (i.e. when the two waves are separated), we can almost always use the ray formulae for head waves.* The more exact formulae, (7.87) and (7.88), are significant only when studying properties of interference reflected head waves inside the interference zone (see sections 7.2.5 and 7.2.7).

The above conclusions may not be precisely valid when the refractive index n is small or when γ_1 is substantially different from γ_2 (for example, for head waves from a liquid–solid interface). Then the approximation (7.79), which has been accepted as satisfactory so far, can lead to errors.

7.2.5 *The Interference Reflected-Head Wave*

In the previous sections we derived expressions for reflected and head waves. These waves interfere with each other beyond the critical point ($r > r^*$). The resultant wave will be called the *interference reflected-head wave* or simply the *interference wave*. We are interested here mainly in the expressions in the critical region because at larger distances it is possible to use the ray expressions (see section 3.8). Again, only the vertical component of the displacement will be discussed in detail, the final expressions for the horizontal component being added later. We obtain, from (7.67) and (7.91),

$$\overline{W}_z^{131} = \frac{\cos\theta_1}{L}\exp[-i\omega(t-\tau_{11})]\Big\{R_{11}-\left(\frac{h+H}{2\Lambda}\right)^{-1/4} g^*(n)[2^{3/4}ig_1(y^0) -2^{3/4}i(y^0)^{1/2}]+(1-n^2)^{1/2}L(1-u_0{}^2)^{-1/2}r^{-1} \times\left(\frac{h+H}{2\Lambda}\right)^{-1/4} g^*(n)ng_2(y^*)2^{-3/4}\exp(-i\bar{y}^2)\Big\}, \tag{7.92}$$

where we have used $\overline{W}_z^{131}$ for $W_z^{11}+W_z^{131}$, and

$$\bar{y} = \omega^{1/2}(\tau_{11}-\tau_{131})^{1/2}. \tag{7.93}$$

The first term in the curly brackets of (7.92) represents the zero-order ray solution for the reflected wave, the second term is the correction to this expression in the critical region, and the third term is the head wave.

Both the second and the third terms have their largest magnitude in the critical region, and with increasing distance from the critical point their significance decreases. Therefore, we can put, approximately, $L/r \sim 1/n$ and $u_0 \sim n$ in the third term. Then

$$\overline{W}_z^{131} = \frac{\cos\theta_1}{L}\exp\{-i\omega(t-\tau_{11})\}\Big\{R_{11}-\left(\frac{h+H}{2\Lambda}\right)^{-1/4} g^*(n)$$
$$\times[2^{3/4}ig_1(y^0)-2^{3/4}i(y^0)^{1/2}-2^{-3/4}g_2(y^*)\exp(-i\bar{y}^2)]\Big\}. \quad (7.94)$$

The function in square brackets in the above expression depends on three parameters: y^0, y^*, and $\bar{y}$. We can easily see, from (7.56), (7.82), and (7.93), that they are close to one another in the neighbourhood of the critical point. They can therefore be replaced by a single parameter, say $\bar{y}$. If we introduce

$$G_2(\bar{y}) = 2^{3/4}ig_1(\bar{y})-2^{3/4}i\bar{y}^{1/2}-2^{-3/4}g_2(\bar{y})\exp(-i\bar{y}^2), \quad (7.95)$$

we can write, from (7.94),

$$\begin{bmatrix}\overline{W}_r^{131}\\ \overline{W}_z^{131}\end{bmatrix} = \frac{1}{L}\exp[-i\omega(t-\tau_{11})]\Big\{R_{11}-\left(\frac{h+H}{2\Lambda}\right)^{-1/4} g^*(n)G_2(\bar{y})\Big\}$$
$$\times\begin{bmatrix}\sin\theta_1\\ \cos\theta_1\end{bmatrix}, \quad (7.96)$$

which is very similar to the expression (7.67) for the reflected wave. Taking (7.68), (7.69), and (7.89) into consideration, we obtain from (7.95)

$$G_2(\bar{y}) = -2^{3/4}\pi^{-1/2}\exp\{7i\pi/8-i\bar{y}^2\}\int_{-\infty}^{\infty} y_1^{1/2}\exp\{-y_1^2$$
$$-2\bar{y}y_1\exp(i\pi/4)\}dy_1$$
$$-2^{7/4}\pi^{-1/2}\exp\{7i\pi/8-i\bar{y}^2\}\int_0^{\infty} y^{1/2}\exp\{-y^2-2\bar{y}y\exp(i\pi/4)\}dy$$
$$-2^{3/4}i\bar{y}^{1/2},$$

where $\arg y_1^{1/2} = \pi$ for $y_1 > 0$. From this we obtain further

$$G_2(\bar{y}) = -2^{3/4}\pi^{-1/2}\exp(7i\pi/8-i\bar{y}^2)$$
$$\times\int_{-\infty}^{+\infty} y^{1/2}\exp\{-y^2-2y\bar{y}\exp(i\pi/4)\}dy-2^{3/4}i\bar{y}^{1/2}, \quad (7.97)$$

where

$$\arg y^{1/2} = 0 \qquad \text{for } y > 0. \quad (7.98)$$

The formula (7.96), with the function $G_2(\bar{y})$ defined by (7.97), gives the final expression for the interference reflected-head wave.

A comparison of (7.97) with (7.69) shows that the expression for $G_2(\bar{y})$ beyond the critical point is the same as for $G_1(y^0)$ before the critical point. We must remember, however, that $y^0 < 0$ before the critical point, whereas $\bar{y} > 0$ beyond it. We can easily prove that the value of y^0 is very near to the value of $-|\bar{y}|$.

In the following, when we use the expression *reflected wave in the critical region* we mean the simple *reflected wave* before the critical point and *interference reflected-head wave* beyond it. The displacement components of the reflected wave in the critical region will be denoted by $\tilde{W}_r^{131}$ and $\tilde{W}_z^{131}$. The appropriate expressions for them are obtained by a combination of equations (7.67) to (7.70) and (7.96) to (7.98), viz.,

$$\begin{bmatrix} \tilde{W}_r^{131} \\ \tilde{W}_z^{131} \end{bmatrix} = \frac{1}{L} \exp[-i\omega(t-\tau_{11})] \left\{ R_{11} - \left(\frac{h+H}{2\Lambda}\right)^{-1/4} g^*(n) G(\bar{y}) \right\} \begin{bmatrix} \sin\theta_1 \\ \cos\theta_1 \end{bmatrix}, \tag{7.99}$$

where

$$G(\bar{y}) = -2^{3/4} \left\{ \pi^{-1/2} \exp(7i\pi/8 - i\bar{y}) \times \int_{-\infty}^{+\infty} y^{1/2} \exp\{-y^2 - 2y\bar{y}\exp(i\pi/4)\} dy - i\bar{y}^{1/2} \right\}, \tag{7.100}$$

$$\arg y^{1/2} = 0 \qquad \text{for } y > 0, \tag{7.101}$$

and

$$\bar{y} = \pm\omega^{1/2} |\tau_{11} - \tau_{131}|^{1/2}. \tag{7.102}$$

In the expression (7.102) for $\bar{y}$, the positive sign applies beyond the critical point ($u_0 > n$) and the negative one before it ($u_0 < n$). It must be pointed out again that the expression (7.99) can be used only in the neighbourhood of the critical point (for $\bar{y}$ small). If we are not near the critical point, it is more convenient to calculate the displacement components of the reflected wave and those of the head wave independently of one another. Beyond the critical point, the resultant interference wave is then obtained as the sum of the two waves.

The displacement components $\tilde{W}_r^{131}$ and $\tilde{W}_z^{131}$ are continuous and smooth functions of the epicentral distance in the neighbourhood of the critical point. They do not have such irregularities as sharp peaks and infinite derivatives, which arise in the case of the zeroth ray approximation.

Considering the identity

$$\int_{-\infty}^{+\infty} t^{1/2} \exp[-t^2 - 2\bar{y}t \exp(i\pi/4)]\mathrm{d}t$$
$$= (2\pi)^{1/2} 2^{-3/4} \exp(\tfrac{1}{2}i\bar{y}^2 + \tfrac{1}{4}i\pi) D_{1/2}(\bar{y}(i-1)), \quad (7.103)$$

where $D_{1/2}(\bar{y}(i-1))$ is a Weber-Hermite function (parabolic cylinder function), we can rewrite (7.100) as

$$G(\bar{y}) = 2^{1/2} \exp(i\pi/8 - i\bar{y}^2/2) D_{1/2}(\bar{y}(i-1)) - 2^{3/4} i\bar{y}^{1/2}. \quad (7.104)$$

Using the tables of the Weber-Hermite functions (see, for example, Kireyeva and Karpov, 1959), we can easily find the values of $|G(\bar{y})|$ and $\arg G(\bar{y})$ which are given in figure 7.7, a and b (see also Červený, 1963a).

Weber-Hermite functions have been used in connection with head waves also (see Brekhovskikh, 1960, section 22). Our integrals g_2, μ_2 in section 7.2.4 can be expressed in terms of $D_{-3/2}$.

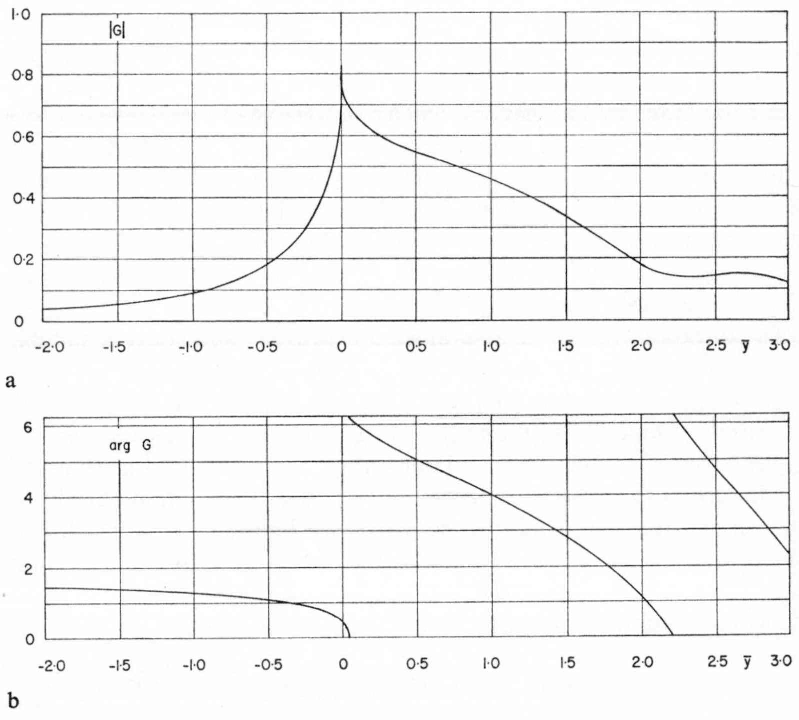

FIGURE 7.7 Modulus and argument (in radians) of G as a function of $\bar{y}$.

7.2.6 *Exact Numerical Integrations*

The formulae for the displacement components of the reflected wave, head wave, and the interference reflected-head wave, given in sections 7.2.3, 7.2.4, and 7.2.5, are very simple and suitable for numerical calculations. Each of these formulae contains only one special function (depending on one parameter). They have no singularities just at the critical point and they are more exact in the neighbourhood of the critical point than the formulae of the ray theory. Their accuracy near the critical point is about the same as in other regions. However, these formulae are not quite exact; a number of simplifying assumptions were made in their derivation. Maximum error is introduced by the approximations (7.48) and (7.79), and if the functions $\tilde{B}_1(y)$ and $\tilde{B}_2(y)$ change very rapidly with y, for y close to zero, these approximations are not valid. The most favourable conditions for the accuracy of the method are when $\gamma_1 \sim \gamma_2$ and when the refractive index n lies in the range $0.75 < n < 0.95$. For the interface with a large velocity contrast ($n < 0.5$) and for a very small velocity contrast ($0.97 < n < 1.0$), the formulae given in the preceding sections cannot be used. Note that for very small velocity contrasts, the assumptions (7.51) and (7.80) are also invalid. Moreover, it is not possible to use these formulae when γ_1 is small (e.g. for a liquid–solid interface). In all cases, the error is larger for lower frequencies (or smaller $(h+H)/2\Lambda$). Even when $0.7 < n < 0.95$, it is not possible to use these formulae when $(h+H)/2\Lambda$ is small.

In some cases, it is possible to generalize the above method to overcome the difficulties mentioned above. This was done by Červený (1959) for the refractive index n very near to unity, and by Smirnova (1962, 1964a, 1966a) for this and other situations. However, all these formulae are considerably more complicated than those given above.

The integrals (7.27) can be evaluated exactly, if we choose a suitable integration contour (for example, the integration contour given in figure 7.4). Along the integration contour Ω_0 the integrand decreases rapidly with increasing distance from the saddle point and does not oscillate to any appreciable extent, and numerical integration can be done by the standard quadrature method. Integration along the part of the contour parallel to the real axis (see figure 7.4) can be performed by the Filon method. It is also possible to choose other alternative contours along which numerical integration can be carried out without difficulty. Some results of exact numerical integrations will be given in the following section.

7.2.7 *The Reflected Wave in the Critical Region: Numerical Examples*

It was shown in section 7.2.4 that beyond the interference zone the ray formulae can be used for head waves, usually with a maximum error of

10 per cent. Properties of head waves on the basis of ray theory were studied in detail in chapter 3. Therefore, we shall consider here only the critical region, where the head wave interferes with the reflected wave. We shall first consider an interface with a small velocity contrast, and then one with a large velocity contrast. In both cases we assume that $\gamma_1 = \gamma_2 = 1/\sqrt{3}$, $\rho_1/\rho_2 = 1.0$. We shall call the amplitude–distance curves of the vertical component of displacement of reflected waves calculated using the ray theory (for infinite frequencies) the *ray amplitude curves*, and those calculated using the methods of this chapter *exact amplitude curves* (or simply *amplitude curves*.)

A Small Velocity Contrast

One example of the exact amplitude curve of a reflected wave was given in figure 3.40. While the ray amplitude curve has a sharp peak at the critical point, where it reaches its maximum value, the exact amplitude curve is quite smooth and continuous in the neighbourhood of the critical point and has a maximum value at a certain distance beyond this point. (This maximum can be found approximately by the ray method also, if we construct the interference reflected-head wave; see section 3.8.) At greater distances, the amplitude curve decreases, while it oscillates around the ray amplitude curve. As soon as the head wave separates from the reflected wave, the amplitude curve stops oscillating and gradually decreases. The shape of the exact amplitude curve and the position of its maximum depend mainly on the frequency (or $(h+H)/2\Lambda$), on the refractive index n, and on the depth of the interface.

(a) *Dependence of amplitude curve on frequency* (or $(h+H)/2\Lambda$) Figure 7.8 shows the exact amplitude curves of the reflected wave, $\tilde{A}_z^{131}$, for $h = H = 30$ km, $\alpha_1 = 6.4$ km/sec, $\alpha_2 = 8$ km/sec, for five different H/Λ: 3, 10, 30, 100, and ∞. These values of H/Λ correspond to frequencies $f = 0.64$, 2.13, 6.4, 21.3, and ∞ cps. The ray amplitude curve corresponds to $H/\Lambda = \infty$. Note that the refractive index n equals 0.8. The amplitude curves for finite frequencies are not given for all epicentral distances, but are always stopped after about two oscillations. The exact amplitude curves are smooth in the neighbourhood of the critical point for finite frequencies, and beyond the critical point, they change appreciably as a function of H/Λ (or frequency). The greatest changes occur in the position of the maximum of the amplitude curves. The greater the frequency (or H/Λ), the closer the maximum lies to the critical point. We also see that the amplitude curve has a wider form and less pronounced peak when the frequency is lower.

(b) *Dependence of amplitude curves on refractive index n* Figure 7.9 gives the exact amplitude curves of the reflected wave, $\tilde{A}_z^{131}$, for $h = H = 30$ km, $f = 6$ cps, $\alpha_2 = 8$ km/sec. The refractive index n changes from

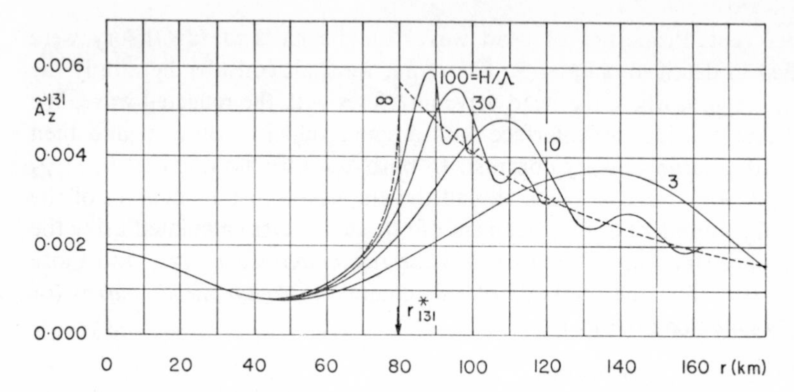

FIGURE 7.8 Amplitudes of the vertical displacement components $\mathbf{A}_z{}^{131}$ vs. distance, for different values of H/Λ. $h = H = 30$ km, $\alpha_1 = 6.4$ km/sec, $\alpha_2 = 8$ km/sec, $\rho_1/\rho_2 = 1$, $\gamma_1 = \gamma_2 = 0.577$.

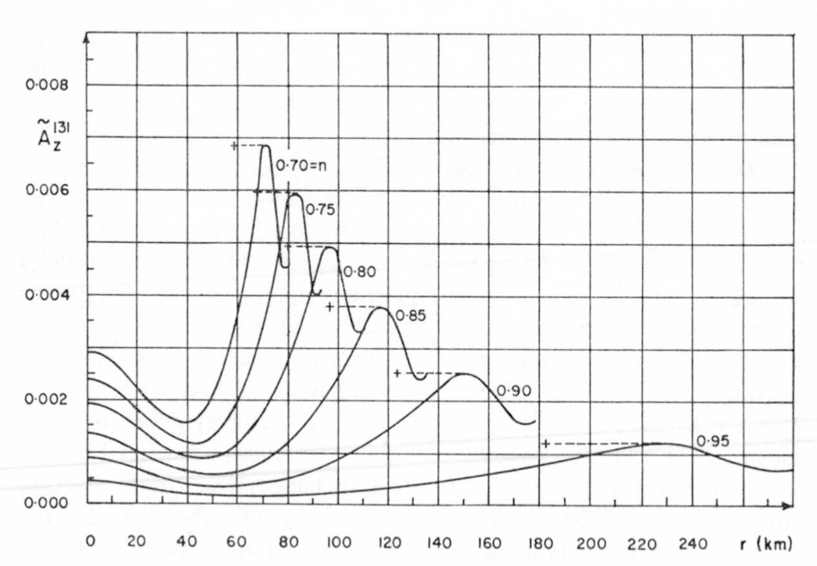

FIGURE 7.9 Amplitudes of the vertical displacement components $\tilde{\mathbf{A}}_z{}^{131}$ vs. distance, for different values of the refractive index n. $H = h = 30$ km, $\alpha_2 = 8$ km/sec, $\rho_1/\rho_2 = 1$, $\gamma_1 = \gamma_2 = 0.577$, $f = 6$ cps. The crosses indicate the corresponding critical distances.

0.7 to 0.95. Note that the velocity α_1 is determined by $\alpha_1 = n\alpha_2$. The crosses indicate the corresponding critical distances (where the ray amplitude curves have a maximum). We can see that the refractive index has a great influence on the amplitude curves of reflected waves. The nearer the refractive index is to one, the farther the maximum of the amplitude curve lies from the source. Moreover, with increasing refractive index the distance between the maximum and the corresponding critical point increases. The closer the refractive index is to one, the flatter and less pronounced is the curve. For smaller n the curve has a sharper peak.

(c) *Dependence of amplitude curves on depth of interface* We shall assume that $h = H$ and call H the depth of the interface. Figure 7.10 gives the amplitude curves of the reflected wave, $\tilde{A}_z^{131}$, for $n = 0.8, f = 6$ cps, $\alpha_1 = 6.4$ km/sec, $\alpha_2 = 8$ km/sec. The depth of the interface changes from

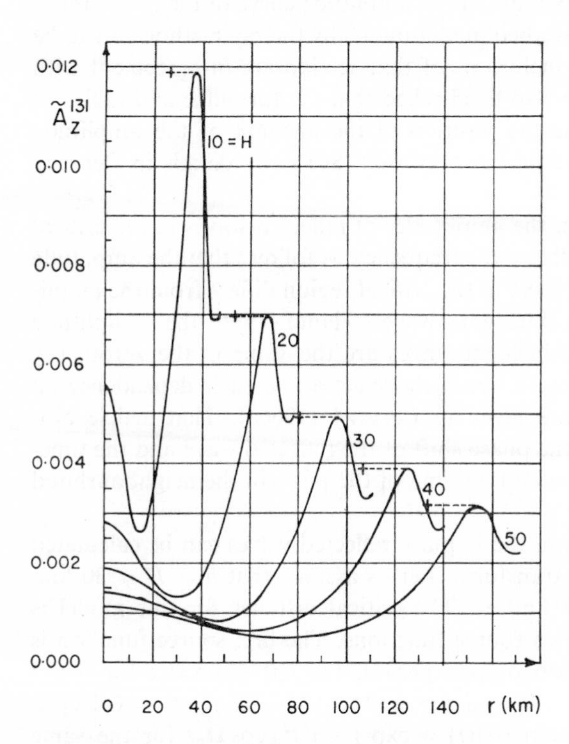

FIGURE 7.10 Amplitudes of the vertical displacement components $\tilde{A}_z^{131}$ vs. distance, for different values of H. $H = h$, $n = 0.8$, $\alpha_1 = 6.4$ km/sec, $\rho_1/\rho_2 = 1.0$, $\gamma_1 = \gamma_2 = 0.577$, $f = 6$ cps. The crosses indicate the corresponding critical distances.

10 km to 50 km. The greater the depth of the interface, the farther the maximum lies from the source and the greater the distance between the maximum of the amplitude curve and the corresponding critical point. For greater depths of the interface, the amplitude curves are less pronounced and flatter; at smaller depths the maximum is much sharper.

Whereas the parameters investigated above (such as refractive index, frequency, and the depth of the interface) have a great influence on the exact amplitude curves of reflected waves in the critical region, the other parameters, viz., ρ_1/ρ_2, γ_1, and γ_2, have only very little influence on them. The amplitude curves for different ρ_1/ρ_2, γ_1, and γ_2 differ mainly at small epicentral distances, but the position of the maximum is not appreciably influenced by changes in these parameters. Numerical examples are given in Červený (1966a).

The position of the maximum of the amplitude curve of the interference reflected-head wave was studied in section 3.8 by the ray method. It can be easily seen that all the conclusions of that section are in agreement with those given above. It can also be checked that the formulae and tables of r_{131}^{M} given there determine the position of the maxima of the amplitude curves given in the above figures very well. See more details in Červený (1962b).

As we have seen above, the amplitudes of reflected waves in the critical region depend substantially on the frequency. It follows that the amplitude spectrum of the reflected wave in the critical region differs from the amplitude spectrum of the incident wave. (Note that the amplitude spectra of reflected and incident waves are the same in the zeroth ray approximation.) More details about the spectra and their dependence on the epicentral distance are given in Červený (1966a). That article also contains a discussion of the phase shift of the reflected wave and the time-distance curves of maxima and minima of the pulse in the neighbourhood of the critical point.

If the spectra are known, the impulse reflected waves can be calculated using numerical Fourier transform. Let us assume that $h = H = 30$ km, $\alpha_1 = 6.4$ km/sec, $\alpha_2 = 8$ km/sec. The critical distance for this model is 80 km. Let us consider two source functions. The first source function is sinusoidal with a duration of one period, i.e. $f(t) = \sin \Omega_0 t$ for $0 \leqslant t \leqslant 2\pi/\Omega_0$, $f(t) = 0$ for $t < 0$, and $t > 2\pi/\Omega_0$ ($\Omega_0 = 2\pi f_0$, $f_0 = 6.4$ cps). The second source function is $f(t) = \exp(-\beta^2 t^2) \cos \Omega_0 t$ for the same Ω_0 as in the previous case, and $\beta/\Omega_0 = 1/4$. Note that the same numerical example for the first source function was calculated by the ray theory in section 3.8 (see figure 3.39). The theoretical seismograms for the first source function are given in figure 7.11 and for the second in figure 7.12. From these figures it is evident that the form of the pulse of reflected wave changes only very little on passing through the critical point ($r^* = 80$ km, in our case). The amplitudes increase monotonically there. It is not, therefore, possible to use the change in pulse shape to identify the position of the critical point. The other conclusions to be drawn from figures 7.11 and 7.12 are the same as in section 3.8, using ray theory.

From all that was said above, we see that the critical point is not as important for finite frequencies as for infinite frequencies. All the characteristics (amplitude, shape of pulse, etc.) of reflected waves change quite smoothly across the critical point, the largest changes occurring only some distance beyond it.

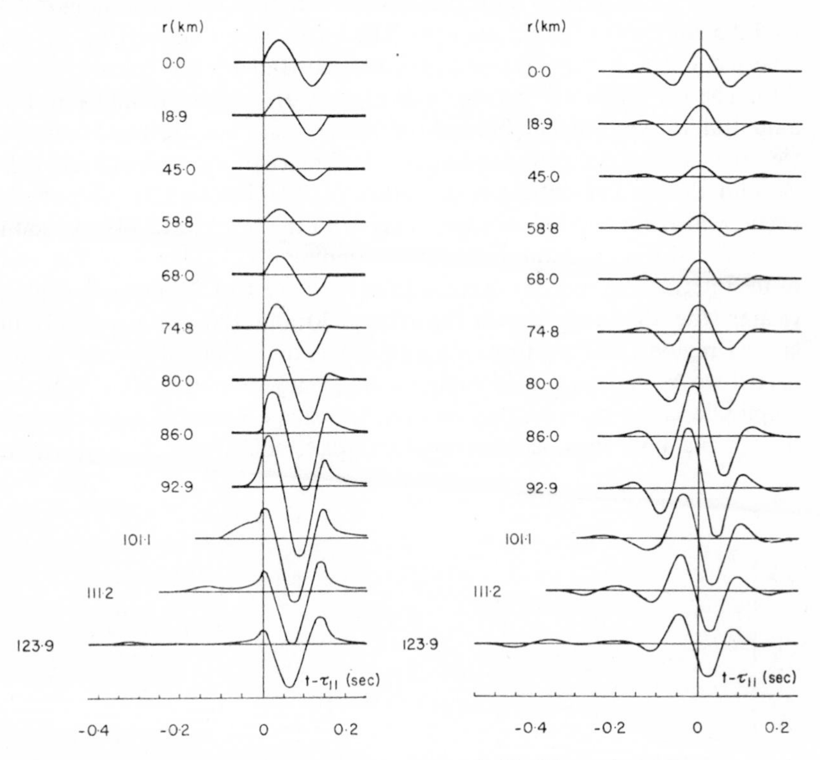

FIGURE 7.11 Theoretical seismograms at different epicentral distances for the source function $f(t) = \sin \Omega_0 t$ for $0 \leqslant t \leqslant 2\pi/\Omega_0$. $H = h = 30$ km, $\alpha_1 = 6.4$ km/sec, $\alpha_2 = 8$ km/sec, $\rho_1/\rho_2 = 1$, $\gamma_1 = \gamma_2 = 0.577$, $f_0 = 6.4$ cps.

FIGURE 7.12 Theoretical seismograms at different epicentral distances for the source function $f(t) = \exp(-\beta^2 t^2) \cos \Omega_0 t$. $\beta/\Omega_0 = 1/4$; other parameters are the same as in figure 7.11.

B Large Velocity Contrast

Let us assume that $n = 0.4$, $\gamma_1 = \gamma_2 = 1/\sqrt{3}$, $\rho_1/\rho_2 = 1$, $h = H = 1$ km. Then the reflected wave of the type 11 has two critical points, $r^*_{131} = 0.873$ km and $r^*_{141} = 1.921$ km. The head wave 131 exists for distances $r > r^*_{131}$ and the head wave 141 for $r > r^*_{141}$. In the zeroth ray approximation (for infinite frequency), the amplitude curve of the reflected wave has the characteristic shape shown in figure 3.31: a sharp maximum at the first critical point, a broad maximum beyond the second critical point, and a deep minimum in between. The situation changes considerably when we consider finite frequencies and interference of the reflected wave with head waves 131 and 141. Note that we must use the exact numerical integrations in this case (see section 7.2.6); the approximate formulae of sections 7.2.3 to 7.2.5 are not accurate enough. The exact amplitude curves of the reflected wave, $\tilde{A}_z$, for $H/\Lambda = 3$, 10, 30, 100, and ∞, are given in figure 7.13. The ray amplitude curve is indicated by $H/\Lambda = \infty$. In all cases, the amplitude curves oscillate beyond the first critical point for finite frequencies, the oscillations corresponding to interference of the waves 11 and 131. Beyond the second critical point, other oscillations with larger periods occur, which correspond to interference with the head wave 141. As soon as any head wave separates, the corresponding oscillations stop. The end of the interference zone is not considered in figure 7.13. To make the figure clearer, the amplitude curves for $H/\Lambda = 30$ and 100 are given only in critical regions. We see from the figure that the amplitude curves of reflected waves depend considerably on frequency. For very high H/Λ, the amplitude curve preserves, in general, the shape resulting from the ray theory, but both maxima shift towards greater distances. At lower H/Λ,

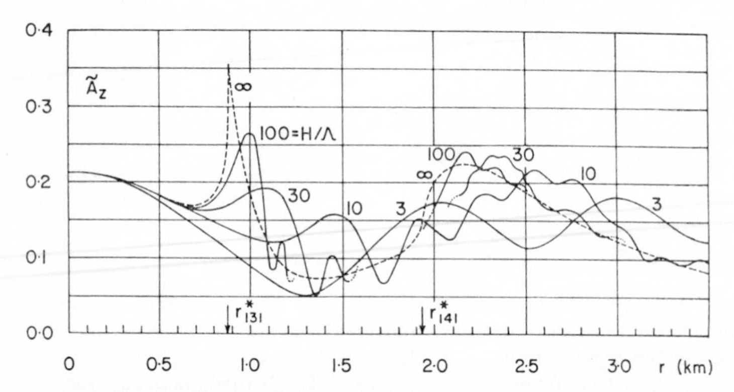

FIGURE 7.13 Amplitudes of the vertical displacement components of the interference reflected wave vs. distance, for different values of H/Λ. $h = H = 1$ km, $\rho_1/\rho_2 = 1$, $\gamma_1 = \gamma_2 = 0.577$, $n = 0.4$.

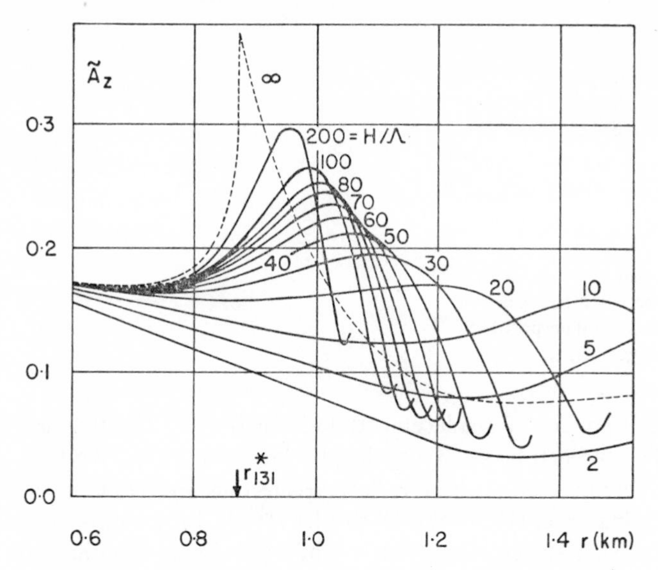

FIGURE 7.14 Details of figure 7.13 near the first critical point r^*_{131}.

the whole character of the amplitude curve changes. The maxima are shifted to considerable distances beyond the critical points, and the sharp peaks are smoothed. The amplitude curve has no peculiarities at the critical points. In some cases, the amplitude–distance curve has a maximum just where the ray amplitude curve has a minimum, and vice versa. In figure 7.14, the region of the first critical point is shown in detail.

7.3
LAYERED MEDIA

The formal solution of the problem of a point source in a horizontally layered elastic medium (with homogeneous layers) can be constructed in the same manner as for the one-interface problem (see section 7.2.1). The impulse response to a point source has the form

$$\mathscr{W}(r, z, t) = \frac{1}{2\pi}\int_{-\infty}^{\infty} \mathrm{e}^{i\omega t} W(r, z, \omega)\mathrm{d}\omega, \tag{7.105}$$

where

$$W(r, z, \omega) = \int_{0}^{\infty} kJ_n(kr)Q(\omega, k, z)\mathrm{d}k \tag{7.106}$$

(see Ewing *et al.*, 1957; Harkrider, 1964; Phinney, 1965). W is the horizon-

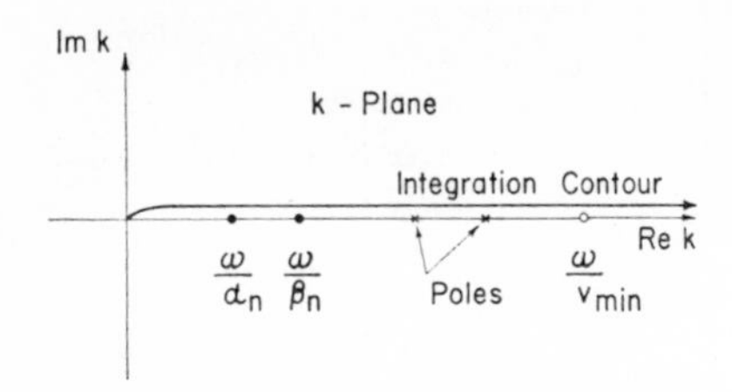

FIGURE 7.15 Position of the poles in the complex k-plane when the real parts of all radicals are taken positive on the sheet of integration.

tal or vertical component of displacement, and n is 0 or 1. The function $Q(\omega, k, z)$ can be obtained by satisfying the interface conditions. We shall not write explicit expressions for it here, however, as they are very cumbersome, but see Ewing *et al.* (1957) and Harkrider (1964), for example. For an n-layered elastic half-space (with a free boundary and $n-1$ interfaces at which the surrounding layers are in welded contact), we can write

$$Q(\omega, k, z) = \Delta_2(\omega, k, z)/\Delta(\omega, k), \tag{7.107}$$

where $\Delta(\omega, k)$ is a determinant of order $4n-2$ and $\Delta_2(\omega, k, z)$ is some combination of determinants of the same order. Direct evaluation of Δ_2 and Δ on a computer is not simple, but a suitable recursive algorithm is given in Harkrider (1964); see also Thomson (1950), Haskell (1953), Knopoff (1964), and Fuchs (1968b).

The formula (7.105) describes the whole wave field at the receiving point (i.e., the whole theoretical seismogram). It is not divided into individual signals, corresponding to R-waves, H-waves, etc. It is generally possible to expand W in a series of integrals each of which represents a ray (see the exact ray theory in the following), but for thin layers and/or a large number of interfaces this approach fails because it becomes necessary to consider a very large number of rays.

Note that the formal solution to the n-layer – point source problem can be written in many different forms. We can, for example, use the operational form (see Pekeris *et al.*, 1965, where many other references are given). In the following, we shall consider the solution in the form of the Fourier integral (7.105).

The inner integral (7.106) gives the frequency response to the point source. If we find $W(r, z, \omega)$ for a series of closely spaced frequencies ω, we can evaluate the impulse response $\mathscr{W}(r, z, t)$ by the standard methods of numerical Fourier analysis (Filon, 1928–9; Cooley and Tukey, 1965).

Therefore, we shall mainly deal with the frequency response $W(r, z, \omega)$.

The integrand of (7.106) contains radicals of the form

$$\nu_{\alpha_i} = [k^2 - (\omega/\alpha_i)^2]^{1/2}, \qquad \nu_{\beta_i} = [k^2 - (\omega/\beta_i)^2]^{1/2}, \tag{7.108}$$

$i = 1, 2, \ldots, n$, where α_i and β_i are compressional and shear velocities in the ith layer. It is shown in Ewing *et al.* (1957, p. 244) that the integrand is an even function of ν_{α_i} and ν_{β_i} for $i = 1, 2, \ldots, n-1$. From this follows the conclusion that for wave propagation from a point source in a half-space formed by parallel layers, the only branch points which have to be considered when we transform the integration contours are those of ω/α_n and ω/β_n. All other expected branch line integrals vanish.

Poles are the other singularities of the integrand of (7.106). The poles lie on the real k-axis if it is required that the real parts of all radicals be taken positive on the sheet of integration. They lie to the right of the branch points ω/α_n and ω/β_n and to the left of the limiting point $\omega/v_{\min}$, where $v_{\min}$ is the least velocity in any layer (see figure 7.15). We shall call these the *real poles.* The other poles can lie on a lower sheet of the Riemann surface (where Re $\nu_{\alpha_n} < 0$ and/or Re $\nu_{\beta_n} < 0$). They do not lie on the real axis, but in other parts of the complex plane, and will therefore be called *complex poles.* The position of the poles was investigated in detail by Gilbert (1964). The complex poles can lie quite close to the branch points.

Numerical evaluation of the integral (7.106) along the original contour from $k = 0$ to $k = \infty$ is very complicated. The integrand does not vary smoothly along the contour, but oscillates very rapidly (especially for large r) and passes over all the singularities. This problem of rapid oscillation can be avoided to some extent by using a suitable numerical algorithm for integration (see, e.g., Filon, 1928–9). The position of the poles along the real axis, however, makes the original contour of integration unsatisfactory.

The methods usually used to evaluate the above integrals are the exact ray expansion and normal mode expansion. If we are not interested in the whole theoretical seismogram, but only in the first arrivals, Phinney's method of λ-spectra is very convenient. We shall describe these methods very briefly in the following. Please note that what we are presenting here is only a very brief outline of these techniques and is not intended to be a full exposition; the reader is referred to the papers mentioned below for details.

7.3.1 *Exact Ray Expansion*

It is obvious that the method of exact ray theory cannot be classified as a wave method. However, we would like to describe it briefly here because of its importance in many problems dealing with head waves. The function $Q(\omega, k, z)$ can be expanded in an infinite series. In formal analogy with the theory of propagation of *plane* waves, we can identify each term of this series with the ray of some multiply reflected and refracted wave. The individual double integrals in the series for $\mathscr{W}(r, z, t)$ are considerably simpler than the original integral (7.105), and can be transformed into single integrals, suitable for numerical evaluation by, for example, Cagniard's method. Each of these integrals is characterized by some time τ_i so that it vanishes for $t < \tau_i$. τ_i is the same as found by ray methods (the travel time). The theoretical seismogram $\mathscr{W}(r, z, t)$ then can be found as a superposition of all the individual waves (integrals). For a given finite time t, there is only a finite number of waves for which $t > \tau_i$; therefore, $\mathscr{W}(r, z, t)$ is given as a superposition of a finite number of individual waves. The number of waves which have to be considered is, of course, higher for larger t. For more details see Pekeris (1948, 1955a, 1955b, 1960), Pekeris and Lifson (1957), Pekeris and Longman (1958), Pekeris *et al.* (1963, 1965), Bortfeld (1967), Müller (1967, 1968), Spencer (1965a), and Helmberger (1968). Many other references are given in Pekeris *et al.* (1965). This approach to the evaluation of $\mathscr{W}(r, z, t)$ is quite exact if the individual integrals are evaluated exactly. When the number of waves is large, the method becomes very cumbersome, especially when we are interested in the whole theoretical seismogram (7.105). For large t, the number of waves which have to be considered is very large even for very simple media (e.g. for $n = 2$). For small t, the situation becomes simpler, but even then the number of waves may be very large when the number of interfaces n is not small or if the thickness of layers is small. It would be very difficult to use this method when thin layers, of thickness of the same order of magnitude as, or less than, the wavelength, are considered, because the number of waves then becomes very high. The exact ray method can be effective if we are interested in the first arrivals in a thick-layered medium with only a few interfaces.

Note that expressions for the individual waves can also be found directly, without expansion of (7.105) (see, e.g., Pod"yapol'skiy, 1959a, 1966b; Brekhovskikh, 1960; Müller, 1967, 1968; Bortfeld, 1967).

As we have shown, the integrals for the individual multiple waves can be evaluated exactly. They can, however, more easily be evaluated approximately. If we use the method of steepest descent, the results correspond to those found by the ray theory (see sections 2.6.2 and 4.4). Various modifications of the saddle-point method enable us to find the approximate formulae for individual waves in regions where the ray approximation (or, the method of steepest descent) cannot be used, as in the critical regions (see, for example, Červený, 1968; Smirnova, 1968). The accuracy of ray expressions for individual waves was investigated by Pod"yapol'skiy (1966b), who also derived formulae for the second term in the ray series (from exact expressions) and found approximate expressions for the displacement of individual R-waves when the angle between the ray and the interface is small (Pod"yapol'skiy, 1959c).

7.3.2 *Normal Mode Expansion*

The original contour of integration of (7.106) can be deformed in the complex k-plane. $W(r, z, \omega)$ is then expressed in terms of the residue contributions (due to real poles of the integrand) and the branch line integral (integral along the branch line). The residues thus obtained are the *normal modes*. Thus, in this method, the wave field from a point source in a layered medium is studied as a whole (unlike in the exact ray expansion where the field is separated into individual waves, corresponding to specific rays). This solution automatically includes all multiple reflections. More details are given in Pekeris (1948) and Ewing *et al.* (1957, p. 134). For large epicentral distances, the relative contribution of the branch line is small and the solution is reasonably well described by the series of normal modes. This approach has been used mainly in the investigation of surface waves, where it has yielded very satisfactory results. When we are interested in the signals which arrive early (body waves), the relative contribution of the branch line integral becomes large. Numerical evaluation of the branch line integral is not easy. The real and the complex poles may lie near the branch line, which complicates the situation considerably, and the contour is usually deformed onto the lower Riemann sheets so that the complex poles must be considered. Details are given in Gilbert (1964), Rosenbaum (1960, 1961, 1964, 1965), and Phinney (1961a, 1961b, 1965). When the number of layers is small, the position of the complex poles can be evaluated and the method can lead to effective results. For example, using this method, Rosenbaum (1961, 1964, 1965) solved the problem of head waves propagating along a thin layer, which is of basic importance in seismic prospecting. Other references to the problem of head waves propagating along a thin layer and to screened head waves are given in Petrashen (1965), Spencer (1965b), Musgrave (1967), and in Krauklis (1968). We can conclude that this method is very effective for the investigation of surface waves, but becomes usually more complicated for the investigation of early arriving signals owing to complications in evaluation of the branch line integral.

A similar method can also be used for inhomogeneous media if the solution in the form (7.106) is known. Nakamura (1964) evaluated the branch line integral and found an expression for head waves from a linear transition layer in a liquid. However, the possible contribution of the poles lying near the branch line integrals was not considered by him. The problem of head waves from a transition layer in a liquid was also investigated by Tsepelev (1961, 1968) and from a transition layer in a solid by Hirasawa and Berry (1969). The effect on head waves of a small velocity gradient in the substratum was studied by Chekin (1964, 1965, 1966); see chapter 6 also.

7.3.3 *Direct Evaluation*

As we have shown, it is very difficult to evaluate the spectrum $W(r, z, \omega)$ numerically directly along the integration contour along the real k-axis, as this contour passes over all the real poles. In order to perform the numerical integration, we must smooth the integrand. For example, we can, by suitable techniques, move the singularity of the integrand away from the contour. By moving the singularity, however, the problem itself changes, as the analytic function is determined by its singularities. We shall describe briefly one possibility of moving the singularities which has a clear physical meaning and consequences. (For more details, see Phinney (1965).)

We introduce a complex-valued frequency ω^* by

$$\omega^* = \omega - i\lambda, \tag{7.109}$$

where λ is a real and positive constant, and substitute ω^* for ω in (7.106). The singularities of the function $Q(\omega^*, k, z)$ will be along a straight line in the fourth quadrant which passes through the origin (see figure 7.16). The higher the value of λ, the farther the singularities lie from the real k-axis. No singularities lie on the real k-axis in this case and it is consider-

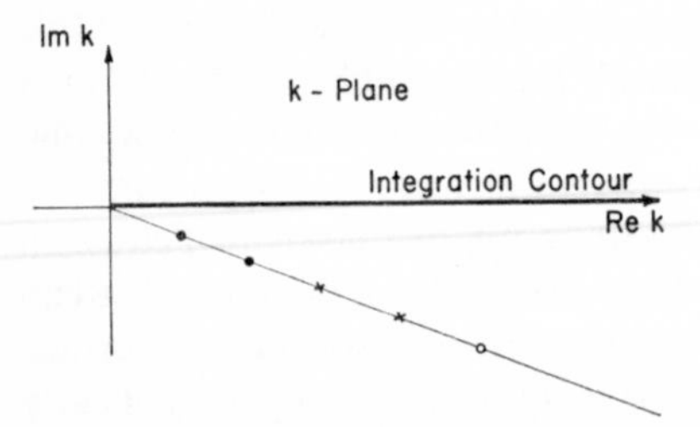

FIGURE 7.16 The singularities of $Q(\omega^*, k, z)$ lie in a straight line in the fourth quadrant of the k-plane.

ably easier to perform the integration (7.106) along the real k-axis. As a result we find the spectrum $W(r, z, \omega^*) = W(r, z, \omega - i\lambda)$. Let us define

$$W_\lambda(r, z, \omega) = W(r, z, \omega - i\lambda), \tag{7.110}$$

and call $W_\lambda(r, z, \omega)$ the λ-spectrum. It is clear that the spectrum $W(r, z, \omega)$ is different from the λ-spectrum $W_\lambda(r, z, \omega)$ when $\lambda \neq 0$. W_λ is considerably smoother and simpler than W, especially for large λ. Now we find the relation between the λ-spectrum W_λ and the impulse response $\mathscr{W}(r, z, t)$. The relation between the frequency response $W(r, z, \omega)$ and impulse response $\mathscr{W}(r, z, t)$ is given by the Fourier integral (see 7.105)

$$W(r, z, \omega) = \int_{-\infty}^{+\infty} e^{-i\omega t}\mathscr{W}(r, z, t)\mathrm{d}t. \tag{7.111}$$

From (7.110) and (7.111) it follows that

$$\begin{aligned} W_\lambda(r, z, \omega) &= \int_{-\infty}^{+\infty} \exp[-i(\omega - i\lambda)t]\mathscr{W}(r, z, t)\mathrm{d}t \\ &= \int_{-\infty}^{+\infty} e^{-i\omega t}\mathscr{W}_\lambda(r, z, t)\mathrm{d}t, \end{aligned} \tag{7.112}$$

where

$$\mathscr{W}_\lambda(r, z, t) = e^{-\lambda t}\mathscr{W}(r, z, t). \tag{7.113}$$

Thus, the λ-spectrum is the spectrum of the function $\mathscr{W}_\lambda(r, z, t) = e^{-\lambda t} \times \mathscr{W}(r, z, t)$. $\mathscr{W}_\lambda(r, z, t)$ is the theoretical seismogram $\mathscr{W}(r, z, t)$ as viewed through the exponentially decaying time window. The exponentially decaying time window emphasizes the early arriving phases in comparison with later arrivals (including surface waves, of course). We are in a position to calculate a spectrum for the first arrival by choosing λ according to the length of the interval we wish to discuss. For large λ, this interval is very short and only the first arriving phase is obtained (other phases are suppressed). The λ-spectrum is very smooth in this case and its numerical evaluation is considerably easier than the evaluation of $W(r, z, \omega)$.

For details of the method of λ-spectra, see Phinney (1965), where some numerical examples are also given. The method seems to be very suitable for the investigation of compressional head waves which appear as first arrivals. For example, it enables us to investigate head waves from transition layers when the transition layer is simulated by a series of homogeneous layers. A computer program for the λ-spectra of head waves from transition layers, based on Phinney's method, has been prepared by

C. Keen (personal communication), who will soon publish her results. Note that the ray theory does not allow us to investigate head waves from transition layers; the interference character of these head waves differs considerably from the simple character of head waves from a sharp interface. Head waves from transition layers simulated by a series of homogeneous layers were also investigated by different methods by Bessonova and Michota (1968) and by Fuchs (1968a). The latter reference has many numerical calculations of practical interest.

APPENDIX: Some useful vector identities

The following vector identities can be found in, or derived from the formulae in, any standard work on vector algebra, such as Schwartz *et al.* (1960).

NOTATION

$$\nabla = \mathbf{i}\frac{\partial}{\partial x} + \mathbf{j}\frac{\partial}{\partial y} + \mathbf{k}\frac{\partial}{\partial z},$$

where **i**, **j**, **k** are unit vectors along the Cartesian coordinates x, y, and z.

$$\nabla a = \text{grad } a, \quad \nabla\cdot\mathbf{a} = \text{div } \mathbf{a}, \quad \nabla\times\mathbf{a} = \text{curl } \mathbf{a}.$$

IDENTITIES

$$\mathbf{a}\times(\mathbf{b}\times\mathbf{c}) = \mathbf{b}(\mathbf{a}\cdot\mathbf{c})-\mathbf{c}(\mathbf{a}\cdot\mathbf{b})$$

$$(\mathbf{a}\times\mathbf{b})\cdot(\mathbf{c}\times\mathbf{d}) = (\mathbf{a}\cdot\mathbf{c})(\mathbf{b}\cdot\mathbf{d})-(\mathbf{b}\cdot\mathbf{c})(\mathbf{a}\cdot\mathbf{d}) \quad \text{(Laplace identity)}$$

$$\nabla\cdot a\mathbf{b} = \mathbf{b}\cdot\nabla a+a(\nabla\cdot\mathbf{b})$$

$$\nabla\times a\mathbf{b} = \nabla a\times\mathbf{b}+a(\nabla\times\mathbf{b})$$

$$\nabla\cdot(\mathbf{a}\times\mathbf{b}) = \mathbf{b}\cdot(\nabla\times\mathbf{a})-\mathbf{a}\cdot(\nabla\times\mathbf{b})$$

$$\nabla\times(\mathbf{a}\times\mathbf{b}) = (\mathbf{b}\cdot\nabla)\mathbf{a}-\mathbf{b}(\nabla\cdot\mathbf{a})+\mathbf{a}(\nabla\cdot\mathbf{b})-(\mathbf{a}\cdot\nabla)\mathbf{b}$$

$$\nabla(\mathbf{a}\cdot\mathbf{b}) = (\mathbf{a}\cdot\nabla)\mathbf{b}+\mathbf{a}\times(\nabla\times\mathbf{b})+(\mathbf{b}\cdot\nabla)\mathbf{a}+\mathbf{b}\times(\nabla\times\mathbf{a})$$

$$\nabla\cdot(\nabla\times\mathbf{a}) = 0$$

$$\nabla\times\nabla a = 0$$

$$\nabla^2\mathbf{a} = \nabla(\nabla\cdot\mathbf{a})-\nabla\times(\nabla\times\mathbf{a})$$

$$\nabla(\mathbf{b}\cdot\nabla a) = (\mathbf{b}\cdot\nabla)\nabla a+(\nabla a\cdot\nabla)\mathbf{b}+\nabla a\times(\nabla\times\mathbf{b})$$

$$(\nabla a\cdot\nabla)\nabla a = \tfrac{1}{2}\nabla(\nabla a)^2$$

$$\nabla(\nabla\cdot a\mathbf{b}) = \nabla(\mathbf{b}\cdot\nabla a)+(\nabla\cdot\mathbf{b})\nabla a+a\nabla(\nabla\cdot\mathbf{b})$$

$$\nabla(\nabla\cdot a\mathbf{b}) = (\mathbf{b}\cdot\nabla)\nabla a+(\nabla a\cdot\nabla)\mathbf{b}+\nabla a\times(\nabla\times\mathbf{b})+(\nabla\cdot\mathbf{b})\nabla a+a\nabla(\nabla\cdot\mathbf{b})$$

$$\nabla\times(\nabla\times a\mathbf{b}) = a(\nabla\times(\nabla\times\mathbf{b}))+\nabla a\times(\nabla\times\mathbf{b})+(\mathbf{b}\cdot\nabla)\nabla a -(\nabla a\cdot\nabla)\mathbf{b}+\nabla a(\nabla\cdot\mathbf{b})-\mathbf{b}(\nabla\cdot\nabla a)$$

$$\nabla^2 a\mathbf{b} = a\nabla^2\mathbf{b}+2(\nabla a\cdot\nabla)\mathbf{b}+\mathbf{b}(\nabla\cdot\nabla a)$$

$(\mathbf{t}\cdot\nabla)\mathbf{a} = \dfrac{\mathrm{d}\mathbf{a}}{\mathrm{d}s}$ (directional derivative of **a** in the direction of a unit vector **t**)

REFERENCES

Alekseyev, A. S., V. M. Babich, and B. Y. Gel'chinskiy. 1961. Ray method for the computation of the intensity of wave fronts. In G. I. Petrashen (ed.), *Problems in the Dynamic Theory of Propagation of Seismic Waves* (Leningrad University Press, Leningrad), 5: 3–24 [in Russian]

Alekseyev, A. C., and B. Y. Gel'chinskiy. 1958. Determination of the intensity of head waves in the theory of elasticity by the ray method. *Doklady Akad. Nauk SSSR*, 118: 661–4 [in Russian]

— 1959. On the ray method of computation of wave fields for inhomogeneous media with curved interfaces. In G. I. Petrashen (ed.), *Problems in the Dynamic Theory of Propagation of Seismic Waves* (Leningrad University Press, Leningrad), 3: 107–60 [in Russian]

— 1961. The ray method of computation of the intensity of head waves. In G. I. Petrashen (ed.), *Problems in the Dynamic Theory of Propagation of Seismic Waves* (Leningrad University Press, Leningrad), 5: 54–72 [in Russian]

Ansel, E. A. 1930. Das Impulsfeld der praktischen Seismik in graphischer Behandlung. *Beitrage zur angewandten Geophysik* (Gerlands), 1: 117–36 [in German]

Babich, V. M. 1961. On the convergence of series in the ray method of calculation of the intensity of wave fronts. In G. I. Petrashen (ed.), *Problems in the Dynamic Theory of Propagation of Seismic Waves* (Leningrad University Press, Leningrad), 5: 25–35 [in Russian]

Babich, V. M., and A. S. Alekseyev. 1958. A ray method of computing wave front intensities. *Bull.* (Izv.) *Acad. Sci. USSR, Geophys. Ser.*, no. 1: 9–15

Berry, M. J., and G. F. West. 1966. Reflected and head wave amplitudes in a medium of several layers. In J. S. Steinhart and T. J. Smith (eds.), *The Earth beneath the Continents* (Geophysical Monograph 10, American Geophysical Union, Washington, D.C.), 464–81

Berzon, I. S., L. I. Ratnikova, and M. I. Rac-Chizgiya. 1966. *Seismic Transformed Reflected Waves* (Nauka, Moscow) [in Russian]

Berzon, I. S., A. M. Yepinat'yeva, G. N. Pariyskaya, and S. P. Starodubrovskaya. 1962. *Dynamic Characteristics of Seismic Waves in Real Media* (Izd. Akad. Nauk SSSR [Acad. Sci. USSR Press], Moscow) [in Russian]

Bessonova, E. N., and G. G. Michota. 1968. On interference head waves. In V. I. Keilis-Borok (ed.), *Computational Seismology* (Nauka, Moscow), 4: 263–74 [in Russian]

Bortfeld, R. 1961. Approximations to the reflection and transmission coefficients of plane longitudinal and transverse waves. *Geophys. Prospect.* 9: 485–502

— 1962a. Exact solution of the reflection and refraction of arbitrary spherical compressional waves at liquid-liquid interfaces and at solid-solid interfaces with equal shear velocities and equal densities. *Geophys. Prospect.* 10: 35–67

— 1962b. Reflection and refraction of spherical compressional waves at arbitrary plane interfaces, *Geophys. Prospect.* 10: 517–38

— 1964. Examples of synthetic refraction arrivals and wide-angle reflections. *Geophys. Prospect.* 12: 100–14

— 1967. Elastic waves in layered media. *Geophys. Prospect.* 15: 644–50

Brekhovskikh, L. M. 1948. Reflection of spherical waves at a plane interface between two media. *Zh. Tekhn. Fiz.* 18: 455 [in Russian]

— 1949. Reflection and refraction of spherical waves. *Uspekhi Fiz. Nauk*, 38: 1–42 [in Russian]

— 1960. *Waves in Layered Media* (Academic Press, New York) [Trans. from Russian (Akad. Nauk SSSR, Moscow, 1957) by D. Lieberman]

Buldyrev, V. S., and A. I. Lanin. 1965. Interfering waves at the surface of an elastic inhomogeneous sphere. *Revs. Geophys.* 3: 49–54

— 1966. On the investigation of the interference wave field at the surface of an elastic sphere. In *Numerical Methods for Solution of Problems of Mathematical Physics* (Nauka, Moscow), 131–43 [in Russian]

Bullen, K. E. 1965. *An Introduction to the Theory of Seismology* (Cambridge University Press, Cambridge)

Cagniard, L. 1962. *Reflection and Refraction of Progressive Seismic Waves* (McGraw-Hill, New York) [Trans. from French (Gauthier-Villars, Paris, 1939) by E. A. Flinn and C. H. Dix]

Červený, V. 1957a. The reflection of spherical elastic waves at a plane boundary (Travaux Inst. Géophys. Acad. Tchécosl. Sci., no. 44). *Geofysikální Sborník* (Čs. Akad. Věd), 4: 343–66

— 1957b. A study of the amplitudes of head waves. *Studia Geophys. et Geod.* 1: 256–84 [in Russian]

— 1959. On the reflection of spherical waves at a plane interface with refractive index near to one, Part 1. *Studia Geophys. et Geod.* 3: 116–34

— 1961. The amplitude curves of reflected harmonic waves around the critical point. *Studia Geophys. et Geod.* 5: 319–51

— 1962a. On the length of the interference zone of a reflected and head wave beyond the critical point and on the amplitudes of head waves. *Studia Geophys. et Geod.* 6: 49–64

— 1962b. On the position of the maximum of the amplitude curves of reflected waves. *Studia Geophys. et Geod.* 6: 215–34

— 1963a. Simplified relations for amplitudes of spherical compressional harmonic waves reflected from plane interface. *Studia Geophys. et Geod.* 7: 337–52

— 1963b. Determination of the position of the critical point from the amplitude curves of reflected waves. *Studia Geophys. et Geod.* 7: 286–7 [in Russian]

— 1966a. The dynamic properties of reflected and head waves around the critical point (Travaux Inst. Géophys. Acad. Tchécosl. Sci., no. 221). *Geofysikální Sborník* (Čs. Akad. Věd), 13: 135–245

— 1966b. On dynamic properties of reflected and head waves in the *n*-layered earth's crust. *Geophys. J.* 11: 139–47

— 1967a. On some kinematic and dynamic properties of reflected and head waves in the case of a layered overburden (Travaux Inst. Géophys. Acad. Tchécosl. Sci., no. 244). *Geofysikální Sborník* (Čs. Akad. Věd), 14: 105–79

— 1967b. The amplitude-distance curves for waves reflected at a plane interface for different frequency ranges. *Geophys. J.* 13: 187–96

— 1968. The theory of reflected and head waves in the case of a layered overburden (Travaux Inst. Géophys. Acad. Tchécosl. Sci., no. 269). *Geofysikální Sborník* (Čs. Akad. Věd), 15: 133–80

Červený, V., F. Hron, and B. Novák. 1964. Reflection coefficients of plane waves of PP-type for weak interfaces (Travaux Inst. Géophys. Acad. Tchécosl. Sci., no. 181). *Geofysikální Sborník* (Čs. Akad. Věd), 12: 79–142

Červený, V., and J. Janský. 1966. Über einige dynamische Eigenschaften der Tauchwelle. *Proc. European Seismological Commission Meeting* (Akademisk Forlag, Copenhagen), 397–402 [in German]

Červený, V., and A. M. Yepinat'yeva. 1967. The influence of the structure of the overburden on the amplitudes of reflected and head waves. *Acta Univ. Carolinae, Math. et Phys.*, no. 1: 55–88 [in Russian]

— 1968. Amplitudes of reflected and head waves in the presence of a stratified overlying rock mass. *Bull.* (Izv.) *Acad. Sci. USSR, Earth Phys. Ser.* 3: 153–60

Chekin, B. S. 1959. Reflection and refraction of seismic waves at a weak interface. *Bull.* (Izv.) *Aacd. Sci. USSR, Geophys. Ser.*, no. 1: 9–13

— 1964. On the reflection of elastic, spherical waves from an inhomogeneous half-space. *Bull.* (Izv.) *Acad. Sci. USSR, Geophys. Ser.*, no. 5: 426–9

— 1965. The effect on a head wave of small inhomogeneities in a refracting medium. *Bull.* (Izv.) *Acad. Sci. USSR, Earth Phys. Ser.*, no. 3: 143–7

— 1966. The effect on head waves of a change of the parameters of an elastic medium with depth. *Bull.* (Izv.) *Acad. Sci. USSR, Earth Phys. Ser.*, no. 10: 623–6

Cooley, J. W., and J. W. Tukey. 1965. An algorithm for machine calculation of complex Fourier series. *Math. Comput.* 19: 297–301

Costain, J. K., K. L. Cook, and S. T. Algermissen. 1963. Amplitude, energy, and phase angles of plane SV waves and their application to earth crustal studies. *Bull. Seismol. Soc. Amer.* 53: 1039–74; and corrigendum in vol. 55 (1965), 567–75

Courant, R., and D. Hilbert. 1962. *Methods of Mathematical Physics*, vol. 2: *Partial Differential Equations* (Interscience, New York)

Dix, C. H. 1939. Refraction and reflection of seismic waves, II: Discussion of the physics of refraction prospecting. *Geophysics*, 4: 238–41

— 1963. Cagniard's method and associated numerical techniques. *J. Geophys. Res.* 68: 1184–5

Donato, R. J. 1963a. The S head wave at a liquid-solid boundary. *Geophys. J.* 8: 17–25

— 1963b. Amplitude of the head wave near the critical angle. *Geophys. J.* 8: 203–16

— 1964. Amplitude of P-head waves. *J. Acoust. Soc. Amer.* 36: 19–25

Ewing, W. M., W. S. Jardetzky, and F. Press. 1957. *Elastic Waves in Layered Media* (McGraw-Hill, New York)

Filon, L. N. G. 1928–9. On a quadrature formula for trigonometric integrals. *Proc. Roy. Soc. Edinburgh*, 49: 38–47

Fock, V. A. 1950. Generalization of the reflection formulas to the case of reflection of an arbitrary wave from a surface of arbitrary form. *J. Exptl. Theoret. Phys.* (ZETF), *USSR*, 20: 961–78 [translation in N. A. Logan, *Diffraction, Refraction and Reflection of Radio Waves* (Air Force Cambridge Research Centre, Bedford, Mass., Rept. no. TN-57-102, 1957), 283–317]

Friedlander, F. G. 1946. Simple progressive solutions of the wave equation. *Proc. Cambridge Phil Soc.* 43: 360–73

— 1958. *Sound Pulses* (Cambridge University Press, London)

Friedrichs, K. O., and J. B. Keller. 1955. Geometrical acoustics, II: Diffraction, reflection and refraction of a weak spherical or cylindrical shock at a plane interface. *J. Appl. Phys.* 26: 961–6

Fuchs, K. 1968a. The reflection of spherical waves from transition zones with arbitrary depth-dependent elastic moduli and density. *J. Phys. Earth*, 16: 27–41

— 1968b. Das Reflexions- und Transmissionsvermögen eines geschichten Mediums mit beliebiger Tiefen-Verteilung der elastischen Moduln und der Dichte für schrägen Einfall ebener Wellen. *Zeitschr. Geophys.* 34: 389–413 [in German]

Gel'chinskiy, B. Y. 1961. An expression for the spreading function. In G. I. Petrashen (ed.), *Problems in the Dynamic Theory of Propagation of Seismic Waves* (Leningrad University Press, Leningrad), 5: 47–53 [in Russian]

Gilbert, F. 1964. Propagation of transient leaking modes in a stratified elastic waveguide. *Revs. Geophys.* 2: 123–53

Grant, F. S., and G. F. West. 1965. *Interpretation Theory in Applied Geophysics* (McGraw-Hill, New York)

Gupta, Ravindra N.* 1965. Reflection of plane waves from a linear transition layer in liquid media. *Geophysics*, 30: 122–32

— 1966a. Reflection of sound waves from transition layers. *J. Acoust. Soc. Amer.* 39: 255–60

— 1966b. Reflection of elastic waves from a linear transition layer. *Bull. Seismol. Soc. Amer.* 56: 511–26

— 1966c. Reflection of plane elastic waves from transition layers with arbitrary variation of velocity and density. *Bull. Seismol. Soc. Amer.* 56: 633–42; and corrigendum in vol. 57 (1967), 302

Gutenberg, B. 1944. Energy ratio of reflected and refracted seismic waves. *Bull. Seismol. Soc. Amer.* 34: 85–102

Harkrider, D. G. 1964. Surface waves in multilayered elastic media, I: Rayleigh and Love waves from buried sources in a multilayered elastic halfspace. *Bull. Seismol. Soc. Amer.* 54: 627–79

Haskell, N. A. 1953. The dispersion of surface waves in multilayered media. *Bull. Seismol. Soc. Amer.* 43: 17–34

Heelan, P. 1953. On the theory of head waves. *Geophysics*, 18: 871–93

Helmberger, D. V. 1968. The crust-mantle transition in the Bering Sea. *Bull. Seismol. Soc. Amer.* 58: 179–214

Hirasawa, T., and M. J. Berry. 1969. Reflected and head waves from a linear transition layer. *Trans. Amer. Geophys. Union*, 50: 251

Honda, H., and K. Nakamura. 1953. On the reflection and refraction of the explosive sounds at the ocean bottom, I. *Sci. Repts. Tohoku Univ., Ser. 5*, 4: 125–33

— 1954. On the reflection and refraction of the explosive sounds at the ocean bottom, II. *Sci. Repts. Tohoku Univ., Ser. 5*, 6: 70–84

Hook, J. F. 1961. Separation of the vector wave equation of elasticity for certain types of inhomogeneous isotropic media. *J. Acoust. Soc. Amer.* 33: 302–13

— 1965. Determination of inhomogeneous media for which the vector wave equation of elasticity is separable. *Bull. Seismol. Soc. Amer.* 55: 975–87

Hron, F. 1962. On the reflection of impulse spherical waves at a plane interface (Travaux Inst. Géophys. Acad. Tchécosl. Sci., no. 151). *Geofysikální Sborník* (Čs. Akad. Věd), 9: 95–121 [in Russian]

*Known as Ravi Ravindra since August 1966. See also under that name.

Jardetzky, W. S. 1952. On the general solution of the wave equation (Part I), On waves generated at an interface (Part II). *Technical Report no. 22, Lamont Geological Observatory* (Palisades, New York)

Jeffreys, H. 1926. On compressional waves in two superposed layers. *Proc. Cambridge Phil. Soc.* 23: 472–81

— 1931. The formation of Love waves in two-layer crust. *Gerlands Beitr. Geophys.* 30: 336–50

Jeffreys, H., and B. S. Jeffreys. 1956. *Methods of Mathematical Physics* (Cambridge University Press)

Joos, G., and J. Teltow. 1939. Zur Deutung der Knallwellenausbreitung an der Trennschicht zweier Medien. *Phys. Zeitschr.* 40: 289–93 [in German]

Karal, F. C., and J. B. Keller. 1959. Elastic wave propagation in homogeneous and inhomogenous media. *J. Acoust. Soc. Amer.* 31: 694–705

Kaufman, H. 1953. Velocity functions in seismic prospecting. *Geophysics*, 18: 289–97

Keilis-Borok, V. I. 1960. *Investigation of the Mechanism of Earthquakes* (Soviet Research in Geophysics, vol. 4; English translation published by American Geophysical Union)

Keller, J. B. 1958. A geometrical theory of diffraction. In L. M. Graves (ed.), *Calculus of Variations and Its Application* (McGraw-Hill, New York), 27–52

— 1963. Geometrical methods and asymptotic expansions in wave propagation. *J. Geophys. Res.* 68: 1182–3

Keller, J. B., R. M. Lewis, and B. D. Seckler. 1956. Asymptotic solution of some diffraction problems. *Commun. Pure Appl. Math.* 9: 207–65

Kireyeva, I. E., and K. A. Karpov. 1959. *Tables of Weber Functions*, vol. 1 (Num. Centrum Akad. Sci. SSSR, Moscow) [in Russian]

Kline, M. 1951. An asymptotic solution of Maxwell's equation. *Commun. Pure Appl. Math.* 4: 225–62

— 1954. Asymptotic solution of linear hyperbolic partial differential equations. *J. Ratl. Mech. Anal.* 3: 315–42

Kline, M. and I. W. Kay. 1965. *Electromagnetic Theory and Geometrical Optics* (Interscience Publishers, New York)

Knopoff, L. 1964. A matrix method for elastic wave problems. *Bull. Seismol. Soc. Amer.* 54: 431–8

Knott, C. G. 1899. Reflection and refraction of seismic waves with seismological applications. *Phil. Mag.* 48: 64–97

Koefoed, O. 1962. Reflection and transmission coefficients for plane longitudinal incident waves. *Geophys. Prospect.* 10: 304–51

Kosminskaya, I. P. 1946. On initial points of time-distance curves of Mintrop waves. *Izv. Akad. Nauk SSSR, Ser. Geograf. i Geofiz.*, no. 5 [in Russian]

Krauklis, P. V. 1968. Head waves in a medium with a high-velocity layer. *Trudy Mat. Inst. Akad. Nauk SSSR*, 95: 98–105 [in Russian]

Lamb, M. 1904. On the propagation of tremors over the surface of an elastic solid. *Phil. Trans. Roy. Soc. London A*, 203: 1–42

Lanin, A. I. 1966. Investigation of interference wave fields in problems of diffraction by a transparent cylinder and an elastic sphere. PH.D. thesis, Leningrad University [in Russian]

— 1968. Calculation of interference waves in problems of diffraction by a cylinder and a sphere. In V. M. Babich (ed.), *Mathematical Problems in the Theory of Propagation of Waves* (Nauka [Leningrad Division], Leningrad), 64–104 [in Russian]

Luneburg, R. K. 1964. *Mathematical Theory of Optics* (University of California Press, Berkeley and Los Angeles)
Malinovskaya, L. N. 1957a. On the methodology of calculating the dynamic characteristics of seismic waves. *Bull.* (Izv.) *Acad. Sci. USSR, Geophys. Ser.*, no. 4: 10–25
— 1957b. On the dynamic characteristics of longitudinal reflected waves beyond the critical angle. *Bull.* (Izv.) *Acad. Sci. USSR, Geophys. Ser.* 5: 22–29
Marcinkovskaya, N. G., and V. G. Krasavin. 1968. Algorithm for the calculation of the field of reflected waves in media with curvilinear interfaces. In G. I. Petrashen (ed.), *Problems in the Dynamic Theory of Propagation of Seismic Waves* (Nauka [Leningrad Division], Leningrad), 9: 135–44 [in Russian]
McCamy, K., R. P. Meyer, and T. J. Smith. 1962. Generally applicable solutions of Zoeppritz' amplitude equations. *Bull. Seismol. Soc. Amer.* 52: 923–55
McConnell, R. K., Ravindra N. Gupta, and J. T. Wilson. 1966. Compilation of deep crustal seismic refraction profiles. *Revs. Geophys.* 4: 41–100
Menzel, D. H. 1955. *Fundamental Formulas of Physics* (Prentice-Hall, New York)
Mohorovičić, A. 1910. Das Beben vom 8 Okt. 1909. *Jahrb. Meteorol. Obs. Zagreb* (*Agram*), *1909*, 9: IV [in German]
Morse, P. M., and H. Feshbach. 1953. *Methods of Theoretical Physics* (McGraw-Hill, New York)
Müller, G. 1967. Theoretische Seismogramme für Punktwellen in geschichteten Medien. PH.D. thesis, Technische Hochschule Clausthal [in German]
— 1968. Theoretical seismograms for some types of point-sources in layered media I, II. *Zeitschr. Geophys.* 34: 15–35, 147–62
Musgrave, A. W. (ed.). 1967. *Seismic Refraction Prospecting* (The Society of Exploration Geophysicists, Tulsa)
Muskat, M. 1933. The theory of refraction shooting. *Physics*, 4: 14–28
Muskat, M., and M. W. Meres. 1940. Reflection and transmission coefficients for plane waves in elastic media. *Geophysics*, 5: 115–148
Nafe, J. E., 1957. Reflection and transmission coefficients at a solid-solid interface of high velocity contrast. *Bull. Seismol. Soc. Amer.* 47: 205–20
Nakamura, Y. 1964. Head waves from a linear transition layer in a liquid. *J. Geophys. Res.* 69: 4349–54
O'Brien, P. N. S. 1957a. The variation with distance of the amplitude of critically refracted waves. *Geophys. Prospect.* 5: 300–16
— 1957b. Multiply reflected refractions in a shallow layer. *Geophys. Prospect.* 5: 371–80
— 1963. A note on the reflection of seismic pulses with application to second event refraction shooting. *Geophys. Prospect.* 11: 59–72
— 1967. The use of amplitudes in seismic refraction survey. In A. W. Musgrave (ed.), *Seismic Refraction Prospecting* (The Society of Exploration Geophysicists, Tulsa), 85–118
Ott, H. 1942. Reflexion und Brechung von Kugelwellen. Effekte 2. Ordnung. *Ann. Physik*, 41: 443–66 [in German]
Pekeris, C. L. 1948. Theory of propagation of explosive sound in shallow water. *Geol. Soc. Amer. Mem.* 27
— 1955a. The seismic surface pulse. *Proc. Natl. Acad. Sci. U.S.* 41: 469–80
— 1955b. The seismic buried pulse. *Proc. Natl. Acad. Sci. U.S.* 41: 629–39
— 1960. Propagation of seismic pulses in layered liquids and solids. In N. Davids

(ed.), *Symposium on Stress Wave Propagation in Materials* (Interscience Publishers, New York), 45–57

Pekeris, C. L., Z. Alterman, and F. Ambramovici. 1963. Propagation of an SH-torque pulse in a layered solid. *Bull. Seismol. Soc. Amer.* 53: 39–57

Pekeris, C. L., Z. Alterman, F. Ambramovici, and H. Jarosch. 1965. Propagation of a compressional pulse in a layered solid. *Revs. Geophys.* 3: 25–47

Pekeris, C. L., and H. Lifson. 1957. Motion of the surface of a uniform elastic halfspace produced by a buried pulse. *J. Acoust. Soc. Amer.* 29: 1233–8

Pekeris, C. L., and I. M. Longman. 1958. Ray-theory solution of the problem of propagation of explosive sound in a layered liquid. *J. Acoust. Soc. Amer.* 30: 323–8

Petrashen, G. I. 1957a. Certain interference phenomena in a two-layer medium. *Bull.* (Izv.) *Acad. Sci. USSR, Geophys. Ser.*, no. 10: 1219–31

— (ed.) 1957b. *Materials for the Quantitative Investigation of the Dynamics of Seismic Waves*, vols. I, II, III (Leningrad University Press, Leningrad) [in Russian]

— 1959. Elements of the dynamic theory of propagation of seismic waves. In G. I. Petrashen (ed.), *Problems in the Dynamic Theory of Propagation of Seismic Waves* (Leningrad University Press, Leningrad), 3: 11–106 [in Russian]

— 1965. The work of the Leningrad school on seismic wave propagation. *Revs. Geophys.* 3: 199–209

Phinney, R. A. 1961a. Leaking modes in the crustal waveguide, Part I: The oceanic PL wave. *J. Geophys. Res.* 66: 1445–69

— 1961b. Propagation of leaking interface waves. *Bull. Seismol. Soc. Amer.* 51: 527–55

— 1965. Theoretical calculation of the spectrum of first arrivals in layered elastic mediums. *J. Geophys. Res.* 70: 5107–23

Pod"yapol'skiy, G. S. 1959a. The propagation of elastic waves in a layered medium, I, II. *Bull.* (Izv.) *Acad Sci. USSR, Geophys. Ser.*, no. 8: 788–93 and no. 9: 913–19

— 1959b. On a certain formula connecting the coefficients of head waves with reflection and refraction coefficients. *Bull.* (Izv.) *Acad. Sci. USSR, Geophys. Ser.*, no. 11: 1108–13

— 1959c. An approximate expression for the displacement in the vicinity of the principal front when the angle between the ray and an interface is small. *Bull.* (Izv.) *Acad. Sci. USSR, Geophys. Ser.*, no. 12: 1238–44

— 1966a. Physics of elastic waves. In *Handbook of Geophysics* (Nedra, Moscow), 28–96 [in Russian]

— 1966b. A ray series expansion for reflected and transmitted waves. *Bull.* (Izv.) *Acad. Sci. USSR, Earth Phys. Ser.*, no. 6: 347–63

Ravindra, R.* 1967. Propagation of S–H waves in inhomogeneous media. *J. Acoust. Soc. Amer.* 41: 1328–29

— 1968. Usual assumptions in the treatment of wave propagation in heterogeneous elastic media. *Pure and Appl. Geophys.* 70: 12–17

— 1969. Dispersion of waves in a continuously stratified elastic medium. *Pure and Appl. Geophys.* 72: 17–21

Riblet, H. J., and C. B. Barker. 1948. A general divergence formula. *J. Appl. Phys.* 19: 63

Richards, T. C. 1961. Motion of the ground on arrival of reflected longitudinal and transverse waves at wide-angle reflection distances. *Geophysics*, 26: 277–97

*See also Gupta, Ravindra N.

Rosenbaum, J. H. 1960. The long-time response of a layered elastic medium to explosive sound. *J. Geophys. Res.* 65: 1577–1613

— 1961. Refraction arrivals along a thin elastic plate surrounded by a fluid medium. *J. Geophys. Res.* 66: 3899–906

— 1964. The response of an elastic plate submerged in a liquid half-space to explosive sound. *Geophysics*, 29: 370–94

— 1965. Refraction arrivals through thin high-velocity layers. *Geophysics*, 30: 204–12

Salm, W. C. 1934. Über die Energie der Mintropwelle. *Beitr. angew. Geophys.* 4: 364–72 [in German]

Schelkunoff, S. A. 1948. *Applied Mathematics for Engineers and Scientists* (D. Van Nostrand, New York)

Schmidt, O. von. 1939. Über Kopfwellen in der Seismik. *Zeitschr. Geophys.* 15: 141–8 [in German]

Schwartz, M., S. Green, and W. A. Rutledge. 1960. *Vector Analysis with Application to Geometry and Physics* (Harper and Brothers, New York)

Skuridin, G. A. 1957. On the theory of scattering of elastic waves from a curvilinear boundary. *Bull.* (Izv.) *Acad. Sci. USSR, Geophys. Ser.*, no. 2: 36–58

Smirnov, V., and S. Sobolev. 1933. On the application of a new method of investigation of the elastic vibrations in a space with axial symmetry. *Trudy Inst. Seism. Akad. Nauk SSSR*, 29 [in Russian]

Smirnova, N. S. 1962. Calculation of wave fields in the neighbourhood of singular points. In G. I. Petrashen (ed.), *Problems in the Dynamic Theory of Propagation of Seismic Waves* (Leningrad University Press, Leningrad), 6: 30–59 [in Russian]

— 1964a. Calculation of wave fields in the neighbourhood of singular points, Part II. In G. I. Petrashen (ed.), *Problems in the Dynamic Theory of Propagation of Seismic Waves* (Leningrad University Press, Leningrad), 7: 77–87 [in Russian]

— 1964b. Some examples of the calculation of theoretical seismogramms. In G. I. Petrashen (ed.), *Problems in the Dynamic Theory of Propagation of Seismic Waves* (Leningrad University Press, Leningrad), 7: 88–103 [in Russian]

— 1966a. On the problem of calculation of wave fields in the region of critical rays. In G. I. Petrashen (ed.), *Problems in the Dynamic Theory of Propagation of Seismic Waves* (Nauka, Moscow-Leningrad), 8: 5–15 [in Russian]

— 1966b. On the character of the field of reflected waves near the critical point. In G. I. Petrashen (ed.), *Problems in the Dynamic Theory of Propagation of Seismic Waves* (Nauka, Moscow-Leningrad), 8: 16–22 [in Russian]

— 1966c. Comparison of theoretical seismograms calculated by two different methods. In G. I. Petrashen (ed.), *Problems in the Dynamic Theory of Propagation of Seismic Waves* (Nauka, Moscow-Leningrad), 8: 23–9 [in Russian]

— 1968. Calculation of wave fields of multiple waves in the vicinity of critical points. In V. M. Babich (ed.), *Mathematical Problems in the Theory of Propagation of Waves* (Nauka [Leningrad Division], Leningrad), 212–40 [in Russian]

Smirnova, N. S., and N. I. Yermilova. 1959. On the construction of theoretical seismograms in the neighbourhood of critical points. In G. I. Petrashen (ed.), *Problems in the Dynamic Theory of Propagation of Seismic Waves* (Leningrad University Press, Leningrad), 3: 161–213 [in Russian]

Sobolev, S. 1932. Application de la théorie des ondes planes à la solution du problème de H. Lamb. *Trudy Inst. Seism. Akad. Nauk SSSR*, 18 [in French]

— 1933. Sur les vibrations d'un demiplan et d'une couch à conditions initiales arbitraires. *Sbornik Mat.* (Moscow), 40: 236–66 [in French]

Sommerfeld, A. 1909 Über die Ausbreitung des Wellen in der Drahtlosen Telegraphie. *Ann. Physik*, 28: 665–736 [in German]

Spencer, T. W. 1965a. Long-time response predicted by exact elastic ray theory. *Geophysics*, 30: 363–8

— 1965b. Refraction along a layer. *Geophysics*, 30: 369–88

Steinhart, J. S., and R. P. Meyer. 1961. *Explosion Studies of Continental Structure* (Carnegie Institute Publ. 622, Washington)

Thomson, W. T. 1950. Transmission of elastic waves through a stratified solid medium. *J. Appl. Phys.* 21: 89–93

Thornburgh, H. R. 1930. Wave-front diagrams in seismic interpretation. *Bull. Amer. Assoc. Petrol. Geol.* 14: 185–200

Tooley, R. D., T. W. Spencer, and H. F. Sagoci. 1963. *Reflection and Transmission of Plane Compressional Waves* (Rept. AD 429797, U.S. Department of Commerce, Washington [Vela Uniform Report])

— 1965. Reflection and transmission of plane compressional waves. *Geophysics*, 33: 552–70

Tsepelev, N. V. 1959. Reflection of elastic waves in a non-homogeneous medium. *Bull.* (Izv.) *Acad. Sci. USSR, Geophys. Ser.*, no. 1: 5–8

— 1961. Propagation of waves in an acoustical medium with a transition layer. In G. I. Petrashen (ed.), *Problems in the Dynamic Theory of Propagation of Seismic Waves* (Leningrad University Press, Leningrad), 5: 169–205 [in Russian]

— 1968. Propagation of waves in a medium with a transition layer. *Trudy Mat. Inst., Akad. Nauk SSSR*, 95: 184–212 [in Russian]

Vasil'yev, Y. I. 1959. Certain consequences of an analysis of reflection and refraction coefficients of elastic waves. *Trudy Inst. Fiziki Zemli, Akad. Nauk SSSR*, 6: 52–80 [in Russian]

Vavilova, T. I., and G. I. Petrashen. 1966. On the calculation of fields of interference multiple reflected waves in multilayered media. In G. I. Petrashen (ed.), *Problems in the Dynamic Theory of Propagation of Seismic Waves* (Nauka, Moscow-Leningrad), 8: 48–54 [in Russian]

Vavilova, T. I., and A. D. Pugach. 1966. On the intensity of interference multiple reflected waves in a three-layered medium. In G. I. Petrashen (ed.), *Problems in the Dynamic Theory of Propagation of Seismic Waves* (Nauka, Moscow-Leningrad), 8: 30–47 [in Russian]

Vlaar, N. J. 1968. Ray theory for an anisotropic inhomogeneous elastic medium. *Bull. Seismol. Soc. Amer.* 58: 2053–72

Werth, G. C. 1967. Method for calculating the amplitude of the refraction arrival. In A. W. Musgrave (ed.), *Seismic Refraction Prospecting* (The Society of Exploration Geophysicists, Tulsa), 119–37.

Wolf, A. 1936. The amplitude and character of refraction waves. *Geophysics*, 1: 319–26

Yanovskaya, T. B. 1966. Algorithm for the calculation of reflection and refraction coefficients and head wave coefficients. In V. I. Keilis-Borok (ed.), *Computational Seismology* (Nauka, Moscow), 1: 107–11 [in Russian]

— n.d. Approximate methods in elastic wave theory. Unpublished notes of lectures delivered at the Department of Applied Mathematics and Theoretical Physics, Cambridge University

Yeliseyevnin, V. A. 1964. Calculation of rays propagating in an inhomogeneous medium. *Akusticheskiy Zh.* 10: 284–8 [in Russian]

Yepinat'yeva, A. M. 1959. Some results of the analysis of formulae for the amplitudes of refracted waves. *Trudy Inst. Fiziki Zemli, Akad. Nauk SSSR*, 6 (173): 7–51 [in Russian]
— 1960. Investigation of compressional seismic waves propagating in certain real layered media. *Trudy Inst. Fiziki Zemli, Akad. Nauk SSSR*, 14 [in Russian]
Yepinat'yeva, A. M., and V. Červený. 1965. Reflected waves in the region of the second critical point. *Studia Geophys. et Geod.*, 9: 259–71 [in Russian]
Zaitsev, L. P. 1959. On the head wave of the surface type. In G. I. Petrashen (ed.), *Problems in the Dynamic Theory of Propagation of Seismic Waves* (Leningrad University Press, Leningrad), 3: 378–83 [in Russian]
Zaitsev, L. P., and N. V. Zvolinskiy. 1951a. Investigation of the head waves generated at the boundary between two elastic liquids. *Izv. Akad. Nauk SSSR, Ser. Geograf. Geofiz.*, no. 1: 20–39 [in Russian]
— 1951b. Investigation of the axisymmetric head wave generated on the plane boundary dividing two elastic liquids. *Izv. Akad. Nauk SSSR, Ser. Geograf. i Geofiz.*, no. 5: 40–50 [in Russian]
Zöppritz, K. 1919. Über Erdbebenwellen, VIIb. *Göttingen Nachrichten*, 66–84
Zvolinskiy, N. V. 1957. The reflected and head wave arising at a plane interface between two elastic media, I. *Bull.* (Izv.) *Acad. Sci. USSR, Geophys. Ser.*, no. 10: 1201–18
— 1958. The reflected and refracted wave arising at a plane interface between two elastic media, II, III. *Bull.* (Izv.) *Acad. Sci. USSR, Geophys. Ser.*, no. 1: 1–7 and no. 2: 97–101
— 1965. Wave problems in the theory of elasticity of the continuum. *Izv. Akad. Nauk SSSR, Ser. Mekhanika*, 1: 109–23 [in Russian]

INDEX

A-wave, 210, 211
accuracy of ray method, 13, 122, 123, 224, 252, 256, 275–6
additional components, 26–38, 44–6, 101, 104
Alekseyev, A. S., 7, 11. 15, 31, 33, 50, 52, 76, 92, 97, 187, 297
Algermissen, S. T., 299
Alterman, F., 303
Ambramovici, F., 303
amplitude coefficients of head waves, 133; *see also* amplitudes of H-waves
amplitude spectra, 249, 250, 284; *see also* theoretical seismograms
amplitudes of H-waves: amplitude–distance curve, 147–63, 174–9, 209, 212–24, 242, 247–50; comparison with R-waves, 150–63, 212–24; curves of constant amplitudes, 150; definition, 121, 213; dependence on various parameters, 146–63; fundamental H-waves, 212–24; interference H-wave, 173–82, 247–50; multiayered media, 197–9, 203, 204; SH-waves, 122, 198, 199
amplitudes of interference waves, 173–9, 181–2, 280–7
amplitudes of R-waves, 82, 86, 150–63, 174–9, 212–24, 248, 249
Ansel, E. A., 8, 297
arrival time, *see* travel time
associated R-wave, 210, 211

B-interface, 188–97, 200–3, 211, 212, 214, 225, 226, 227, 231, 232, 236
B-wave, 191
Babich, V. M., 7, 11, 15, 16. 31, 297, 301, 304
Barker, C. B., 55, 303
basic head waves, 127, 133, 135–41, 159, 182, 183
Berry, M. J., 6, 146, 213, 292, 297, 300
Berzon, I. S., 8, 69, 297
Bessonova, E. N., 6, 294, 297
Bortfeld, R., 5, 68, 252, 290, 297, 298
boundary layer method, 7, 13
Brekhovskikh, L. M., 4, 8, 256, 259, 262, 279, 290, 298
Buldyrev, V. S., 8, 182, 239, 298
Bullen, K. E., 85, 298

C-wave, 239–45, 246–50 *passim*
C^{+}-wave, 239, 245–50
C_s-wave, 218, 237–50
Cagniard, L., 4, 8, 252, 298
Cagniard, method of, 4, 290
caustic, 13, 76, 84, 94, 207
Červený, V., 4, 6, 68, 69, 95, 96, 102, 131, 140, 146, 168, 175. 213, 250, 256. 257, 259, 261, 262, 264, 267, 279, 280, 284, 291, 298, 299, 306
Chekin, B. S., 7, 8, 11, 40, 182, 239, 245, 292, 299
classification of H-waves, 123–7
coefficients: of conversion, 67, 68, 69, 84, 198, 199, 203, 234; of head waves, 133, *see also* H-coefficients; of formation of H-waves, 133; of reflection and refraction, *see* R-coefficients
compressional velocity, 20, 189, 204, 210, 235
compressional wave, 20, 27, 34, 38, 101; distinguished from longitudinal wave, 20
conical wave, 8
Conrad discontinuity, 218, 219, 221, 222, 223
constant velocity gradient, 86–8, 205–6; *see also* linear variation of velocity
continuity of displacement and stress, 39, 42, 45, 254

conversion: coefficients, 67, 68, 69, 84, 198, 199, 203, 234; vector, 67, 74
Cook, K. L., 299
Cooley, J. W., 288, 299
Costain, J. K., 68, 299
coupling: of P and S waves, 19; of principal components of shear waves, 35–8; of SV and SH waves, 57, 58, 226
Courant, R., 21, 299
critical: angle, 10, 94, 100, 101, 182, 190, 236, 237; distance, 10, 101, 110, 121, 123, 147–63 *passim*, 192, 196, 201, 202, 205, 207, 213, 216, 236, 272; point, 109, 110, 111, 122, 123, 147–63 *passim*, 166, 182, 208, 210, 214, 216, 217, 224, 237 245, 249, 251, 264, 268–91 *passim*; ray, 13, 100, 101, 123, 187, 196, 245, 264; region, 7, 13, 100, 166, 213, 249, 252, 256, 259, 264, 273–9 *passim*, 279–87, 290
curvature, radii of, 32, 35, 41, 52, 54, 89, 186, 187, 225–34 *passim*
curve: of double curvature, 41, 57; plane, 24, 36, 37, 38, 57, 89, 226; space, 41, 89, 226
curved: interface, 40, 51, 56, 88–93, 163, 179, 181, 182, 225–34, 239, 251; wave front, 51, 179, 187, 210

damped head waves, 8, 181, 182, 225, 239
Davids, N., 303
differential equations for rays, 22, 23, 25
diffracted waves, 182, 219
dipping interface, 88–93, 225–34, 251
directional characteristic of a source, 75
displacement H-coefficients, 132, 133; *see also* H-coefficients
displacement potentials, 63, 252
displacement vector: H-wave, 102, 108–11, 114–16, 117, 121, 122, 184, 189, 191–7, 208, 209, 211, 231, 232, 233; impulse wave, 15, 93–6, 163–6, 233, 240; incident wave, 59, 93, 98, 103, 183, 231, 232, 233; R-wave, 60, 65, 70, 77, 82, 88, 91, 92, 94, 99, 184, 210, 211
displacements, 63
diving wave, 180, 237
Dix, C. H., 8, 252, 299
Donato, R. J., 8, 299
dynamic: analogues, 208–11; properties, 172, 242

eikonal equations, 19, 21, 27, 34, 38, 61, 65, 102
elementary ray tube, 30, 80, 110, 185, 194
epicentral distance, 8–10, 78–95 *passim*, 98, 99, 109, 147–63 *passim*, 187, 195, 208–24 *passim*, 236–50 *passim*, 278
Euler's equations, 23; theorem, 54, 186
Ewing, W. M., 4, 11, 19, 45, 252, 253, 255, 287, 288, 289, 291, 299
exact amplitude curve, 174, 175, 281
exact numerical intergrations, 280–7
exact ray expansion, 5, 289–91
expansion coefficients, 31
exponent coefficient, 147, 148

Fermat's principle, 23
Feshbach, H., 23, 256, 302
Filon, L. N. G., 280, 288, 289, 299
first approximation, 14
first-order interface, 39, 40, 51, 59, 88, 179, 217, 219, 225
first-order waves, 242; *see also* higher-order waves
Fock, V. A., 55, 299
Fourier transform 4, 15, 93, 94, 284, 293
free interface, *see* free surface
free surface: H-waves from, 113, 121, 130, 145, 198, 199, 200, 234; R-coefficients, 66–9; receiver on, 66–8, 84, 198, 199, 200, 234
Frenet's formulae, 35
Friedlander, F. G., 11, 15, 97, 299
Friedrichs, K. O., 7, 97, 300
Fuchs, K., 6, 288, 294, 300
fundamental head waves, 203–4, 205, 207, 212–24

Gel'chinskiy, B. Y., 7, 33, 53, 55, 76, 92, 97, 187, 226, 297, 300
geometrical optics, 11, 14, 34, 38
Gilbert, F., 289, 291, 300
Golovnyye volny, 8
Grant, F. S., 4, 11, 19, 25, 69, 300
Graves, L. M., 301
Green, S., 304
Gupta, Ravindra N., 11, 21, 40, 207, 300, 302
Gutenberg, B., 68, 300

H-coefficients, 97, 109, 127–45, 193–203 *passim*, 232, 265; connections between different H-coefficients, 132; displacement, 132, 133; free interface, 113, 130, 145; liquid-liquid interface, 113, 130, 144–5, 163; liquid-solid interface, 113, 129, 143–4; numerical values, 133–45; potential, 132, 133; R-coefficients, connection with, 131; SH-wave, 115, 116, 131, 145–6, 163; solid–liquid interface, 113, 129–30, 144; solid–solid interface, 109, 127–9, 133–43
H-waves, 97; *see also* head waves
Harkrider, D. G., 287, 288, 300
Haskell, N. A., 288, 300
head waves: accuracy of ray method, 122, 123, 224, 252, 256, 275–6; amplitude coefficients of, 133; amplitudes of, 121, 122, 146–63, *see also* amplitudes of H-waves; basic, 127, 133, 135–41, 159, 182, 183; classification of, 123–7; coefficients of, 5, *see also* H-coefficients; curved interfaces, 225–34; damped, 8, 181, 182, 225, 239; dipping interfaces, 225–34; displacement vector for, 102, 108–11, 114–16, 117, 121, 122, 184, 189, 191–7, 208, 209, 211, 231, 232, 233; of first kind, 124–6, 135–41, *see also* basic head waves; of fourth kind, 124–7, 142–3, 159, 163, 183; free interface, 113, 121, 130, 145, 198, 199, 200, 234; fundamental, 203–4, 205, 207, 212–24; general expressions for, 179–87; homogeneous, 102, 103; impossible, 209–10; incident wave, connection with, 185–7; inhomogeneous, 102, 108, 111, 262, 263; interference, 5, 8, 123, 181, 182, 224, 225, 235–50, 275, 276–9; interference reflected, 177–9, 264, 276–9; liquid–liquid interface, 8–10, 112–13, 121, 130, 144–5, 163; liquid–solid interface, 113, 121, 129, 143–4 177, 276, 280; multifold, 208, 210–11; multilayered media, 188–224, 287–94; P-wave source, 98–113, 116–29, 151–62, 189, 198, 203, 208, 209, 252; partially symmetric, 201–3, 204, 205, 207; pure, 123, 169, 179–82, 235, 239–50 *passim*; R-wave, connection with, 183–5, 190, 191; for refractive index approaching unity, 261, 280, 283; regular, 102, 262; schematic representation, 4, 117, 168, 180, 214, 236; screened, 6, 291; of second kind, 124–6, 140–1, 159, 183; SH-wave source, 113–16, 122, 131, 145–6, 163, 189, 198, 199, 200; from single interface, ray method, 97–187, 239–45; from single interface, wave method, 245–50, 252–87; solid–liquid interface, 113, 129–30, 144; solid–solid interface, 99–112, 127–9, 133–43, *see also* from single interface *in this entry*; of surface type, 102; SV-wave source, 98–113, 116–29, 153–62, 189, 198, 208; symmetric, 200–1, 204, 205, 207, 214; in thin layers, 6, 180, 217, 291; of third kind, 124–7, 141–2, 159, 183; time–distance curve, 3, 123, 124, 168, 172, 181, 197, 208, 210, 220, 223, 247; in transition layers, 180, 292, 293, 294; travel time for, 110, 111, 121, 123, 166–9, 196, 201, 202, 203, 205, 207, 224, 235–6, 237, 290; wave front picture of, 8–10, 124, 125, 126
Heelan, P., 4, 131, 132, 300
Helmberger, D. V., 290, 300
high-velocity layer, 213–17
higher-order waves, 51, 53
higher transport equation, 29, 35
Hilbert, D., 21, 299
Hilbert transform, 95
Hirasawa, T., 6, 292, 300
homogeneous head waves, 102, 103
Honda, H., 4, 255, 256, 300
Hook, J. F., 11, 300
Hron, F., 256, 299, 300
Huyghens principle, 8

impossible head waves, 209–10
impulse waves, 15–17, 93–6, 163–79, 199, 200, 204, 233, 240, 249, 284, 287, 288, 293
incident wave, 45, 46, 60, 93, 98, 101, 147, 185–7, 231, 253, 257
inhomogeneous head waves, 102, 108, 111, 262, 263
inhomogeneous media, 11, 19, 23–6, 74, 77–85, 86–8, 163, 179, 181, 183–7, 188–204, 205–8, 217–23, 235–50, 251, 291; coupling of P and S waves, 19; velocity in, 20–1

Integral method, *see* wave methods
interface: of first order, 39, 40, 51, 59, 88, 179, 217, 219, 225; of second order, 195, 197, 207, 218, 219, 222; with large velocity contrast, 158–63, 177, 178, 280, 286–7; with small velocity contrast, 150–63, 178, 280, 281–5
interface conditions, 39–53, 101, 103, 254, 288
interface critical distance, 100, 192
interface generated wave, 8
interfaces: types of, 39, 40; weak, 40
interference: head waves, 5, 8, 123, 177, 179, 224, 225, 235–50, 264, 275, 276–9; reflected head waves, 177–9, 264, 276–9; reflected wave, 177, 179, 264, 275, 276–9; waves, 166–79; zone, 123, 166–79, 275, 276, 286

Janský, J., 250, 299
Jardetzky, W. S., 8, 299, 301
Jarosch, H., 303
Jeffreys, B. S., 256, 301
Jeffreys, H., 4, 256, 301
Joos, G., 4, 301

Karal, F. C., 11, 20, 38, 301
Karpov, K. A., 279, 301
Kaufman, H., 81, 207, 301
Kay, I. W., 11, 12, 23, 29, 31, 32, 301
Keen, C., 6, 294
Keilis-Borok, V. I., 133, 297, 301, 305
Keller, J. B., 7, 11, 20, 38, 97, 300, 301
kinematic: analogues, 208–11; properties, 21, 207, 221
Kireyeva, I. E., 279, 301
Kline, M., 11, 12, 23, 29, 31, 32, 301
Knopoff, L., 288, 301
Knott, C. G., 69, 301
Koefoed, O., 68, 301
Kopfwellen, 8
Kosmininskaya, I. P., 166, 301
Krasavin, V. G., 92, 302
Krauklis, P. V., 6, 291, 301

Lamb, M., 4, 301
λ–spectra, method of, 6, 289, 292–4
Lanin, A. I., 8, 182, 239, 246, 298, 301
Laplace: identity, 32, 295; transform, 4
lateral wave, 8
Lewis, R. M., 301
Lifson, H., 5, 290, 303
linear approximation, 207, 221
linear variation of velocity, 86–8, 205–6, 218–23, 235–50
local velocity, 19
Logan, N. A., 299
longitudinal wave, 19, 20, 27
Longman, I. M., 5, 290, 303
low-velocity layer, 213–17
Luneburg, R. K., 11, 31, 302

Malinovskaya, L. N., 95, 133, 302
Marcinkovskaya, N. G., 92, 302
maxima of amplitude–distance curve, 175–7, 281–7
McCamy, K., 68, 302
McConnell, R. K., 3, 302
Menzel, D. H., 259, 302
Meres, M. W., 68, 302
Merten, 8
Meyer, R. P., 68, 302, 305
Michota, G. G., 6, 294, 297
Mintrop wave, 8
Moho, *see* Mohorovičić discontinuity
Mohorovičić, A., 3, 302
Mohorovičić discontinuity, 151, 212, 218, 219, 222, 223, 248, 249
Molotkov, L. A., 6
Morse, P. M., 23, 256, 302
Müller, G., 4, 5, 290, 302
multifold head wave, 208, 210–11
Musgrave, A. W., 3, 8, 291, 302, 305
Muskat, M., 4, 68, 131, 302

Nafe, J. E., 68, 302
Nakamura, K., 4, 255, 256, 300
Nakamura, Y., 6, 292, 302
non-transformed refracted wave, 4, 52, 53
normal mode expansion, 289, 291–2
normal section, 54

O'Brien, P. N. S., 8, 95, 133, 302
Ott, H., 4, 302

P-waves, 19, 20, 50, 51, 67, 74, 75, 84, 97; R-coefficients for, 58–64; reflection and refraction of, 58–64, 67, 82
parabolic equation method, 7, 13, 181
parameters of the ray, 30
Pariyskaya, G. N., 297

partially symmetric head waves, 201–3, 204, 205, 207
Pekeris, C. L., 5, 225, 288, 290, 291, 302, 303
Petrashen, G. I., 4, 5, 6, 8, 68, 69, 95, 124, 127, 132, 133, 187, 210, 213, 297, 300, 301, 303, 304, 305
Petrashen's method, 5
phase shifts, 83, 86, 121, 122, 173, 197–9, 203, 204, 284, 291
Phinney, R. A., 6, 287, 291, 292, 293, 303
Phinney's method of λ-spectra, 6, 289, 292–4
plane curve, 24, 36, 37, 38, 57, 89, 226
plane of incidence, 41
Pod"yapol'skiy, G. S., 4, 5, 6, 68, 131, 132, 133, 290, 291, 303
point source, 74–6, 97, 251–94 *passim*
potential H-coefficients, 132, 133
potentials of compressional and shear waves, 252, 253
Press, F., 299
principal: components, 26–38, 44–6, 51, 101, 104; normal sections, 53, 89, 186, 227; radii of curvature, 32, 41, 51, 53, 54, 59, 71, 88, 89, 186, 225–34 *passim*
principle: Fermat's, 23; Huyghens, 8; of isolated element, 40, 51; of local field character, 40
Pugach, A. D., 210, 305
pulse, *see* impulse waves
pure head wave, 123, 169, 179–82, 235, 239–50 *passim*

R-coefficients, 51, 53, 62–9, 74, 86, 92, 94, 99, 107, 108, 191–203 *passim*, 231, 258, 260, 266; connections between, 64, 69; displacement potentials, 63; free surface, 66, 67; H-coefficients, relation with, 131, 260; P and SV waves, 62–4; SH-waves, 64–6
R-waves, 14, 46, 51, 63, 64, 76, 95, 123–6, 150–63 *passim*, 183–5, 191, 210–11, 291; displacement vector for, 64, 65, 70, 77, 82, 88, 91, 92, 94, 99, 184, 210, 211; travel time for, 70, 74, 76, 81, 82, 83, 85, 87, 91, 99, 123, 166–9, 224
Rac-Chizgiya, M. I., 297
radiation condition, 43
radii of curvature, 32, 35, 41, 52, 54, 89, 186, 187, 225–34 *passim*
radius of torsion, 35
Ratnikova, L. I., 297
Ravindra, R., 21, 207, 303
ray amplitude curve, 281; *see also* amplitudes
ray: coordinates, 29; critical, 13, 100, 101, 123, 187, 196, 245, 264; method, 7, 11–96; parameters of, 30; plane, 72; series, 7, 11–17; system of differential equations, 22, 23, 25; tube, 30, 80, 110, 185, 194
ray path parameter, 24, 84
Rayleigh function, 66
reflected waves, 39, 42, 51, 53, 102, 103, 107, 124, 212–24 *passim*, 255, 265–71, 276–9, *see also* R-waves; multiple reflections, 71–4, 179; P and SV waves, 60–4; SH-waves, 64–6
reflection coefficients, *see* R-coefficients
refracted waves, 3, 39, 42, 51, 124, 180, 218–24 *passim*, 237, 255, *see also* R-waves; multiple interfaces, 71–4, 179; non-transformed, 52, 53; P and SV waves, 60–4; SH-waves, 64–6
refraction: arrival, 3, 4; method, 3
regular head wave, 102, 262
Riblet, H. J., 55, 303
Richards, T. C., 68, 304
Rosenbaum, J. H., 6, 291, 304
Runge-Kutta method, 25
Rutledge, W. A., 304

S-waves, 19, 20, 67, 75
saddle point, 259–65, 290
Sagoci, H. F., 305
Salm, W. C., 8, 304
Schelkunoff, S. A., 29, 304
Schmidt, O. von, 8, 304
Schwartz, M., 30, 35, 295, 304
screened head waves, 6, 291
Seckler, B. D., 301
second-order discontinuity, 195, 197, 207, 218, 219, 222
separability of vector wave equation, 11, 19
SH-waves, 50, 51, 57, 58, 67, 74, 75, 84, 85, 86, 97; H-waves, *see* head waves; R-coefficients for, 64–6; reflection and refraction of, 64–6, 82
shadow zone, 181, 182, 219, 221
shear velocity, 20, 189, 210

shear wave, 20, 27, 34, 38, 101; distinguished from transverse wave, 20
Skuridin, G. A., 40, 304
Smirnov, V., 5, 304
Smirnova, N. S., 4, 6, 133, 224, 256, 280, 291, 304
Smith, T. J., 297, 302
Snell's law, 24, 43, 62, 81, 83, 102, 240
Sobolev, S., 5, 304, 305
Sobolev and Smirnov, method of, 5
Sommerfeld, A., 4, 253, 305
Sommerfeld integral method, 4, 131, 253
source: directional characteristics of, 75; point, 74–6, 97, 251–94 *passim*; problem, 74; symmetric, 76, 77, 97, 189, 213
space curve, 41, 89, 226
Spencer, T. W., 5, 290, 291, 305
spreading, 76, 110
spreading function, 76–93
Starodubrovskaya, S. P., 297
steepest descent, method of, 251, 252, 256, 259–65, 290
Steinhart, J. S., 68, 297, 305
step approximation, 207
Stoneley wave, 255, 262
subcritical reflected wave, 217
supercritical reflected wave, 217
surface type head waves, 102
SV-waves, 50, 51, 57, 74, 75, 84, 85, 97; R-coefficients for, 58–64; reflection and refraction of, 58–64, 82
symmetric head waves, 200–1, 204, 205, 207, 214

Teltow, J., 4, 301
theoretical seismograms, 163–6, 171, 172, 223, 224, 284, 285, 288, 289, 290, 293; *see also* amplitude spectra
thin layers, head waves in, 6, 180, 217
Thomson, W. T., 288, 305
Thornburgh, H. R., 8, 305
time-distance curve, 3, 84, 123, 124, 168, 172, 197, 207, 208, 210, 220, 223, 247
Tooley, R. D., 68, 305
torsion of a ray, 36, 57
torsion, radius of, 35
transformed refracted wave, 52, 53, 156, 159, 237
transient waves, *see* impulse waves
transition layers, head waves in, 180, 292, 293, 294
transport equation, 29, 35
transverse wave, 19, 20, 27
travel time: H-waves, 110, 111, 121, 123, 166, 169, 196, 201, 202, 203, 205, 207, 224, 235–6, 290; R-waves, 70, 74, 76, 81, 82, 83, 85, 87, 91, 99, 123, 166–9, 224
travel time curves, *see* time–distance curve
Tsepelev, N. V., 6, 40, 292, 305
Tukey, J. W., 288, 299
turning point of a ray, 83, 84, 87, 88, 180, 189, 207, 208, 218, 219, 236, 247

Vasil'yev, Y. I., 68, 305
Vavilova, T. I., 210, 305
vector indentities, 295–6
velocity gradient, 87, 181, 182, 208, 218, 219, 221, 235, 237, 248, 250, 292; *see also* inhomogeneous media
velocity in heterogeneous media, 20, 21
vertically inhomogeneous media, *see* inhomogeneous media
Vlaar, N. J., 31, 305

wave front diagram, 8, 124, 125, 126
wave methods, 3, 131, 181, 245–50, 251–94
wave potential, 252
weak interfaces, 40, 52
Werth, G. C., 146, 305
West, G. F., 4, 11, 19, 25, 69, 146, 213, 297, 300
Wilson, J. T., 302
Wolf, A., 4, 305

Yanovskaya, T. B., 7, 11, 12, 13, 15, 20, 31, 76, 97, 132, 180, 305
Yeliseyevnin, V. A., 26, 306
Yepinat'yeva, A. M., 8, 69, 146, 168, 213, 297, 299, 306
Yermilova, N. I., 4, 256, 304

Zaitsev, L. P., 5, 102, 306
zero-order ray theory, *see* zeroth approximation
zero-order wave, 242
zeroth approximation, 14, 33, 34, 51, 57–74, 76, 100, 187, 238, 239, 278, 284, 286
Zöppritz, K., 67, 306
Zvolinskiy, N. V., 4, 5, 132, 133, 306